# The Great Fires

Indian Burning
and Catastrophic Forest Fire Patterns of
the Oregon Coast Range, 1491-1951

July 8, 2003

PhD Defense
Bob Zybach

To the Reader:

This is the first of a planned series of books to be published that feature my written works, maps, and photographs that have accumulated through the years. It is also the first book to be published by my family business. NW Maps Co. was created in 1994 to conduct historical research in the fields of forest and fire history, reforestation planning, local and regional human history, and the management of native plants and animals. One result has been the publication of several thousand of my reports, articles, editorials, maps, comments, correspondence, and photographs on the Internet, in monographs and reports, magazines, journals, and newspapers since then; but this remains the first book.

*The Great Fires: Indian Burning and Catastrophic Forest Fire Patterns of the Oregon Coast Range, 1491-1951* is the defintive fire history of Oregon Coast Range forests and grasslands for the past 500 years. Its comprehensive research methods, references, and recommendations serve as a model for other landscape-scale fire histories and is primarily why it is being updated and reprinted at this time. To serve a wider audience of students, scholars, and resource managers.

The following pages contain a slightly revised version of my 2003 PhD dissertation at Oregon State University. Principal differences are the removal of several academic-related forms and signatures at the beginning, and a complete repagination of the contents. This latter change was caused by a change in font sizes that resulted in the final manuscript being 104 pages shorter than the original. Future books are planned tthat will contain selections of articles and editorials, book-length reports, photographs, and historical research. Some may be selections by other authors that are little known or difficult to obtain, and are out of print and in the public domain.

The title of "Great Fires" to reference the major Oregon Coast Range forest fires that took place between 1845 and 1910, was first used in 1912 by colorful Theosophist and US Geographer, Fred G. Plummer, in his "Bulletin 117 of the Forest Service," *Forest Fires, Their Causes, Extent, and Effects, with a Summary of Recorded Destruction and Loss.* Of the 15 total "greatest forest fires in the United States since the year 1800," Plummer lists three (the Yaquina, Nestucca, and Coos) that are mapped and described here, as well as the "Historic" 1868, 1902, and 1910 wildfire years he also identifies.

Bob Zybach
Cottage Grove, Oregon
January 14, 2018

# The Great Fires:
# Indian Burning and Catastrophic Forest Fire Patterns of the Oregon Coast Range, 1491-1951

by
Bob Zybach

A DISSERTATION

submitted to

Oregon State University

In partial fulfillment of
The requirements for the
degree of

Doctor of Philosophy

Presented July 8, 2003
Commencement June 13, 2004

Est. 1994

2nd Edition


Citation: Zybach, Bob 2018. *The Great Fires: Indian Burning and Catastrophic Forest Fire Patterns of the Oregon Coast Range, 1491-1951.* 2nd Edition, NW Maps Co., Cottage Grove, Oregon: 347 pp.

Keywords: Afforestation, Catastrophic Fires; Columbia River; Douglas Fir Region; Fire History; Fire Season; Forest Fires; Forest History; Indian Burning; Oak Savannah; Oregon Coast Range; Oregon History; Oregon Indians; Reforestation, Wildfire; Willamette Valley

AN ABSTRACT OF THE DISSERTATION OF

Bob Zybach for the degree of Doctor of Philosophy in Environmental Sciences presented on July 8, 2003.

Title: The Great Fires: Indian Burning and Catastrophic Forest Fire Patterns of the Oregon Coast Range, 1491-1951.

Abstract approved: Kurt M. Peters

This study was designed to examine the relationship between land management practices of Indian communities prior to contact with Europeans and the nature or character of subsequent catastrophic forest fires in the Oregon Coast Range. The research focused on spatial and temporal patterns of Indian burning across the landscape from 1491 until 1848, and corresponding patterns of catastrophic fire events from 1849 until 1951. Archival and anthropological research methods were used to obtain early surveys, maps, drawings, photographs, interviews, Geographic Information Systems (GIS) inventories, eyewitness accounts and other sources of evidence that document fire history. Data were tabulated, mapped, and digitized as new GIS layers for purposes of comparative analysis. An abundance of useful historical evidence was found for reconstructing precontact vegetation patterns and human burning practices in western Oregon. The data also proved useful for documenting local and regional forest fire histories. Precontact Indians used fire to produce landscape patterns of trails, patches, fields, woodlands, forests and grasslands that varied from time to time and place to place, partly due to demographic, cultural, topographic, and climatic differences that existed throughout the Coast Range. Native plants were systematically managed by local Indian families in even-aged stands, usually dominated by a single species, throughout all river basins of the study area. Oak, filberts, camas, wapato, tarweed, yampah, strawberries, huckleberries, brackenfern, nettles, and other plants were raised in select areas by all known tribes, over long periods of time. However, current scientific and policy assumptions regarding the abundance and extent of precontact western Oregon old-growth forests may have been erroneous. This study demonstrates a high rate of coincidence between the land management practices of precontact Indian communities, and the causes, timing, boundaries, severity, and extent of subsequent catastrophic forest fires in the same areas. Information developed from this study will be of value to researchers, wildlife managers, forest landowners, and others with an interest in the history and resources of the Oregon Coast Range.

## ACKNOWLEDGEMENTS

This dissertation is the culmination of more than twenty years' research and study. It was made possible through the efforts of many people. Unfortunately, there is not room enough to list every individual or institution that provided significant help or assistance during that time, so I ask for understanding from those I may have inadvertently overlooked or otherwise failed to acknowledge.

First and foremost, I must acknowledge the advice, support, and other resources provided by Wayne Giesy, Ralph Hull, Wilma Hull, and Betty Joe Giesy. This degree would not have been undertaken and could not have been achieved without their help and encouragement.

Next, I would like to thank Kurt M. Peters for accepting me as a student for my doctoral studies. His advice, enthusiasm, and guidance were critical to the orderly completion of this program. Committee members George Taylor, Kermit Cromack, Cynthia Twohy, and Court Smith must also be thanked for their time and assistance, as must former committee members Jesse Ford and David Brauner. All provided valuable direction and encouragement at key points in time. William Winner, founder and former Head of the OSU Environmental Sciences Program, also deserves recognition for his timely encouragement and assistance in helping me to undertake and complete this research at Oregon State University.

In addition to my committee, significant critical review and technical assistance were provided by the following individuals: Amanda Matzke, mapping and GIS; Albert Gruen, research strategy, patience, and website design; Nana Lapham, editorial assistance and encouragement; Scott Byram, text review and editing; Frank Lake, discussion, research, and review; and Jenine Filipelli, "spiritual guidance."

Next, I would like to acknowledge the individuals that first encouraged and directed me along the lines of fire history research, forest resources management, and the history of the Indians of the Oregon Coast Range: Rex Wakefield, Richard Hermann, Henry P. Hansen, Kenneth Munford, Bill Atkinson, Royal Jackson, Robert Buckman, John McWade, Jerry Phillips, Md. Shahid Islam, Benjamin Stout, Angela Sondenaa, Larry Fick, George Martin, Kevin Sherer, and Harold

Sandstrom. Several of these people are no longer living, and others are retired, yet I hope their work and ideas are credibly presented and continued through this document.

Other individuals and institutions that provided key support or services at critical times during this research include: Judy Juntanen, Benton County Historical Museum Library; Cheryl Oakes, Forest History Society Archives; Elizabeth Winroth, Oregon Historical Society Library; Sue Montgomery, OSU Environmental Sciences Program; Evelyn Reynolds, OSU Department of Ethnic Studies; Joey Heimberg and Renee Windsor, OSU Graduate School; and Megan Moholt, Weyerhaeuser Timber Company Archives.

Finally, I would like to thank Ralph and Wilma Hull of Corvallis, Oregon, Wayne and Betty Jo Giesy of Philomath, Oregon, the Ralph Hull Foundation, the Board of Oregon Websites and Watersheds Project, Inc., George Taylor of the OSU Oregon Climate Services, Bernard Bormann of the USDA Pacific Northwest Forest and Experiment Station, and Kermit Cromack of the OSU Department of Forest Sciences, for providing funding, employment, encouragement, and other critical resources needed to complete this research in a timely and professional manner.

# TABLE OF ACRONYMS AND ABBRV.'S

The following acronyms and abbreviations are used throughout this dissertation. The list does not include such standard abbreviations as p., pg., pp., ibid., et al., US, USA, and etc.

| | |
|---|---|
| BLM | [USDI] Bureau of Land Management. |
| BP | Years before the present time. |
| CCC | Civilian Conservation Corps. |
| d.b.h. | diameter (in inches) at breast height (4 1/2 feet above ground). |
| DLC | [Oregon] Donation Land Claim [Act of 1850]. |
| E. | East of the Willamette Meridian. |
| GIS | Geographic Information Systems. |
| GLO | [USDI] General Land Office. |
| N. | North of the Willamette Baseline. |
| NF | [USDA] National Forest. |
| ODF | Oregon Department of Forestry. |
| OSU | Oregon State University (formerly OSC, OAC, and others). |
| PLS | [US] Public Land Survey. |
| PNW | [USDA] Pacific Northwest [Forest and Range Experiment Station]. |
| R. | Range. |
| Rng. | Range. |
| S. (1) | South of the Willamette Baseline. |
| S. (2) | Section. |
| Sec. | Section. |
| T. | Township. |
| T&E | [USDA Fish & Wildlife] Threatened & Endangered species. |
| Tsp. | Township. |
| USDA | United States Department of Agriculture. |
| USDI | United States Department of the Interior. |
| USGS | United States Geological Survey. |
| W. | West of the Willamette Meridian. |
| WPA | Works Progress Administration. |
| W.W.M. | West of the Willamette Meridian. |

**The Great Fires:**
**Indian Burning and Catastrophic Forest Fire Patterns of the Oregon Coast Range, 1491-1951**

TABLE OF CONTENTS

TABLE OF CONTENTS (Continued) Page

TABLE OF CONTENTS (Continued) Page

## LIST OF FIGURES

LIST OF MAPS

## LIST OF MAPS (Continued)

Map | Page



## LIST OF TABLES

Table | Page

## LIST OF APPENDICES



## LIST OF APPENDIX FIGURES

## LIST OF APPENDIX MAPS



## LIST OF APPENDIX TABLES

# The Great Fires:
# Indian Burning and Catastrophic Forest Fire Patterns of the Oregon Coast Range, 1491-1951

## 1. INTRODUCTION: HYPOTHESIS AND SETTING

> *It would be difficult to find a reason why the Indians should care one way or another if the forest burned. It is quite something else again to contend that the Indians used fire systematically to "improve" the forest. Improve it for what purpose? Yet this fantastic idea has been and still is put forth time and again because somebody's grandfather said that is what happened.*
>
> --C. Raymond Clar (1959: 7)

> *When the forest burned, fires were often of high intensity and uncontrollable. The tribes of coastal Oregon were the victims of some of these fires, having been driven to the waters of the Pacific Ocean to survive.*
>
> --James Agee (1993: 56)

Between 1840 and 1850, several American Indian nations comprised of dozens of tribes and communities in the Oregon Coast Range were all but replaced by a comparatively homogenous population of white European American immigrants. This abrupt change in land ownership, cultures, and technology caused immediate changes to the region's forest and grassland environments. The transition from Indian to European land management practices resulted in large-scale changes to landscape patterns that persist to the present time. Many wildlife species were decimated and extirpated in favor of domesticated plants and animals: California condors, grizzly bears and whitetail deer gave way to chickens, cattle, and swine; fields of camas, brackenfern, and tarweed were transformed to corn, potatoes, and wheat; geese and grouse gave way to Chinese pheasants and turkeys. Even fire was affected. Expansive grasslands that were annually fired to produce and harvest food crops were plowed and grazed instead. Interior forestland trails, prairies, meadows, brakes, and berry patches—created and maintained by fire--were abandoned and began converting to trees. Near the end of the decade, probably in 1849 or 1850, the first of a century-long series of catastrophic forest fires took place in the region. These wildfires were so large and notable they became known as the "Great Fires" and acquired individual names: the Yaquina; the Coos; the Nestucca; the Tillamook. It is ironic that each Great Fire was named for the people who formerly owned and lived in the lands in which these catastrophic events took place: the Yakona; the Koos; the Nestucca; the Killamox.

This thesis examines the change in vegetation patterns of the Coast Range of western Oregon (see Map 1.1) from the late 15th century, through the beginning of historical time in the late 18th century, until 1951: the year of the last significant Tillamook fire. It is the first study to place the region's precontact cultural landscapes in context to the ignition patterns, fuel histories, and boundaries of subsequent historical wildfires.

A certain amount of history has been written and documented about the Oregon Coast Range ("Coast Range" or "the Range") forest fires of ca. 1849 to 1951 (e.g., Gannett 1902; Morris 1934b; Juday 1976; Pyne 1982). In the eastern Coast Range the times, locations, and results of Indian burning practices were documented by David Douglas from 1825 to 1827 (Douglas 1904; 1905), John Work from 1832 to 1834 (Scott 1923; Maloney 1943), by members of the 1841 Wilkes Expedition (Wilkes 1845a; 1845b), and others (see Table 2.01). However, with the notable exception of the Willamette Valley and the eastern slope (Williams 2003), very little has been written about precontact Indian burning practices in other areas of the Coast Range. Early landscape drawings and paintings--such as those by Warre in 1845 (Warre 1976: 87; Towle 1982: 72) and Kane in 1847 (Kane 1925), also depicted eastern Coast Range foothills of the Willamette Valley (see Figure 3.04) and Public Land Surveys in the 1850s and thereafter (e.g., Freeman 1852; Hyde 1852a; 1852b; Hathorn 1854a; 1854b) added significant detail to the written descriptions of the journalists. In the 1860s, landscape photographs became popular, and helped to add detail and certainty to landscape patterns at precise moments in time. Drawing from these resources, subsequent writers and geographers (e.g., Morris 1934b; Thilenius 1968; Johannessen et al 1971) were able to construct reasonably accurate maps and accounts of the burning practices and results of Kalapuyans who occupied the territory in early historical time (Collins 1951; Mackey 1974; Gilsen 1989). By comparison, very little has been documented regarding the burning practices of late prehistoric and early historical people living in the northern, western, and southern parts of the region (see Map 1.01).

LaLande and Pullen (1999: 267) term Indian fires in the southern Coast Range "limited and localized" in the "mid-elevation, mixed conifer forest stands" which characterize the "vast" majority of the area. Whitlock and Knox (2002: 224) go even further, claiming the presence of early historical prairies, savannah, and oak woodlands in the Coast Range were a direct result of prehistoric climate change and lightning-caused fires ("which were probably more abundant . . .

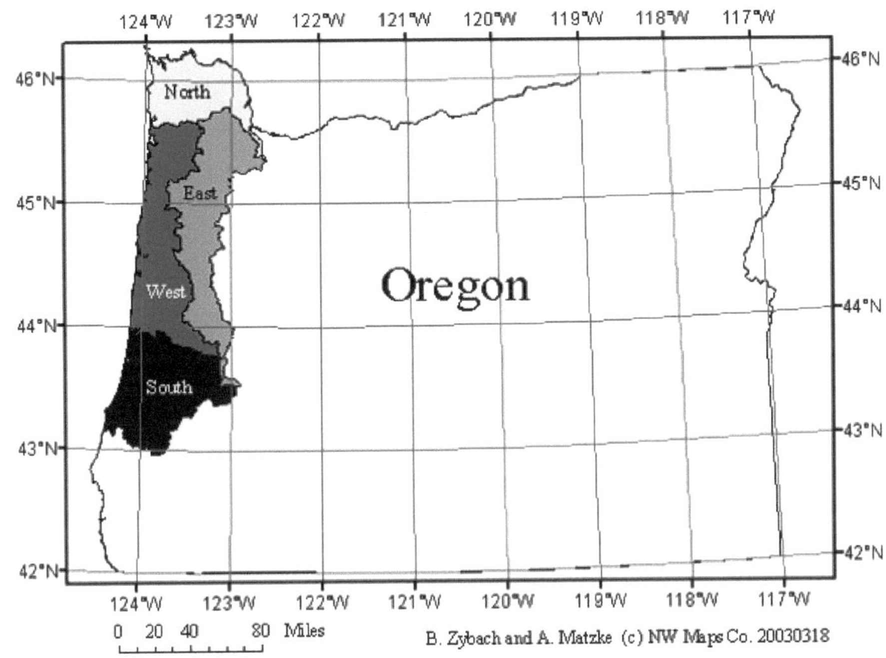


Map 1.01 Location of the Oregon Coast Range study area.

in the early Holocene": p. 206), and had relatively little or nothing to do with human burning practices. Through the use of maps, tables, eyewitness accounts, drawings, and photographs, this dissertation documents the uses of fire by American Indian people living in the Coast Range at the time of contact with white Europeans and Americans in the late 1700s. The same methods are used to describe the subsequent Great Fire events of 1849 to 1951 and the roles Indian fires may have played in their timing, severity, and boundaries.

## 1.1 Hypothesis

The hypothesis of this dissertation is that western Oregon Indian burning practices of the 16th to mid-19th centuries had a direct effect on spatial and temporal patterns of catastrophic forest fires that took place from 1849 to 1951 in the Oregon Coast Range.

The Oregon Coast Range was selected as an area for this research for several reasons: 1) the current study was initiated in 1988, when it was determined that a fire history had not been completed for the Coast Range since Morris (1934b) and "it is an important topic that needs to be researched" (Hermann 1988: personal communication); 2) the incidence of lightning is extremely low in the Coast Range, leaving people as the principal sources of ignition and fire patterning, and allowing for a generally clear division of wildfire causes and seasonality (Keeley 2002); 3) the Coast Range is a bounded portion of the Douglas-fir Region that contains some of the most productive timber growing conditions and important wildlife populations in the world (Heilman et al 1981; FEMAT 1993), therefore providing a value to resource managers that can benefit with a better understanding of fire history and wildfire risk; and 4) the scale of general patterns of Indian burning and wildfire history allow for the display of significant detail on a standard-sized sheet of thesis paper (e.g., Map 4.13).

Maps, tables, and figures are used to show differences in cultural landscape patterns resulting from Indian burning practices. Cultural landscape patterns are landscape-scale designs created and maintained by systematic human burning (and/or by other land management processes), due to their origin and appearance (Winkler and Bailey 2002). Landscape patterns, for purposes of this dissertation, are considered at regional, basin, and local scales (see Chapter 2.3). These patterns are shown to vary between northern, eastern, western, and southern parts of the Coast Range due to differences in national and tribal traditions, weather patterns, demographics, physical setting, and distance from the ocean. There are two primary reasons for dividing the Coast Range into these four subregional areas (Map 1.01) for this study: first, there are significant topographical (Map 1.02), settlement (Table 3.01), climatic (Map 1.04; Table 1.02), and vegetational (Map 1.09) differences that have effects on land use history and wildfire behavior and that should be considered (Table 1.03); second, the use of subregional scales allows for greater mapping detail in the depiction of trails, landmarks, and vegetation patterns that are central to discussion and analysis (e.g., Map 3.02).

Indian burning practices are defined as those uses of fire in pre-European American contact time ("precontact time") and early historical time that resulted in changed or stabilized landscape-scale vegetation patterns. Three principal categories of these practices are recognized: firewood gathering and burning, patch burning, and broadcast burning (Lake and Zybach: In Review). Firewood gathering and burning ("firewood burning") involves the movement of fuels to specific locations before burning, resulting in areas that contained relatively little (or, conversely, stockpiled) large, woody debris and designated spots of intense, repeated, and prolonged heat. Patch burning is defined as having a specific purpose and involving fuels within a bounded area, such as burning an older huckleberry patch, a segment of trail, or a field of weeds. Broadcast burning is the practice of setting fire to the landscape for multiple purposes and with general boundaries, such as burning a prairie to cure tarweed seeds, eliminate Douglas-fir seedlings, expose reptiles and burrowing mammals, and to harvest insects (see Chapter 3.2).

The "cultural legacy" of combined burning actions is shown to have a direct effect on subsequent patterns of catastrophic forest fires in the same areas of the Coast Range. Catastrophic fires are defined as wildfire events greater than 100,000 acres in size. Cultural legacy, in this instance, is used to denote the evidence of trails, savannah, prairies, fields, berry patches, brakes, balds and other environmental indications of Indian land uses that persist through time. Others have used similar terms, typically to denote evidence of past agrarian activities in the landscape. Patterns of burning and wildfire include similarities and differences in sources and locations of ignition; locations and extent of fire boundaries; timing, frequency, seasonality, severity (Feller 1998), and intensity of fires; and effects of fire on local human and wildlife populations.

The terms "Indian" and "American Indian" will be used interchangeably to denote people who lived in the Oregon Coast Range in precontact and early historical time, in accordance with current and accepted use of these terms by the peers and descendants of these people. "Tribes" will be used in the same manner as currently used by the Confederated Tribes of Siletz, Confederated Tribes of Grand Ronde, Coquille Indian Tribe, and the Confederated Tribes of Coos, Lower Umpqua, and Siuslaw. "Nation" will be used to designate adjacent Indian tribes associated by proximity and a shared language, such as the Chinook Nation of the 1940s that became Chinook Indian Tribe, Inc., in 1953: specifically, the Chinook and Athapaskan nations on the Columbia, the Kalapuya and Athapaskan nations of the Willamette, south Umpqua, and upper Coquille valleys, and the Koos, Siuslaw, Yakona, and Salish nations of the coast (see Map 3.01). Tribal and national names will attempt to use the earliest generally accepted historical spellings, rather than modernized spellings or European terms. "White" will be used to denote the predominantly European and American Caucasian immigrants and visitors to the Coast Range in early historical time, with the acknowledgement that many of these people were of Iroquois, Hawaiian, Chinese, or African ancestry. In this instance, the term "white" is intended to represent people of western European culture, rather than a particular race or skin color. Plants and animals are considered "native" to the Oregon Coast Range if they were present in the environment before 1770 (Wilson et al 1991: 17). Species are referenced by accepted local names rather than "common" names (see Zybach 1999: 259-274) -- for example: boomer vs. mountain beaver, filbert vs. hazelnut, or chittam vs. Cascara buckthorn--and are identified by their current scientific names in Appendix A.

## 1.2 Time Period and Scale

The amount of detail that can be considered in this study is constrained by time and space. This section describes and briefly discusses the temporal and spatial scales used to examine landscape patterns created by Indian burning practices and catastrophic fires in the Oregon Coast Range.

The years 1491 and 1951 were selected as beginning and ending points of the time period for this study because they represent significant years in the fire history of the Coast Range. The nearly 500-year time-span is an appropriate scale for considering regional changes in forest composition and migration (Hansen 1947), climate (Bradley and Jones 1995), human demographics (Boyd 1999a), cultural values (Raup 1966), technical complexity (Zybach 1999), and landscape-scale fire patterns (Teensma et al 1993).

1491 predated Columbus' voyage by a year. It is a time that represents a possible apogee of cultural, social, and technical life for North America (Fritz et al 1992: 146-150); at least, a date known to have preceded meaningful contact with Europe, Africa, and Asia, and a point of time for which much recent research has been focused (Crosby 1972; Denevan 1992; Mann 2002). An additional reason for selecting 1491 as a beginning point for this study, is that it represents a point in time during which early historical old-growth conifer forests and oak savannah patterns began to become established in the Coast Range; very few trees, living or dead, have ever been recorded in the region that began growing before 1492 (see Chapter 4.1.1; Appendix G; Weisberg and Swanson 2003).

Until recently, the year 1491 in North America has represented a nearly uninhabited Eden to many scholars: a concept idealized by the Wilderness Act of 1964 as "an area where the earth and its community of life are untrammeled by man, where man himself is a visitor who does not remain" (US Public Law 88-577). Mann (2002) cites Morison's 1974 depiction of precontact North American Indians as living "short and brutish lives, void of any hope for the future" and a 1987 high school textbook (*American History: A Survey*) description of pre-1492 Americas as "empty of mankind and its works," to support his assertion that historians and politicians have minimized the numbers and capabilities of precontact peoples. The surmised reason for these perceptions is justification for the methods in which Europeans claimed and occupied Indian lands throughout North America without just compensation or consideration to their rightful owners and occupants. This study is not concerned with the politics associated with land claims, natural resources management, or with public education policies resulting from misconceptions about precontact populations; nor is it concerned with better estimates of those numbers. The selection of 1491 as an important date in North American cultural and natural history has been established and discussed by others--but something happened in 1492, or shortly thereafter, that had permanent effects on the biology and ecology of American peoples, forests, and grasslands, and--according to the ages of relict Coast Range oaks and conifers--those effects were extended or coincidental to western Oregon (including the Coast Range) as well (see chapters 4.1.1 and 5.3).

The selection of 1951 as an ending point of this research is more obvious. That was the final year (following 1933, 1939, and 1945) of the famous "6-year jinx" of catastrophic fires that shaped the Tillamook Burn (see Chapter 4.1.6 and 4.3.3). It also marked the abrupt ending of catastrophic fires throughout the entire Coast Range from that time until the present (October, 2003), a span of more than 50 years (see Table 4.02). There has been little or no research as to why or how the "Great Fires" ended so suddenly and completely, but one possibility is the large amount of road building and clearcut logging that took place in the Coast Range following World War II (Zybach 2002a). The 460-year time period of this study, therefore, begins with the seeding and juvenile growth of vast conifer forest lands throughout the entire Range and ends, largely, with the completion of their destruction and harvest by wildfire and logging.

Landscape patterns, for purposes of this dissertation, are considered at regional (hundreds of thousands or millions of acres), basin (thousands or tens of thousands of acres), and local (tens or hundreds of acres) scales (see Chapter 2.3). The entire Range and its subregions (see Map 1.01) are millions of acres in size. For the most part, this is sufficient to depict coarse-scale landscape patterns of culture, prescribed burning, and catastrophic wildfire (e.g., Maps 3.01 and 4.01). Basin and subbasin scale maps present more detail, such as major trails, peaks, and creeks, at a finer scale (e.g., Maps 2.12 and 3.03). Local scale landscape patterns can illustrate the processes that result in the broader patterns, depicting roads, plowed fields, and fence lines (e.g., Figure

2.03), small meadows and buildings (e.g., Map 2.04), and even individual trees and snags (e.g., Figure 4.06).

Finally, it should be noted that although this study covers nearly five centuries in time and millions of acres of forests and grasslands, each landscape pattern that is shown represents only a point in time for a selected area. A landscape is altered dramatically between the day before a fire and the day after a fire; during the time of a fire it can change even more dramatically from minute to minute. Patterns for a given area, therefore, represent changes from time to time, or cumulative changes over time, rather than static conditions through time. The degree of those changes depends on the scale in which they are considered.

## 1.3 Physical and Political Setting

This dissertation concerns people, wildfire, and the use of fire by people in a particular forested region of western North America. People first entered the Oregon Coast Range more than 10,000 years ago, probably by foot or watercraft (Schobinger 1994: 20-21). The use of fire by people in the region was coincidental with their arrival; if people didn't arrive with fire, they probably manufactured it within a few hours or days (Pyne 1982: 66-70). The environment began to be transformed immediately. For the first time ever, firewood was gathered, forests were purposefully fired, and grasslands created and regularly burned. As human settlements became more permanent, so did patterns of human use upon the land. This section briefly describes the history of the land these people found, where they settled and why, and how the shape of the land helped to shape the patterns of people and their fires across the landscape.

### 1.3.1 Geological history

A region's topography is a result largely of its geological history. The formation of the Coast Range is relatively young compared to other areas of the earth and of the Pacific Northwest. Its soils and elevations can be traced principally to the erosion of older landmasses, organic decomposition, undersea volcanic eruptions, tectonic plate uplift, lava flows, and a series of cataclysmic floods. According to Orr, et al (1992: 167-202), the beginnings of the Coast Range are thought to have been about 66 million years ago, with underwater eruptions of basaltic pillow lavas that can now be seen on the shoreline of Depoe Bay, throughout the Siletz River Gorge, and as far east as Coffin Butte, along Highway 99 on the eastern boundary of Soap Creek Valley (see Map 2.01). During the same period, the Klamath Mountains in present-day southwest Oregon were steadily eroding, filling the shallow ocean to their north with sediments that would ultimately become the Tyee soils and sandstones of today. Ash and pyroclastics from infant Cascade volcanoes to the east were added to the mix, and the steady collision of the Juan de Fuca tectonic plate with the North American continental shelf forced the mass to rise above sea level and moved the Pacific shoreline west. River valleys and peaks were formed by erosion over millions of years, while rising and lowering seas carved the western boundary and upland terraces. Lava flows, emanating from eastern Oregon 15 million years, ago added further definition to the Range. From 12,800 to 15,000 years ago, a series of ice age floods coursed down the Columbia River, shaping the bluffs and islands along the northern boundary of the Coast Range, and leveling the floodplain of the Willamette Valley from Eugene to Oregon City with soils from eastern Washington, Idaho, and Canada (Allen and Burns 1986). About 5,000 years ago the region achieved its current general configuration, as melting ice caused a worldwide rise in sea levels, flooding coastal river valleys and creating the bays, estuaries, and beaches found today (Orr, et al 1992: 181-182). These events all contributed to the topography of the Coast Range, directly affecting patterns of human settlement and land use (including fire) from the time of discovery, more than 10,000 years ago, to the present.

### 1.3.2 Topography

Fire is said to behave according to ignition patterns constrained by three sides of a "triangle," defined as weather, fuels, and topography (e.g., Martin 1990: 58). Figure 1.01 (Zybach, B. and A. Gruen 2003) is a stylized version of the triangle, emphasizing the relationship between fuel and topography in establishing fire direction and velocity. In general, shrubs, trees, and other forest fuels will burn more readily and completely in hot, dry weather driven by wind, rather than in cool, moist weather or in still air (Martin 1990: 59). Likewise, fire travels faster and hotter moving uphill rather than down, covers the landscape more completely over rolling terrain rather than "skipping" across flatlands, and tends to burn drier fuels on sunny south slopes rather than moister fuels on shady north slopes (ibid: 58). Therefore, the topography of an area has a controlling effect on local fire behaviors and helps to define resulting patterns of burned and surviving vegetation. Topography includes slope (steepness of terrain), aspect (which direction the terrain faces), and elevation above sea level.

Figure 1.01 Fire Triangle: Topography, Weather, and Fuel (Zybach and Gruen 2003).

Map 1.02 shows the Coast Range as sloping steeply east and west from a central north-south ridgeline that runs nearly its entire length. It is bordered at sea level on its western edge by the Pacific Ocean; the Columbia River is subject to tidal influence along its entire northern border; the Willamette Valley floodplain extends for miles on a nearly flat plane along many portions of its eastern boundary; and the Middle Fork Coquille River travels through highly dissected rocky terrain on its southern boundary. The highest point of the Range is Marys Peak, near its center, where it reaches more than 4,000 feet elevation.

### 1.3.3 Rivers and counties

Throughout history, people have preferred to live near the mouths of rivers and creeks and to reside around bays and in valleys in the Oregon Coast Range (see Map 1.03) for the same reasons they have always done so: freshwater; ready transport; access to terrestrial, marine and riverine foods; strategic defense and trade; recreational opportunities; and aesthetics. Currently, most people who live in the Coast Range reside near the mouth of the Willamette River, near the

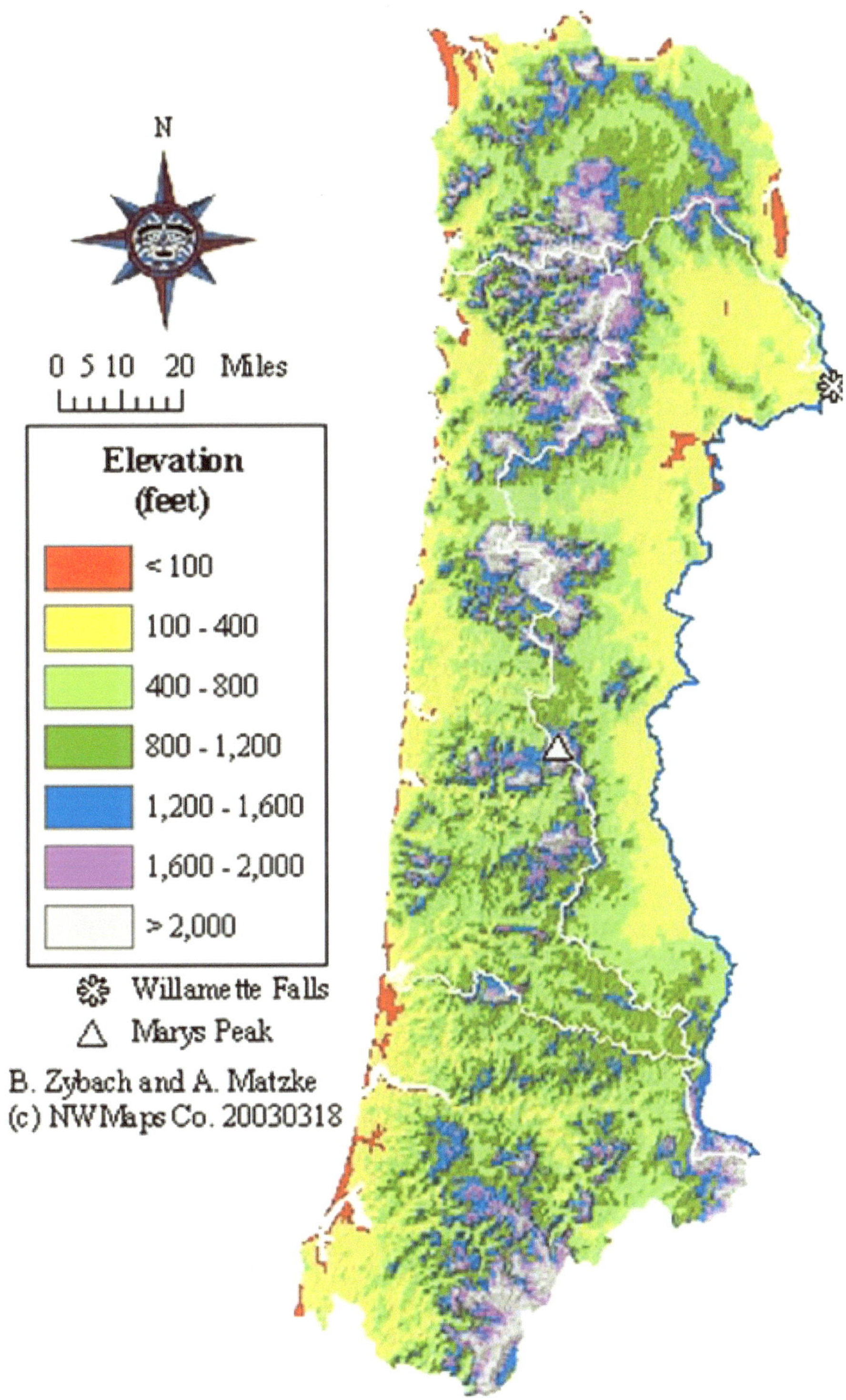


Map 1.02 Topography of the Oregon Coast Range.

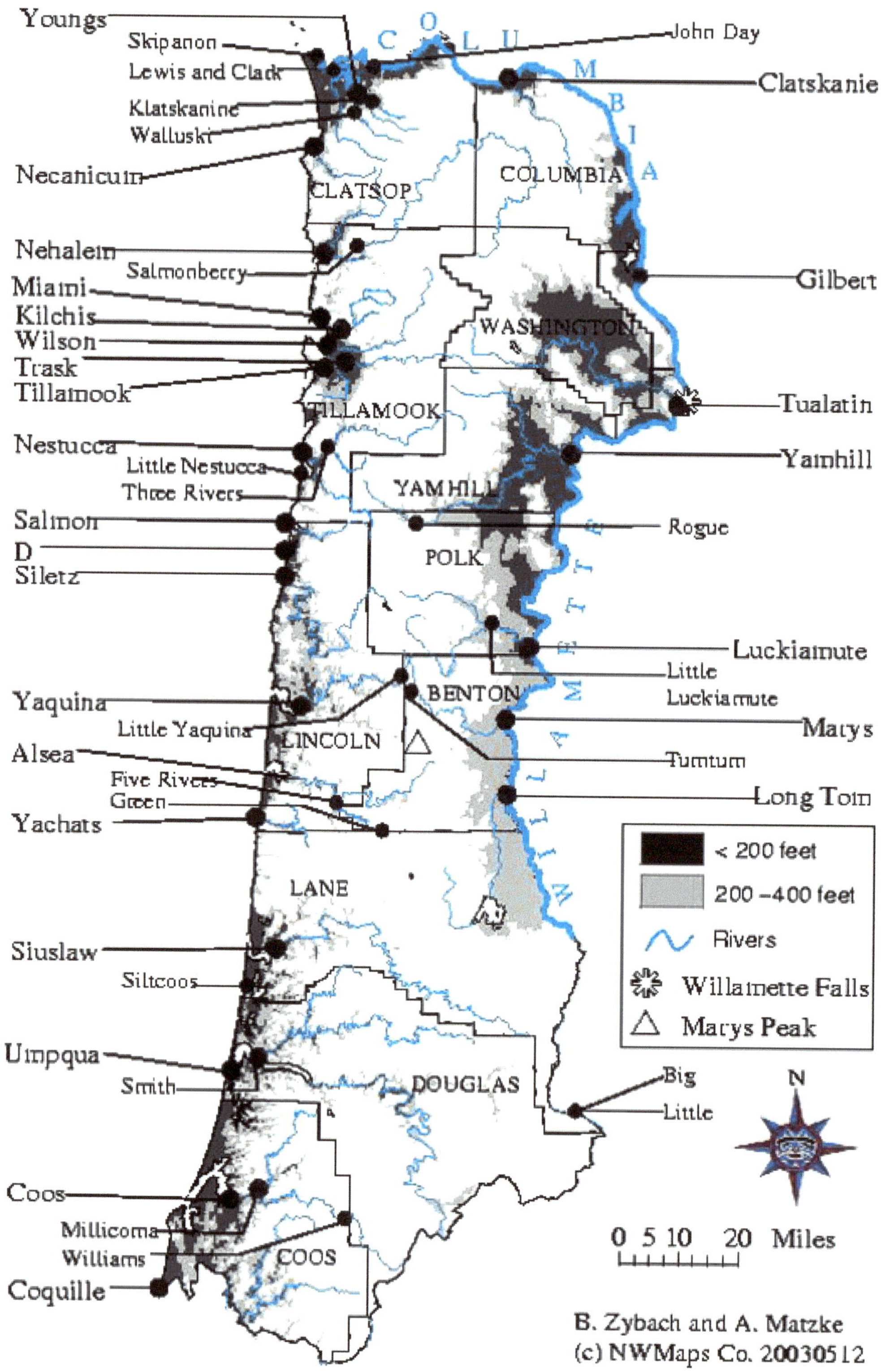


Map 1.03 Rivers and counties of the Oregon Coast Range.

mouths of its tributaries, near the mouth of the Columbia River, or along the bays and estuaries of coastal rivers. Early journalists, such as Haswell in 1788 (Elliott 1928), Lewis and Clark in 1805 and 1806 (Thwaites 1959), McLeod in 1826 (Davies 1961), and Wilkes in 1841 (Wilkes 1845), have reported similar distributions of people in early historical time; archaeologists (e.g., Strong 1960; Aikens 1975; Hall 2001) have found the same patterns for precontact time as well. Until recent times, when people began to heat their homes with electricity, steam, coal, gas, and oil instead of wood, and began to use automobiles instead of feeding herds of ungulates for food or transportation, areas of Coast Range settlement were quartered in grasslands, had relatively few trees, and most dead wood was found gathered or stacked for later use in ovens, stoves, fireplaces, or fire pits. The transition of forests from precontact Indian times continues to the present, with a legacy of older patterns still apparent in the landscape.

The fact that people become associated with the rivers and valleys they occupy is demonstrated by the names on the land of the Oregon Coast Range. Rivers entering the Pacific Ocean from the east are named Coquille, Coos, Umpqua, Siuslaw, Alsea, Yaquina, Nestucca, Tillamook, and Nehalem; after the precontact communities that lived along their bays and traveled their lengths by foot and canoe. Rivers that enter the Columbia from the south are named Clatsop, Klaskanie, and Clatskanie for the same reason. Rivers entering the Willamette from the west are named Long Tom, Luckiamute, Yamhill, and Tualatin, also for the same reason. Even the original historical name for the Willamette--Multnomah--was for a people that lived on Sauvies Island, near the river's mouth. It is reasonable to assume that most intensive firewood gathering, firewood use, trail development, and patch burning in precontact time took place near the mouths of these streams and along the low gradient areas most amenable to foot and canoe traffic. Early historical records (see Table 2.02) and known archaeological sites (e.g., Strong 1960; Aikens 1975; Hall 2001) also support this assumption.

## 1.4 Climate and Weather

The second side of the "fire behavior triangle," after topography, is weather. Weather is the combination of temperature, precipitation, humidity, airflow, and related phenomena (such as fog, cyclones and lightning) that is occurring now or at another given day or point in time (Taylor and Hatton 1999: xii, 7-37); climate is the combination of averages, seasonality, and extremes of these elements over time (Taylor and Hannan 1999: ix, 7-41).

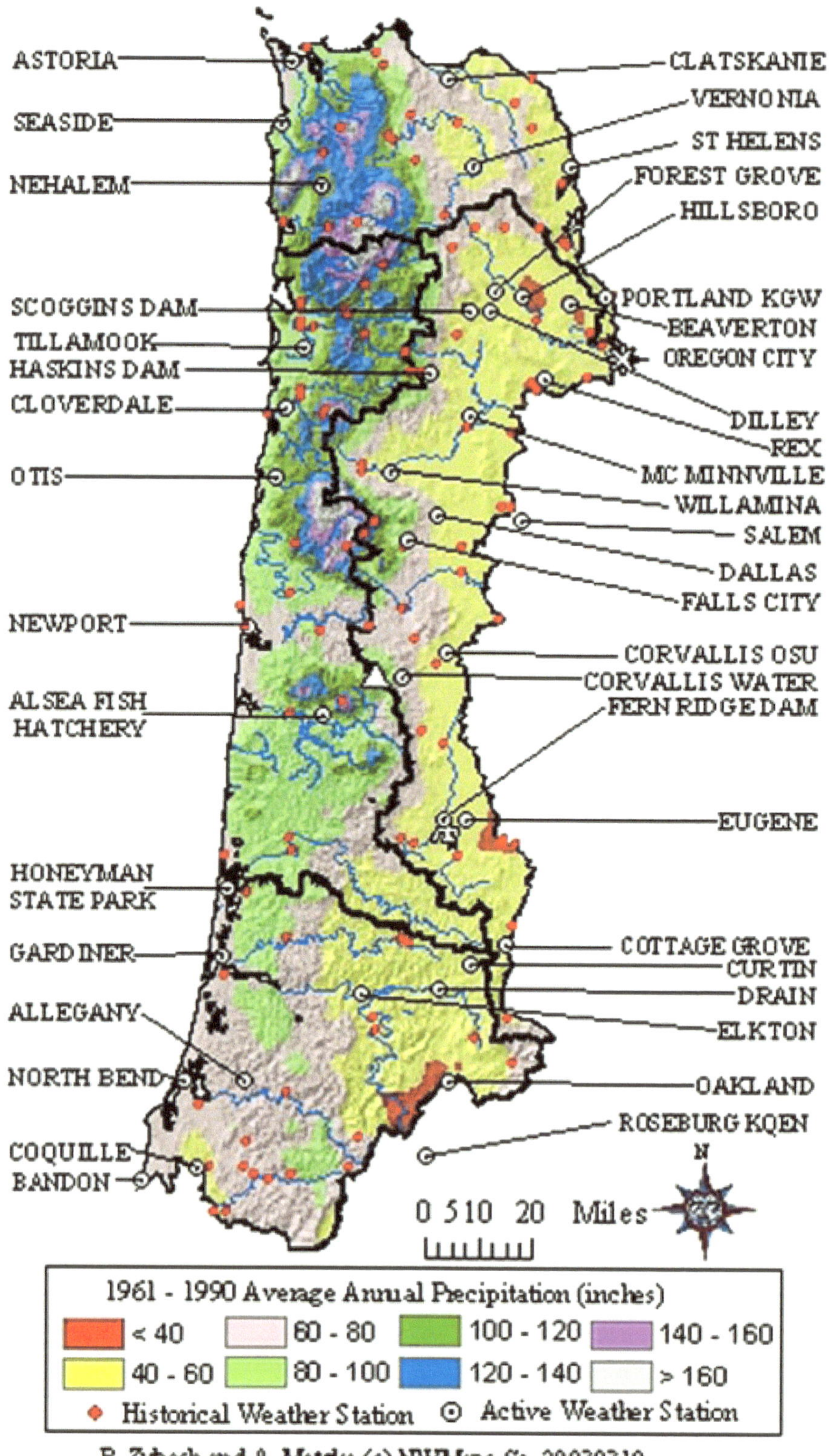


Map 1.04 Precipitation of the Oregon Coast Range, 1961-1990.

Oregon Coast Range climate is classified as a northern extension of the Mediterranean climate that characterizes coastal California, with similar seasonal distributions but cooler temperatures and a longer rainy season (Taylor 2003: personal communication). This means the year has two general seasons: a long, mild, wet winter and a warm, generally very dry summer. There is little snow most years, except for the highest peaks, and that is usually melted by late spring. Likewise, most years have few--if any--days that reach a 100° F. temperature (Redmond and Taylor 1997: 34). Most precipitation falls in the form of rain during the "wet" season from October to March, during which time most days are cloudy and moist (ibid: 28).

### 1.4.1 Precipitation and temperature

Map 1.04 shows average annual rainfall for the Oregon Coast Range for a 30-year period from 1961 to 1990. Note that most precipitation occurs toward the northernmost peaks and the highest elevations. The driest land is in the eastern Coast Range, and coastal beaches and tidelands have less measurable rainfall (but far more fog), than lands a few miles inland and a few hundred feet higher elevation. This coastal "fog belt" has significantly different vegetation patterns and wildlife species than other areas of the Coast Range, and is typically less prone to wildfire because of generally moist, humid conditions (Ruth and Harris 1979: 5).

Table 1.01 was assembled from weather data gathered between 1961 and 1991 at stations marked with white dots and their names on Map 1.04. Locations marked with red dots are former weather station sites for which historical records are available (e.g., Martin and Corbin, ca. 1931; Redmond 1985), but which do not gather data at the present time. In general, there is good conformance with seasonal climate patterns between historical records and more recent data. Heavy precipitation can affect forest vegetation patterns in direct and indirect ways: torrential rains, for example, can trigger landscape-changing floods and landslides (see figures 3.03 and 4.06; Benda 1990: 458; Benner and Sedell 1997: 37-42) and extended freezes and snowfalls can kill massive numbers of livestock that might otherwise browse and trample grasses, wood shrubs, and tree seedlings, thereby resulting in rapid forestation of abandoned pastures and grazing lands (Zybach 1999: 84-87). The snows of 1861-62 (Taylor and Hatton 1999: 118), 1881-82 (Lee and Jackson 1984: 42), and October 1936 (Starker 1939: 47) were all noted, among other things, for the vast amounts of livestock they killed on the ranges of the Willamette Valley and eastern Coast

Table 1.01 Oregon Coast Range seasonal climate markers, 1961-1990.

| **Map 1.04 Location** | **Killing Frosts** | | **Rain (7.0+"/mo.)** | | **Yearly Ave (")** | **Warm (>65°/mo.)** | | **Dry (<3'/mo.)** | | **Frost Free Days** |
|---|---|---|---|---|---|---|---|---|---|---|
| **NORTH** | ***First*** | ***Last*** | ***First*** | ***Last*** | | ***First*** | ***Last*** | ***First*** | ***Last*** | |
| Astoria | Dec. 11 | Mar. 3 | Nov. | Feb. | 66.42 | July | Sep. | June | Sep. | 283 |
| Nehalem | | | Oct. | Apr. | 121.74 | | | July | Aug. | |
| Oregon City | Dec. 6 | Feb. 21 | Dec. | Jan. | 47.06 | May | Sep. | May | Sep. | 290 |
| Portland KGW | Dec. 20 | Feb. 3 | Dec. | --- | 43.15 | May | Sep. | Apr. | Sep. | 320 |
| Seaside | Nov. 3 | Feb. 19 | Nov. | Mar. | 74.46 | June | Sep. | June | Aug. | 284 |
| St. Helens | Nov. 27 | Feb. 20 | --- | --- | 42.76 | May | Oct. | Apr. | Sep. | |
| **EAST** | ***First*** | ***Last*** | ***First*** | ***Last*** | | ***First*** | ***Last*** | ***First*** | ***Last*** | |
| Beaverton | Nov. 26 | Mar. 9 | --- | --- | 39.77 | May | Sep. | Apr. | Oct. | 262 |
| Corvallis | Nov. 26 | Mar. 8 | Dec. | --- | 42.67 | May | Sep. | Apr. | Sep. | 263 |
| Corvallis W.B. | Nov. 26 | Feb. 16 | Nov. | Mar. | 66.13 | May | Sep. | May | Sep. | 284 |
| Cottage Grove | Nov. 10 | Apr. 14 | Nov. | Dec. | 45.54 | May | Oct. | May | Sep. | 210 |
| Dallas | Nov. 12 | Mar. 23 | Nov. | Jan. | 48.42 | May | Oct. | Apr. | Sep. | 235 |
| Dilley | | | Dec. | Jan. | 44.2 | | | Apr. | Sep. | |
| Eugene WSO | | | Nov. | Jan. | 49.25 | May | Sep. | May | Sep. | 259 |
| Fern Ridge | Dec. 1 | Feb. 19 | Dec. | --- | 41.63 | May | Sep. | Apr. | Oct. | 286 |
| Forest Grove | Nov. 12 | Mar. 13 | Dec. | Jan. | 43.86 | May | Oct. | Apr. | Sep. | 245 |
| Haskins | | | Nov. | Mar. | 73.52 | | | May | Sep. | |
| Hillsboro | Nov. 16 | Mar. 6 | --- | --- | 37.57 | May | Sep. | Apr. | Oct. | 255 |
| N. Willamette | Nov. 14 | Mar. 13 | --- | --- | 40.78 | May | Sep. | Apr. | Oct. | 246 |
| Noti | | | Nov. | Mar. | 60.65 | | | May | Sep. | |
| Rex | | | Dec. | --- | 41.37 | | | Apr. | Sep. | |
| Salem WSO | Nov. 2 | Apr. 12 | --- | --- | 39.24 | May | Sep. | Apr. | Oct. | 204 |
| Willamina | | | Nov. | Jan. | 49.96 | | | May | Sep. | |
| **WEST** | ***First*** | ***Last*** | ***First*** | ***Last*** | | ***First*** | ***Last*** | ***First*** | ***Last*** | |
| Alsea | | | Nov. | Mar. | 92 | | | June | Aug. | |
| Cloverdale | Dec. 18 | Feb. 3 | Nov. | Mar. | 84.04 | June | Sep. | July | Aug. | 319 |
| Newport | Dec. 19 | Feb. 9 | Nov. | Mar. | 71.72 | Aug. | Sep. | July | Sep. | 314 |
| Otis | Dec. 12 | Feb. 6 | Oct. | Mar. | 97.27 | June | Sep. | July | Aug. | 310 |
| Summit | | | Nov. | Mar. | 68.86 | | | June | Sep. | |
| Tidewater | Dec. 12 | Feb. 9 | Nov. | Mar. | 91.42 | May | Oct. | June | Sep. | 307 |
| Tillamook | Nov. 8 | Apr. 6 | Oct. | Mar. | 88.65 | July | Sep. | July | Aug. | 216 |

| Map 1.04 Location | Killing Frosts | | Rain (7.0+"/mo.) | | Yearly Ave (") | Warm (>65°/mo.) | | Dry (<3'/mo) | | Frost Free Days |
|---|---|---|---|---|---|---|---|---|---|---|
| **SOUTH** | ***First*** | ***Last*** | ***First*** | ***Last*** | | ***First*** | ***Last*** | ***First*** | ***Last*** | |
| Bandon | Dec.15 | Feb. 9 | Nov. | Mar. | 58.91 | July | Sep. | May | Sep. | 310 |
| Coquille | Nov. 27 | Feb. 10 | | | | | | | | 292 |
| Curtin | | | Nov. | Dec. | 49.82 | | | May | Sep. | |
| Dora | Nov. 28 | Feb. 20 | | | | | | | | 283 |
| Drain | Dec. 4 | Mar. 9 | Nov. | Jan. | 46.17 | May | Oct. | May | Sep. | 270 |
| Elkton | Dec. 17 | Feb. 9 | Nov. | Jan. | 52.17 | May | Oct. | May | Sep. | 312 |
| Honeyman | Dec. 19 | Feb. 2 | Nov. | Mar. | 76.01 | June | Sep. | June | Sep. | 321 |
| Idleyld | Nov. 10 | Apr. 5 | Nov. | Mar. | 63.54 | May | Oct. | May | Sep. | 219 |
| North Bend | Dec. 28 | Jan. 29 | Nov. | Mar. | 63.48 | July | Sep. | May | Sep. | 334 |
| Riddle | Dec. 11 | Mar. 6 | --- | --- | 30.11 | May | Oct. | Apr. | Oct. | 281 |
| Roseburg | Nov. 25 | Mar. 1 | --- | --- | 32.44 | May | Oct. | Apr. | Oct. | 269 |
| Sutherlin | | | --- | --- | 41.13 | | | May | Oct. | |

Range (Oliphant 1932). Prairie lands that had been cleared and maintained by Indian fires in the 1830s had subsequently been kept clear of tree growth by the grazing cattle, hogs, horses, sheep, and goats of immigrant European American ranchers and farmers. There is evidence that reductions in grazing associated with the livestock mortality of those years resulted in the spreading of Douglas-fir to many former hillside pasturages (see Map 4.07; Zybach 1999: 84-87). Starker (1939: 47) also mentions "a similar wet snow about twenty years previous" to the 1936 event that apparently broke the tops and limbs off a number of trees in the Soap Creek area (see Map 2.01) and covered the ground with debris from the breakage. Weather events that result in the spread of conifer trees and other large forest fuels or that result in widespread mortality or breakage of living plants to produce dry fuels, are two ways extreme snows have affected fire and fuel conditions in the Coast Range during historical time. There is no evidence of snow-related fire or fuel patterns in prehistoric time, but their effects must have been episodic and of relatively minor consequence. The cumulative effects of these events and processes, however, have likely had a significant impact on the topography and vegetation patterns of the Coast Range, particularly at a basin or subbasin scale.

### 1.4.2 Wind and humidity

Coast Range winds are dominated by large-scale pressure patterns over the North Pacific and onshore. During winter (and, to a lesser extent, autumn and spring), frequent cyclonic storms reach the area from the west, greatly influencing winds and other weather elements. Summer months see fewer strong storms, and are more typically characterized by sea-land breeze regimes. During summer, Pacific storms tend to be diverted to the north. Typical summer winds along the coast are from the north. Map 1.05 illustrates the prevailing seasonal wind patterns of the Coast Range and episodic easterlies and hurricane-force winds from the south. Several times each year winds exceeding hurricane strength (74 mph) strike the Oregon coast. Damage is confined primarily to vegetation and structures and loss of life is rare. But occasional high intensity windstorms affect virtually all of Oregon (Taylor 2003: personal communication). The Columbus Day storm of 1962, for example, is the most destructive windstorm in Oregon's history. Gusts exceeded 140 mph in some parts of the state, and topped 100 mph in the Willamette Valley. During the course of this event, an estimated 11 billion feet of commercial timber were blown down (Lucia, ca. 1963: 50).

The presence of the Pacific Ocean, combined with generally mild regional temperatures, causes average relative humidity along the coast to be quite high, especially during the cool winter season. As one moves inland, humidities decrease. Table 1.02 are average relative humidities measured at 4:00 PM -- in general, the minimum values for the day -- at key locations along the Oregon coast and eastern Willamette Valley (Taylor 2003: personal communication). Note, in particular, the seasonal differences evident between the coast (Astoria in the north and North Bend in the south) and the Willamette Valley (Portland on the north and Eugene to the south) that occur from April through September. The seasonal droughts of the Pacific Northwest, including the Coast Range, have been noted as one of the primary factors favoring conifer forests over hardwood forests in the region (Franklin and Dyrness ca. 1988: 53-54). Many deciduous trees simply cannot survive long, dry summers without periodic irrigation. Graumlich (1987) analyzed tree rings to arrive at a 300-year precipitation pattern that identified specific years (1717, 1721, 1739, 1839, 1899, 1929, and 1973) and at least one decade per century (1790s, 1840s, 1860s, 1920s, and 1930s) of prolonged regional drought. The annual events don't seem to have a significant relationship to Coast Range fire history, but the prolonged droughts correlate closely with major forest fire events (see Chapter 4). Figure 1.02, the "Palmer Drought Index" for western Oregon (Taylor 2003: personal communication), conforms fairly well with Graumlich's regional findings--particularly for the prolonged drought of the 1920s and 30s--but doesn't indicate unusual conditions for the 1860s; a time of widespread catastrophic fires throughout the Oregon Coast Range and the entire Pacific Northwest.

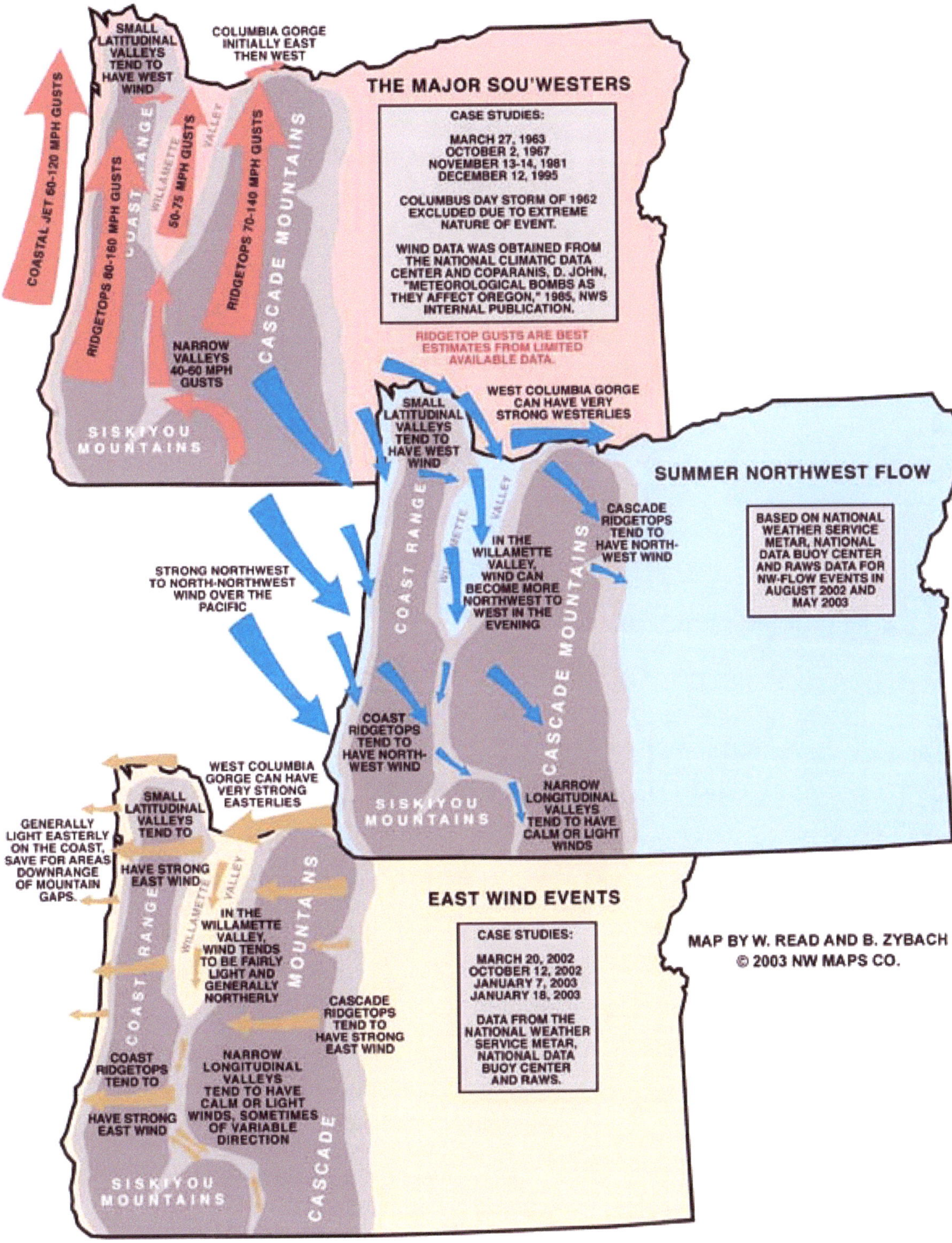


Map 1.05 Wind patterns of the Oregon Coast Range.

Table 1.02 Average relative humidity percentages at 4:00 PM.

| **Month** | **Astoria** | **Portland** | **Eugene** | **North Bend** |
|---|---|---|---|---|
| January | 78 | 77 | 81 | 80 |
| February | 76 | 69 | 73 | 79 |
| March | 69 | 61 | 66 | 74 |
| April | 70 | 55 | 55 | 73 |
| May | 69 | 54 | 53 | 73 |
| June | 71 | 49 | 49 | 74 |
| July | 69 | 46 | 37 | 73 |
| August | 71 | 48 | 39 | 74 |
| September | 70 | 50 | 43 | 75 |
| October | 74 | 66 | 65 | 79 |
| November | 78 | 74 | 78 | 80 |
| December | 81 | 79 | 83 | 82 |
| **Annual** | **73** | **61** | **60** | **76** |

### 1.4.3 Lightning and lightning caused fires

The Oregon Coast Range has remarkably few lightning strikes or storms compared with the remainder of the Pacific Northwest (Morris 1934a; Kirkpatrick 1939: 28) or the entire US (Towle 1974: 37), and the first reported lightning caused fire didn't occur in the region until 1927 (Kirkpatrick 1939: 31). This is an important consideration for this study because of the area's extensive history of forest fires coupled with the existence of numerous early historical oak savannahs, grassy prairies, meadows, brakes, and balds owing their existence to regular occurrences of fire (Keeley 2002: 304-305). These landscape features could have only been made and maintained through sustained and systematic Indian burning--no other cause can be identified. The relative absence of lightning in the Coast Range was recognized in early historical time. Clyman, for example, observed on June 4, 1845:

> Passed over some beautiful farming lands low grumbling thunder heard at a distance and I think this is the third time I have heard thunder in the Territory as thunder and Lightening is varry rare From what cause I cannot tell it may be possibly on account of the lowness of the clouds which rest on the mountains and in fact on the earth even in the vallies (Clyman 1984: 175-176)

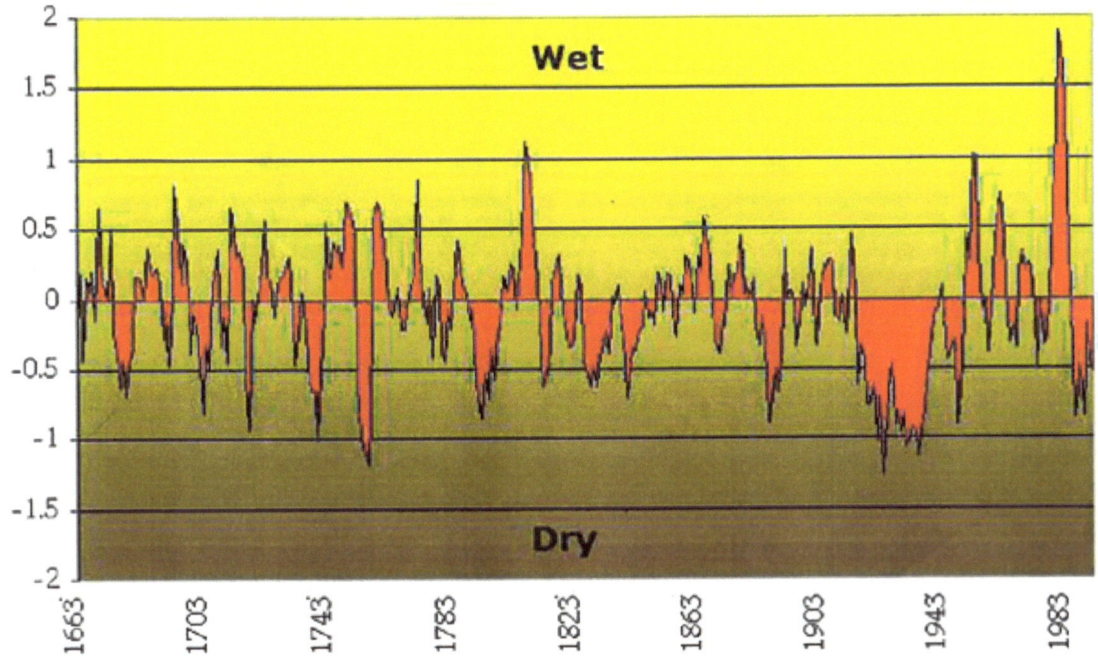


Figure 1.02 Palmer Drought Index, western Oregon, 1663-1983.

Long, et al (1998: 775) cite Agee (1993) when they state: "In the 19th and 20th century, logging activities have often provided an ignition source, but prior to that time fires in the Coast Range were probably started by lightning storms." Morris (1934a), however, depicts lightning strikes, storms, and fires as being extremely rare in the Coast Range from 1925 to 1931, when compared to other areas of the Pacific Northwest (see maps 1.06 and 1.07). Agee (1993: 28) agrees with Morris and cites evidence that these patterns remained similar from 1945 through 1966; the Oregon Department of Forestry shows a similar pattern of lightning strikes from 1992-1996 (Map 1.08).

Map 1.06 shows the locations of more than 2,600 lightning storms recorded over a seven-year period in Oregon, from 1925 to 1931. Note the pattern from southwest, in the 2002 Biscuit Fire area of the Klamath-Siskiyou Mountains, to the northeast, along the crest of the Cascade Range through the area of the 2003 B and B Complex Fire, and into Washington, thereby almost entirely missing the Coast Range.

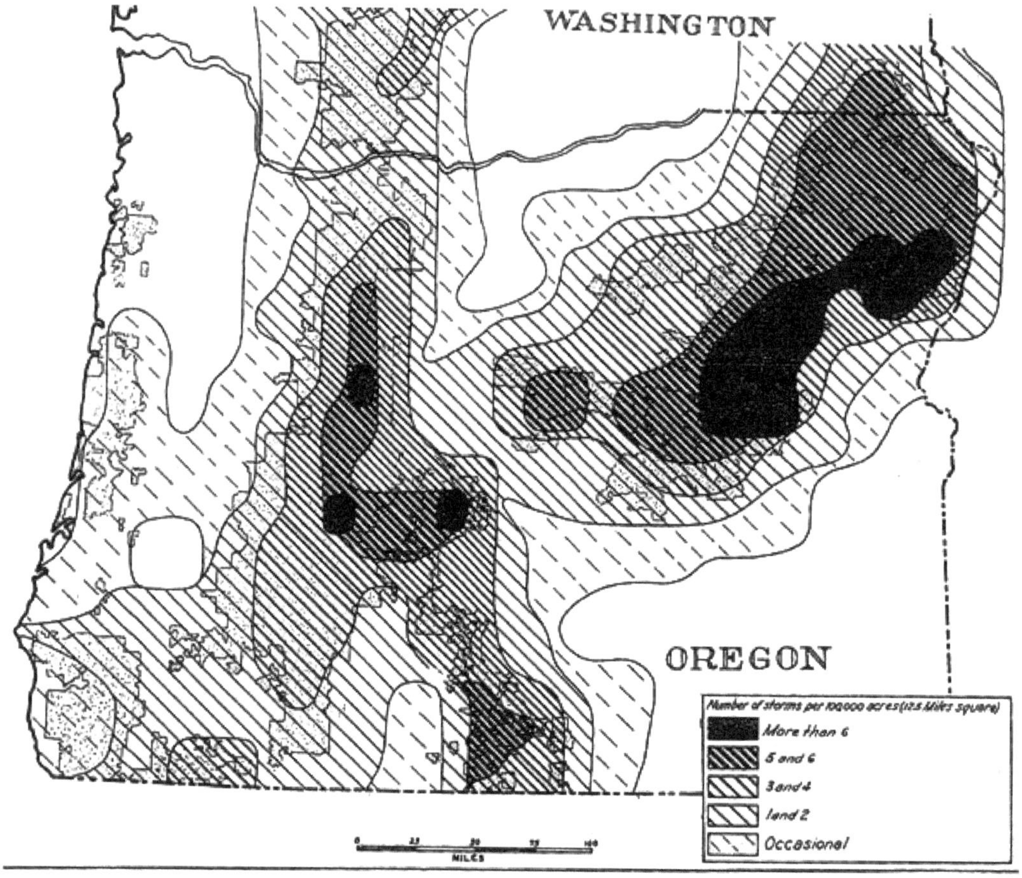


FIGURE 13.--Zones of average yearly lightning storm distribution in the vicinity c the national forests of Oregon and Washington as determined from more than 2600 storms reported by national forest fire lookouts during the 7-year period from 1925 to 1931.

Map 1.06 Lightning storms in Oregon, 1925-1931.

Map 1.07 shows the location of 5,300 lightning fires in Oregon mapped over the same seven-year period, from 1926 to 1931. Note the relatively large number of fires in the Klamath Siskiyou Mountains and along the crest of the Cascade Range, coupled with the almost complete absence of lightning caused fires during the same time period in the Coast Range.

Map 1.08 shows the same basic pattern of lightning activity in Oregon as maps 1.06 and 1.07, but for an entirely different time period: from 1992 to 1996. (It is important to note that these strikes are for the entire year, including the rainy season, and include lightning strikes associated with thunderstorms that typically produce so much rain they will extinguish -- rather than start -- wildfires; e.g., Crombie 2003: C1). Thus, thousands of reports of lightning activity in western

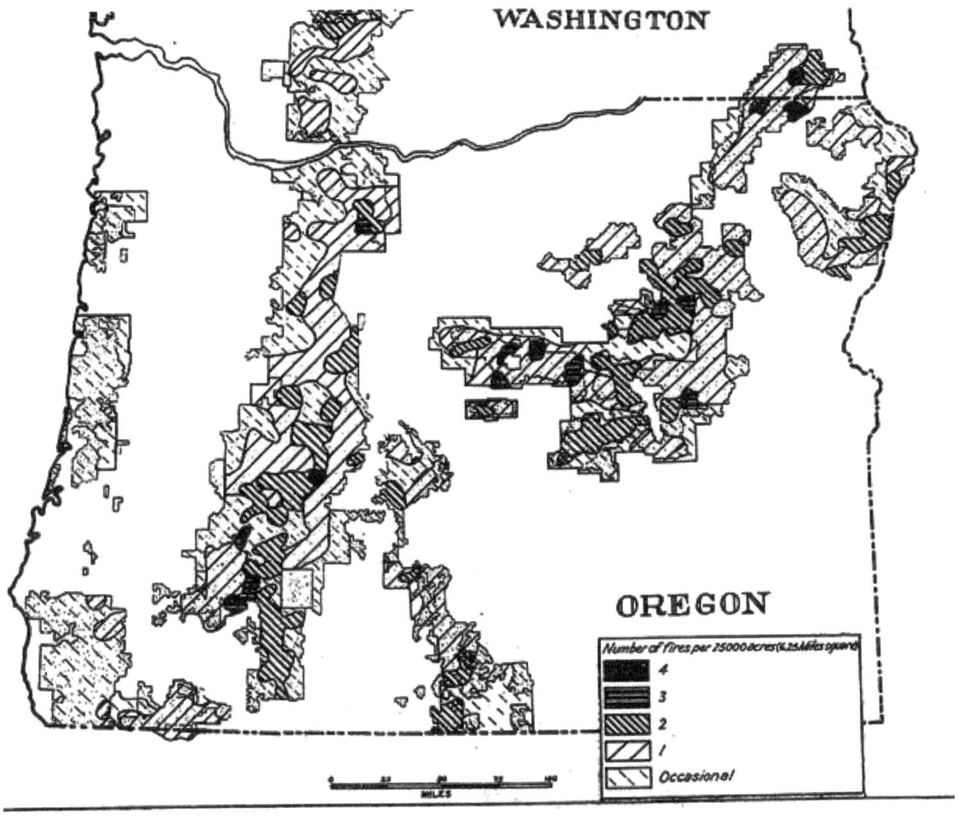


FIGURE 14.--Zones of average yearly lightning fire distribution on the national forests of Oregon and Washington obtained by plotting the locations of the 5300 lightning fires reported from 1925 to 1931.

Map 1.07 Lightning fires in Oregon, 1925-1931.

Oregon--whether strikes, storms, or fires--made over more than 150 years of time, come to the same conclusions: lightning strikes and storms are rare in the Oregon Coast Range, and lightning caused fires are not common or regular events.

### 1.4.4 Seasonal fire weather

There are two basic types of "fire weather": 1) the episodic and seasonal climatic conditions that make wildfire and prescribed burning possible (Pyne 1982: 314-317), and 2) the fire-induced weather patterns that occur during the course of a major fire. The National Weather Service maintains an office in Portland, Oregon that includes a "Fire Weather Program Manager" whose job, among other duties, is to provide a "five day probability forecast" for wildfire events during the late summer/early fall fire season (Taylor 2003: personal communication). Until recently,

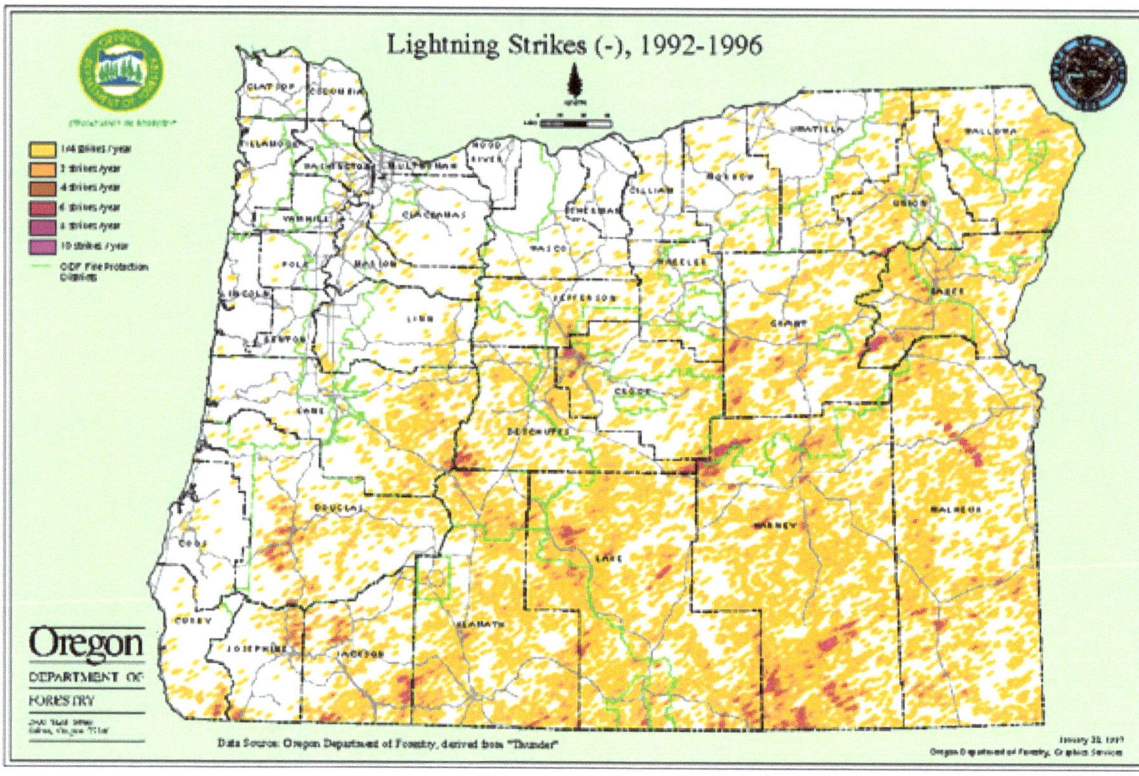


Map 1.08 Lightning strikes in Oregon, 1992-1996.

probability was based on four criteria: 1) one tenth of an inch or more rain measured from at least half of the weather stations within an area (low probability of fire); 2) any reported lightning; 3) low humidity (25% or less minimum relative humidity from at least half of the reporting stations); and 4) 15 mph or greater wind speed averages measured at 20 feet above the ground or canopy, whichever is appropriate, also recorded from at least half the stations. When at least two of the latter three conditions exist--lightning and wind, lightning and low humidity, or wind and low humidity--coupled with low rainfall, a "Red Flag Warning" condition was said to exist. Because lightning is rare in the Coast Range, and because westerly and southerly winds tend to carry large amounts of moisture, "fire weather" conditions for the Coast Range are typically most severe during east wind events occurring in the late summer or early fall (see Table 1.03). This pattern--late summer east winds, following drought--has been long recognized as typifying peak wildfire conditions. The August 1933 "Great Tillamook Fire" (see Chapter 4.1.6; Figure 4.07), for example, "the greatest forest fire ever fought in the Northwest," was blamed on seasonal fire weather conditions: "low relative humidities, fresh to strong easterly winds, and high temperatures were responsible for this huge fire" (Dague 1934: 227). Relative humidities were "generally low, 35 percent and lower" during much of the event, with temperatures occasionally reaching more than 100° F., with east wind velocities "generally fresh too strong in force (19 to 38 miles an hour), [and] probably of gale force (39 to 54 miles an hour) at times" (ibid: 227-229).

Although fire weather is usually associated with late summer and early fall drought, it can also occur during late winter and early spring "dry spells." These periods were favored for "fern burning" projects by Coast Range farmers and ranchers in the early 1900s and could be triggered "with but two or three windy, clear days" (Tillamook Headlight Herald 1947). By the early 1940s "fern burners" were viewed as a threat to Douglas-fir regeneration by the Oregon Department of Forestry (e.g., ODF 1941: 5) and the Keep Oregon Green Association began a "Don't Burn Fern" campaign, complete with posted signs and news releases (Hagenstein 2003: personal communication).

### 1.4.5 Fire-induced weather

Fire-induced weather effects have been noted for decades, but a terminology for these events has only recently been undertaken. Fire-induced weather effects can be said to take place when "fire storms"--"mass fires" that become a "holocaust [that is] a synergistic phenomenon of extreme burning conditions" and is also "stationary"--begin to produce "upper convective columns" (Pyne 1982: 24). Pyne traces the phrase "fire storm" to 1871, a term he uses to describe wildfires that include generalized descriptions of high winds and uncontrollable fire spread with unpredictable burn patterns (Pyne 1982: 204; Haines 1982). Fire storms (Pyne separates these from "conflagrations," which he defines as "mass fires" that would be "storms," except they "travel" rather than remain "stationary") can include the formation of "pyrocumulus clouds" (UCAR 2002: 1; see Figure 4.07), heavy winds (Pyne 1982: 204-205; UCAR 2002: 1), lightning (Latham 1991), and even thunderstorms (Taylor 2003: personal communication; UCAR 2002: 1; personal observations). Fire-generated winds are often the most severe fire-induced effect, and have been measured at speeds as great as 100 mph (UCAR 2002: 1). Winds generated by wildfire can also serve to spread the fire that causes them. This condition was described in the Oregon Department of Forestry's August, 1945 "Forest Log," published during the fire events of that year:

> Firefighters have been faced with conditions that makes it practically impossible to control the fire . . . This is particularly true of the old Tillamook burn area where previous fires have left vast areas covered with snags. The snags continually throw fire across the line in the slightest breeze . . . Drafts and thermals created by the fire in this concentration of fuel have been so great that snags three feet in diameter have been twisted off at the ground line. These same winds have carried the fire for miles ahead of the crews (ODF 1945: 5).

King (1964) describes a filmed "fire-induced tornado" in which:

> . . . a flame rose in the core to a height of 260 feet, that the core velocities were up to 205 m.p.h. vertically, at least 20-30 m.p.h. horizontally and 15-30 revolutions per minute rotationally, [and] that nearby parcels of air rose at up to 100 m.p.h. without spinning and that despite these conditions the ground winds were not strong enough to damage trees (King 1964: 1)

King's description of a fire-induced tornado (Peterson 2000: 8) would border between "severe" and "devastating" if applied to a normal tornado formation (Taylor and Hatton 1999: 137). Lesser whirlwind effects, known as "dust devils" (ibid: 227), can reasonably be termed "fire devils" if fire-induced. Fire devils is a term apparently coined by Zaffino (2003: personal communication) to publicly describe documented occurrences of this behavior during the California wildfires of October, 2003.

Radiant heat (Pyne 1982: 205), of course, and episodic cooling caused by smoke and clouds shading the sun during daylight hours are other fire-related phenomena affecting local weather conditions. Robock (1988) measured daytime temperatures for three weeks in a northern California valley filled with smoke from the 1987 Silver Complex fires in southwest Oregon. During the first week, daily maximum temperatures were nearly 30 degrees F. below normal as a result of shading, and during the second and third weeks temperatures were nearly 10 degrees F. below normal.

Given the inconsistency of terms related to fire-induced weather and the apparent need for such terminology to describe conditions that have become commonly documented via film, videotape, and eyewitness observations, the following terms and definitions are proposed:

> *Blow-up.* When a wildfire suddenly escalates in size, begins to generate its own weather, and transforms into a firestorm.

*Fire devils.* Small whirlwinds of fire that result from wildfire or prescribed fire conditions.

*Fire shadow.* The area of land and water shaded from the sun as a result of flame, smoke, and clouds formed by a wildfire or prescribed burn.

*Firestorm.* A wildfire forming upper convective columns that result in the formation of pyrocumulus clouds, fire-induced tornadoes, gales, or hurricane-force (greater than 75 mph) winds.

*Pyrocanes.* Sustained (measured in minutes) gale-force or hurricane-force winds that result from a major wildfire.

*Pyrocumulus.* Cumulus clouds formed as a result of wildfire or prescribed fire that may also produce lightning or rain.

*Pyronadoes.* Tornadoes formed as a result of wildfire or prescribed fire.

Because wildfires are generally seasonal in nature, due to periodic climatic conditions noted above, fire-induced weather conditions are also--by definition--generally seasonal. The use of these terms would help distinguish weather patterns that might result *in* wildfires or planned burns ("fire weather") from patterns that result *from* such fires ("fire-induced weather").

## 1.5 Vegetation and Fuel Patterns

The final side of the "fire behavior triangle," after topography and weather, is fuel. For forested areas, trees and shrubs form the bulk of available fuel. For grasslands, shrub lands, and prairies, grasses, ferns, and brush species comprise most available fuel. Map 1.09 shows the north-south distribution patterns of the principal wildfire fuel types of the Oregon Coast Range. Three types of north-south patterns dominate: along the coastal "fog belt," lodgepole pine, Sitka spruce, western hemlock, and redcedar are the primary conifer trees (Hansen 1944: 629; Ruth 1954: 1-3)) and red alder and bigleaf maple are the major broadleaf species; the eastern Coast Range is dominated by the cities, towns, and farms of the Willamette and Umpqua river lowlands, with native and exotic grasses, ferns, white oak, black cottonwood, and Oregon ash forming most of the available fuel in unmanaged areas; the remaining central part of the Range is dominated by some of the largest and fastest growing mixed and "pure" stands of Douglas-fir in the world.

The total area of the Coast Range is, according to GIS computations, nearly 7.2 million acres in size. As shown on Map 1.09, about 44,000 acres (less than 1%) is sand or rock; 105,000 acres is

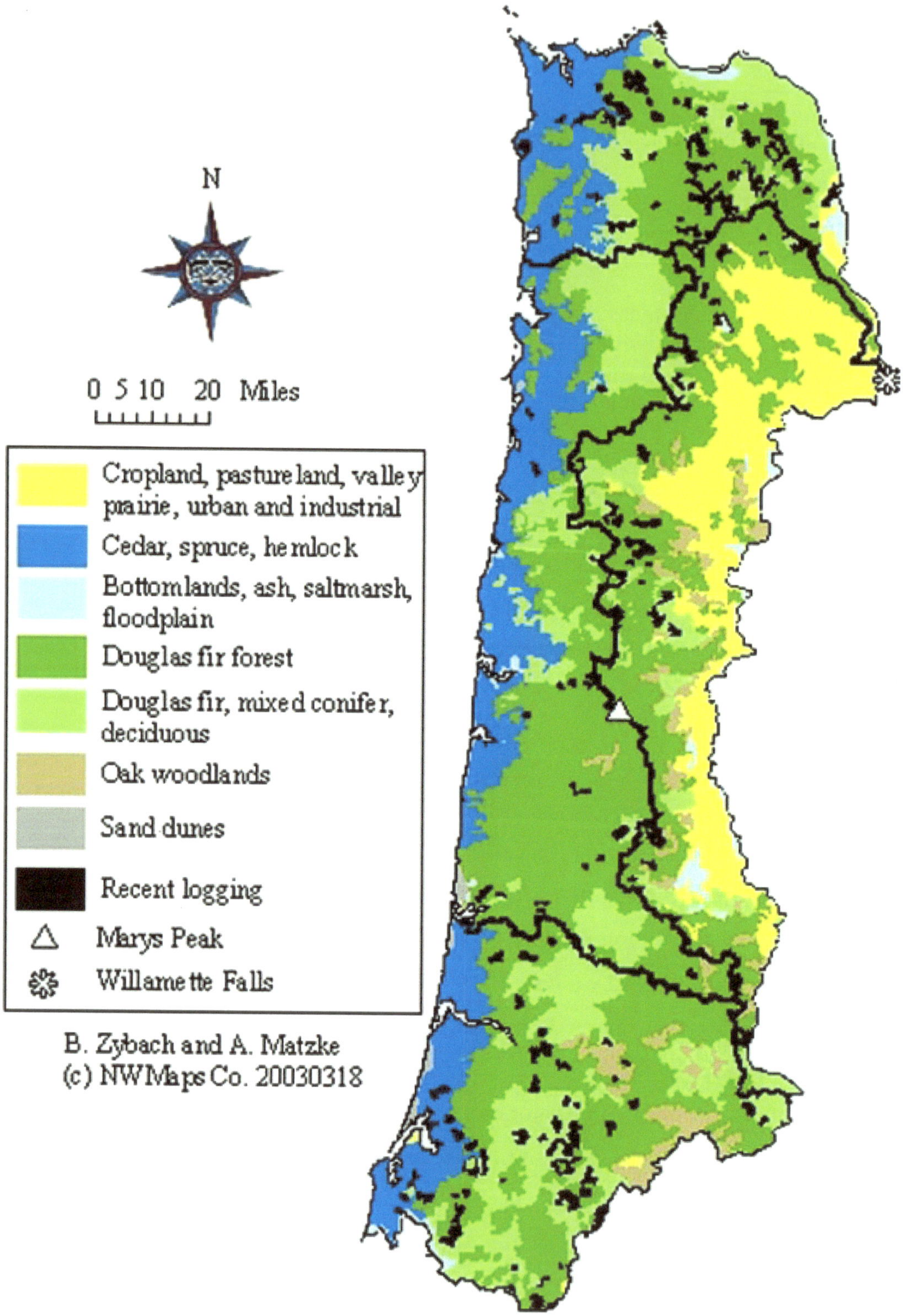


Map 1.09 Current vegetation patterns of the Oregon Coast Range.

wetlands; 221,000 acres are deciduous woodlands; and 908,000 acres are urban or agricultural developments. The rest is forest; about 5.876 million acres, or over 80% of the total area. Of the forested amount, a little more than one million acres is fog belt forest and the remainder is Douglas-fir: 3.223 million acres of nearly pure Douglas-fir stands; 1.403 million acres of stands also containing a mix of other conifer species or hardwoods, and about 204,000 acres "recently logged."

### 1.5.1 Coastal fog belt

The dominant tree species in the coastal fog belt is Sitka spruce, whose "range is limited to a relatively narrow strip along the ocean, characterized by mild winters, cool summers, and abundant moisture" because it "requires abundant moisture throughout the year and will not tolerate prolonged summer drought" (Ruth and Harris 1979: 17-18). Although summer rainfall is slight along the coast (see Chapter 1.4.1), fog is common and spruce is able to obtain sufficient moisture from it:

> During the summer fog commonly forms, covering coastal lowlands and following river drainages . . . As daytime temperatures rise fog is dissipated, often by midmorning in interior areas. It may remain throughout the day along the coast. Fog collects on tree crowns and drips to the ground, adding as much as 25 percent to the total precipitation reaching the ground . . . This "fog drip" is an important factor in reducing fire danger and providing soil moisture during the growing season (Ruth and Harris 1979: 4-5).

Figure 1.03 (Ruth and Harris 1979: 33) illustrates the density and rapidity with which Coast Range fog belt conifers can produce fuels following a major disturbance: whether the disturbance is by fire, logging, windstorm, or other means. Figure 1.04 shows the great size the same species can attain if left undisturbed for a few hundred years. The combination of these two figures also indicate the greater unlikelihood of fire--and the possible greater likelihood of logging--that can accompany such stands over time. Because these fog belt forests exist in a damp and cool environment dominated by the Pacific Ocean and moist, westerly winds, they experience far fewer wildfires than their Douglas-fir neighbors to the east.

*Figure 19.--Natural regeneration of western hemlock-Sitka spruce is often too dense; 10 years after clearcutting, this stand contained approximately 15,000 trees per acre (37,000/ha).*

Figure 1.03 Sitka spruce regeneration: 15,000 trees/acre at 10 years.

### 1.5.2 Douglas-fir forestland

The Douglas-fir forest stretches the entire length of the Coast Range through some of the wettest and steepest lands in the region; from the Middle Fork Coquille on the south to the Columbia River on the north (Snavely and MacLeod 1981; Heilman et al 1981). This forest pattern, combined with seasonal drought and east winds, contains all of the critical elements needed for catastrophic wildfires: a nearly unbroken canopy of massive, living and dead pitchy fuels, millions of acres of rolling and sloping topography, and regular seasonal fire weather patterns, characterized by seasonal droughts, often accompanied by low humidity, warm temperatures, and dry, east winds. In fact, this belt of Douglas-fir is the principal source of fuels for each of the Great Fires (see Chapter 4). Figure 1.05 (Franklin and Dyrness ca. 1988: 87) shows an even-aged 66-year old stand of Douglas-fir near Cottage Grove. Note the general lack of understory vegetation, lower tree limbs, or evidence of an earlier forest.

Figure 1.04 Sitka spruce old-growth, near Waldport, OR, 1923.

Figure 1.05 Douglas-fir 2nd-growth, near Cottage Grove, OR, 66 years.

The lack of older stumps, snags, and logs may indicate that this area was an old field or meadow that seeded in to trees, or that the site had most evidence of an earlier forest removed in preparation for the new stand. The close proximity of the trees makes a stand-replacement crown fire a possibility, but the lack of ladder fuels leading to the crown, makes starting such a fire difficult. The even-aged nature of the stand is typical for the species and for the region. This condition has been attributed to a history of regular disturbance that may be directly related to precontact Indian burning fires (Stewart 2002: 250-255; Moravets 1932).

Grigg and Whitlock (1998: 297) date an "increase in fire frequency" to between 14,850 BP and 14,500 BP for Little Lake, in Lane County; an occurrence that coincides with the "establishment of *Pseudotsuga*" in the area (ibid: 297), and that may also coincidentally indicate the arrival or local settlement of people (Pyne 1982). Hansen's study of a peat bog in nearly the same area of the Coast Range described by Grigg and Whitlock, encompassed as much as 13,000 years' time to the present, and led him to conclude:

> The pollen profiles of this study, however, tend to support the theory that the persistence of Douglas fir as the chief dominant has been largely a result of extensive fires occurring at relatively brief intervals (Hansen 1941: 207).

Hansen's theory is consistent with the findings and thoughts of other noted scientists in the Douglas-fir Region. Munger, for example, made the following conjecture:

> It is somewhat an anomaly that over much of the Douglas-fir region—southern British Columbia, western Washington, and western Oregon—Douglas fir (*Pseudotsuga taxifolia*) is not the climax type. Forests of Douglas fir which now cover millions of acres would not persist as pure stands even if spared by fire and axe, but would make a gradual metamorphosis to forests of the shade-enduring species like western hemlock (*Tsuga heterophylla*), western red cedar (*Thuja plicata*), and balsam firs (*Abies* spp.) (Munger 1940: 451).

Climax vegetation has also been called "potential vegetation," and considers an ultimate, stable plant community that would likely develop in an environment absent of disturbance:

> If all disturbance stopped, plants that had colonized disturbed sites would be replaced by later seral species until climax communities prevailed everywhere. Succession would cease.
>
> There is no great risk of this happening. Disturbance happens on all spatial and temporal scales . . . Because disturbances vary so much in space and time, natural landscapes are not uniform climax communities. Rather, they more closely resemble quilts or mosaics whose irregular-shaped patches are communities with differing structure, species composition, and ecological functioning (Norse 1990: 40-41).

The "mosaic" of Douglas-fir stands throughout the vast forest of the Oregon Coast Range is one of "age-classes" (Braun 1950: 15; Andrews and Cowlin 1940), in which most trees are of about the same (or "even") age, rather than one of "climax" species or "communities" (FEMAT 1993). Even-aged stands typically reflect patterns of regular disturbance. During historical time, disturbance is often related to logging; in precontact time it was most likely the result of regular Indian burning practices.

### 1.5.3 Oak savannah grasslands

The plants, animals, people, and human constructs of the Willamette Valley dominate the eastern slope of the Coast Range. In late precontact time the Valley was maintained largely as a grassy oak savannah by local Kalapuyan families (Habeck 1961; Johannessen et al 1971; Boyd 1986). Now it is the most heavily populated area of the State, and is dominated by farms, ranches, cities and towns. There is no record of a catastrophic forest fire taking place in the eastern Coast Range during precontact or historical times for the simple reason that the land has generally contained too few trees to constitute a forest, much less to fuel a forest fire. Thus, the "cultural legacy" of the Kalapuyans and their predecessors is a vast protein-rich landscape of diverse plants and animals that has been virtually "fire-proofed" of catastrophic wildfires for hundreds or thousands of years.

## 1.6 Discussion and Summary: Subregions and Seasons

This dissertation examines the relationship between patterns of purposeful human burning practices and catastrophic forest wildfires on the Oregon Coast Range. The three principal "fire triangle" components of wildfire and controlled burns are fuel, weather, and topography. All else that is needed is a source of ignition. The Coast Range has abundant forest and grassland fuel, regular seasonal droughts and wind patterns that are ideal for landscape-scale burning practices and wildfire events, and a varied topography that includes large expanses each of flat, sloping, steep, and heavily dissected terrain. Every hour, day, and year for thousands of years, people have started or maintained thousands of fires in a systematic arrangement that borders and crosses the entire Coast Range landscape; typically in patterns that reflect the spatial demographics of the population at any given point in time.

From the late 1400s until the late 1840s, and for a period of time likely extending more than 10,000 years before then, the Coast Range was populated by a wide variety and large number of American Indian cultures, all of whom used fire skillfully on a daily and constant basis. The history of forest wildfires for this period is incomplete, but a great deal can be inferred by the trails systems, expansive prairies, oak savannahs, berry patches, and other cultural legacy left behind by these people.

From 1849 through 1951, a series of catastrophic forest fires took place in the Coast Range that transformed hundreds of thousands of forested acres to charred snags. Wildlife habitat was often burned at a rate of tens of thousands of acres a day during the course of these Great Fires. (From 1952 until 2003, only a few Coast Range fires occurred that could be termed "large," and none approached 100,000 acres in size.)

The combination of wildfires and human burning practices has helped to create the forested landscapes of today. The temporal division between Indian burning practices and historical wildfire events allows for a comparison of the effects each of these fire-related disturbances has had in establishing and maintaining native plant and animal habitat patterns that persist today.

### 1.6.1 Oregon Coast Range subregions

This thesis addresses the question of whether the use of fire across the landscape by Indian people over the span of 350+ years prior to 1849, had any relationship to a subsequent series of catastrophic forest fires, that lasted until 1951. Specifically, were there differences or similarities in the seasonal timing of large-scale fires? Common or differing causes and locations of ignition? Common or differing burn boundaries? Different or similar types and amount of fuels consumed?

To better answer this question, particularly on a spatial basis, the entire region was divided into four subregions: North, East, West, and South for comparison. Each has distinct drainage patterns, weather patterns, vegetation patterns, and human use and settlement patterns; both in precontact and historical time. Subregional similarities and differences in Coast Range topography, fuel loads, climate, and human population densities (sources of ignition) are summarized in Table 1.03.

This table identifies similarities and differences among the four subregions. Perhaps the two most dissimilar subregions are North (the smallest subregion) and East (the largest); also, coincidentally, the two most heavily and densely populated areas. North is dominated by steep, broken, heavily dissected terrain, while East is mostly flat or rolling agricultural lands; North is heavily forested with spruce and hemlock along the coast and mouth of the Columbia, and Douglas-fir almost everywhere else, while East is mostly urban subdivisions or grassland, with scattered woodlands and patches and high elevation belts of conifer forest; North has the shortest

Table 1.03 Oregon Coast Range subregional comparisons.

| Subregion | North | East | West | South |
|---|---|---|---|---|
| Acreage | 4-1,200,000 | 1-2,068,000 | 2-2,045,000 | 3-1,851,000 |
| Topography | Steep/ broken | Flat/ sloping | Steep/ sloping | Mixed |
| Demographics | Urban/ forestry | Urban/ agriculture | Coastal/ forestry | Mixed |
| Population | 1-Portland | 2-Eugene | 3-Newport | 4-Coos Bay |
| Vegetation (fuel) | Douglas-fir/ fog belt | Grassland/ Douglas-fir | Douglas-fir/ fog belt | Douglas-fir/ grassland |
| Fuels (ton/A.) | 1-Heavy | 4-Light | 2-Heavy | 3-Mixed |
| Killing Frosts | Nov3-Mar3 | Nov10-Apr14 | Nov8-Apr6 | Nov10-Apr5 |
| Difference | 1-121 days | 4-156 days | 3-150 days | 2-147 days |
| Frost-free | 1-301 days | 4-245 days | 3-267 days | 2-276 days |
| Rainfall | 1-122/43 in. | 4-74/38 in. | 2-97/69 in. | 3-76/30 in. |
| Difference | 1-79 in. | 3-36 in. | 4-28 in. | 2-46 in. |
| Dry | 3-Apr.-Sep. | 1-Apr.-Oct. | 4-Jun.-Sep. | 1-Apr.-Oct. |
| Warm | May-Oct. | May-Oct. | May-Oct. | May-Oct. |

period of killing frosts and the longest frost-free growing season; East has the longest period of killing frosts and the shortest growing season; North has the heaviest rainfall, and the greatest variation of annual amounts of precipitation, East has the least rainfall and the third least variation of precipitation reports. West and South attributes vary between the extremes of North and East.

North is dominated by the Columbia River, the Willamette River from Willamette Falls to its mouth, the heavily dissected terrain of the Nehalem River basin, and Clatsop Spit. Of the four Coast Range subregions, it is the most heavily influenced by an urban population (Portland), and the area with the greatest extremes in rainfall. It is mostly forested with conifers: spruce, hemlock, cedar and shorepine on the western edge, and Douglas-fir dominated forests throughout the remainder, with oak appearing in groves and scatterings as far west as Oak Point, at Fanny's Bottom, an old Klaskani town site (Ross 1986: 117). Annual spring floods, caused by the melting snows of Rocky Mountain and Cascade Range glaciers and snow pack, regularly inundate and erode the banks of the Columbia. Daily tides reach the entire northern length of the Coast Range, from Astoria to Portland, and extend along its entire western, Pacific Ocean, boundary. Its southern boundary is dominated by Saddle Mountain, one of the wettest areas of the Coast Range; its eastern boundary is dominated by Sauvies Island, one of the driest (rainfall) and most regularly flooded areas of the Range.

East is dominated by the floodplains and upland prairies of the Willamette River, with oak, cottonwood, willow, and other deciduous trees being the most widespread species. Douglas-fir and some true fir forests dominate the western peaks and valleys of the subregion; which are entirely drained by a north-south series of west to east flowing rivers that empty into the Willamette River, which forms the eastern boundary of the region. Elevation ranges from the grassy balds of Marys Peak, at over 4,000 feet elevation, to Willamette Falls, whose pool is not too much above tidewater. East is the driest, flattest, and most generally low in elevation (below 400 feet) of the subregions; it also has the least amount of conifer forest and some of the poorest soils and weather for Douglas-fir and hemlock.

West is dominated by the Pacific Ocean along its entire western boundary; Marys Peak along its common ridgeline boundary with the eastern subregion; Tillamook Bay to the north and Table Mountain to the south. Tideland is usually either rocky or sandy, often lined with a thin strip of shorepine, usually only a few hundred yards wide and rarely a mile wide. To the immediate east of the shorepine is another north-south strip of spruce-hemlock forest that extends as far inland as the fog belt climate it requires for dominance. Inland from the coastal fog belt forest, and extending throughout the remainder of the subregion, is a great north-south swath of Douglas-fir forest that is the setting for some of the fastest growing conifer forests in the world and some of the largest and most memorable forest fires in history. Grassy balds, brackenfern prairies, and occasional stands of true fir trees often characterize the highest peaks and ridgelines.

South is defined by the mainstem Umpqua River Valley, the coastal cranberry bogs, and the Port Orford and myrtlewood forests of the Coos and Coquille river basins. In the early 1850s Coos Bay became the scene of some of the first extensive clearcutting and sawmilling in the Coast Range. Those industries were undertaken to help satisfy the rebuilding of San Francisco and the regional expansion of gold mining. In 1868, the Coos Fire consumed more than 100,000 acres of prime timberland in just a few weeks' time. South is dryer, warmer, and lower in elevation than West or North, but wetter, cooler, and generally higher in elevation than East. The Umpqua Valley area was inhabited by Kalapuyan people who maintained an oak savannah similar to the savannah maintained by other Kalapuyans in the Willamette Valley. Black oak and tanoak are more common in the southern region, though, while white oak is the most widespread tree species in the East; and no species of oak is common throughout most of the western and northern subregions.

### 1.6.2 Oregon Coast Range climatic seasons

Table 1.04 summarizes the weather patterns shown in tables 1.01 and 1.02. It uses seasonal indicators of temperature and precipitation derived from the 1961-1991 time period (Taylor and Hannan 1999) to identify the principal months and seasons in which landscape burning and wildfires are most likely to occur. The rows in Table 1.04 are arranged by month, so seasonal designations are constrained by a lack of specific dates, such as solstices, equinoxes, or lunar cycles. Temporal patterns are further modified from the four standard three-month periods of spring, summer, winter, and fall to better reflect local "wet and dry" weather patterns and to identify key "transitional periods" (when weather was most likely to fluctuate from one seasonal condition to another). The seasonal indicators selected to identify recurring weather patterns were: 1) recorded temperatures of 28° F. or less ("killing frosts"), to signal the first day of fall and the last day of winter; 2) a monthly average of seven inches or more rain to signal the beginning and ending months of the "rainy season"; 3) a monthly average of three inches or less rain to signal the beginning and ending months of the "dry season"; 4) a monthly average of 65° F. or more for daily high temperature, to signal the beginning and ending months of summer.

Table 1.04 Oregon Coast Range seasonal climate patterns.

| Month | Season | Weather | Temperature |
|---|---|---|---|
| January | Winter | Wet | Freezes |
| February | Winter | Wet | Freezes |
| March | Spring | Wet | Freezes |
| April | Spring | Mixed | Warming |
| *May* | *Transition* | *Mixed* | *Warming* |
| June | Summer | Dry | Warm |
| July | Summer | Dry | Warmest |
| August | Late Summer | Dry | Warmest |
| September | Late Summer | Dry | Warm |
| *October* | *Transition* | *Mixed* | *Cooling* |
| November | Fall | Wet | Freezes |
| December | Fall | Wet | Freezes |

An analysis of Table 1.04 shows the year can be divided into two or four general seasons in the Coast Range, or seven fairly specific seasons (winter, spring, spring transition, early summer, late summer, fall, and fall transition). The two general seasons are typically Mediterranean: wet (November to April) and dry (May to October). The wet season, however, is punctuated with occasional killing frosts and freezes, while the dry season includes the warmest months of the year.

The year can also be divided into four general seasons that have a bearing on fire patterns: 1) a warm, wet spring/early summer of low fire hazard due to air moisture, soil moisture, and fast growing trees pulling massive amounts of water from the ground and transferring it to the air through transpiration; 2) a warmer, dry late summer/early fall with moisture stressed plants, dry soils, desiccated grasslands with high fire hazard potential, especially on an east wind; 3) heavy fall rains, knocking desiccated plants and leaves to the ground, filling rivers and creeks with water and fish, with low wildfire potential; and 4) killing frosts, desiccating brackenfern prairies and woodland shrubs and trees, mostly wet or frozen fuels, but fires can spread on occasional east winds, particularly through brakes, across unburned balds, brush piles, and south-sloping berry patches that have had a few days to dry.

# 2. METHODOLOGY

*He had bought a large map representing the sea,*
*Without the least vestige of land:*
*And the crew were much pleased when they found it to be*
*A map they could all understand.*
-Lewis Carroll (1876)

The basic premise of this thesis is to compare changing landscape patterns over time, in order to test the hypothesis that traditional Indian burning practices contributed to subsequent size, location, seasonality and severity of historical Oregon Coast Range forest fires. Maps and tables created from Geographic Information Systems (GIS) software are the primary formats used to compare spatial and temporal patterns. Figures and text are used to illustrate and describe these patterns in context to the hypothesis and to consider such patterns at a finer, more local scale.

Standard archival and anthropological research methods were used to obtain early surveys, maps, drawings, photographs, transcribed interviews, GIS inventories, eyewitness accounts, and other sources of evidence that document fire history for the Oregon Coast Range. Focus was on spatial and temporal patterns of Indian burning across the landscape from 1491 until 1848, and corresponding patterns of catastrophic fire events from 1849 until 1951. Data were inventoried, tabulated, digitized, and mapped as GIS layers for comparison.

The research in this thesis was performed in accordance with the "method of multiple working hypotheses," first described by Chamberlin in 1890 (Chamberlin 1965): "the effort is to bring up into view every rational explanation of new phenomena, and to develop every tangible hypothesis respecting their cause and history." Two key features of this methodology are that research questions are answered based on the "weight of evidence" from multiple sources, and research findings include the formal identification of additional "tangible" questions as they present themselves. Chapters 5 and 6 include a series of such questions developed through this research. The preponderance of available tangible evidence is used to select likely answers, given the likelihood that such answers are intended to change or be modified as more information is made available.

This chapter identifies study boundaries, describes the types of information used for research, and demonstrates the methods used to construct and compare GIS layers, maps and tables in order to test the study hypothesis. It is composed of three parts. The first part describes principal exterior and subregional boundaries of the study area, including brief descriptions of four research focal points, and provides references to key historical maps of those areas. Part 2.2 is a listing and general description of the types of information that were used to conduct research, and shows how data were organized into GIS layers, maps and tables for purposes of display and analysis. Part 2.3 defines and compares the three basic spatial scales (region, basin, local) used for GIS map displays of fire and vegetation patterns for given points in time.

## 2.1 Research Boundaries and Focal Points

The Oregon Coast Range is bounded by the Pacific Ocean on the west, the Columbia River on the north, the Willamette and Umpqua rivers on the east, and the middle fork of the Coquille River on the south (Orr, et al 1992: 167). For purposes of this study, the Coast Range was divided into four subregions: labeled North, East, West, and South (see Map 1.01). The primary reasons for these divisions were to create a finer scale to display and consider research data, and because significant differences were found to exist between each of the subregions. Chapter 1.6.1 identified subregional differences in drainage patterns, topography, weather, and vegetation types (see Table 1.03). Chapters 3 and 4 will show that subregional scale differences also exist for precontact cultural landscape patterns, early historical land use patterns, and historic catastrophic forest fire patterns.

Map 2.01 is an index for the entire study. In addition to the four subregions, locations for various areas and landmarks will be in accordance with standard survey terminology: township, range, and section (see Appendix D), which are designated by the numbering grid shown on Map 2.01. Shaded townships show those areas that have been mapped from original land survey notes and converted to GIS layers by Natural Heritage Program and Oregon State University (OSU) researchers (Christy and Alverson 2003). Marys Peak and Willamette Falls will be used as reference point landmarks throughout this thesis, representing the highest peak in the Coast Range, and the abrupt change to tidewater taken by the Willamette River, nearby to the Willamette Stone survey marker (McArthur 1982: 798). The four spatial and temporal focal points of each subregion are also shown: Tsp. 4 N., Clatsop County (1913); Soap Creek Valley (1826); "Alseya" Valley (1850); and the Coos Bay Quadrangle (1895).

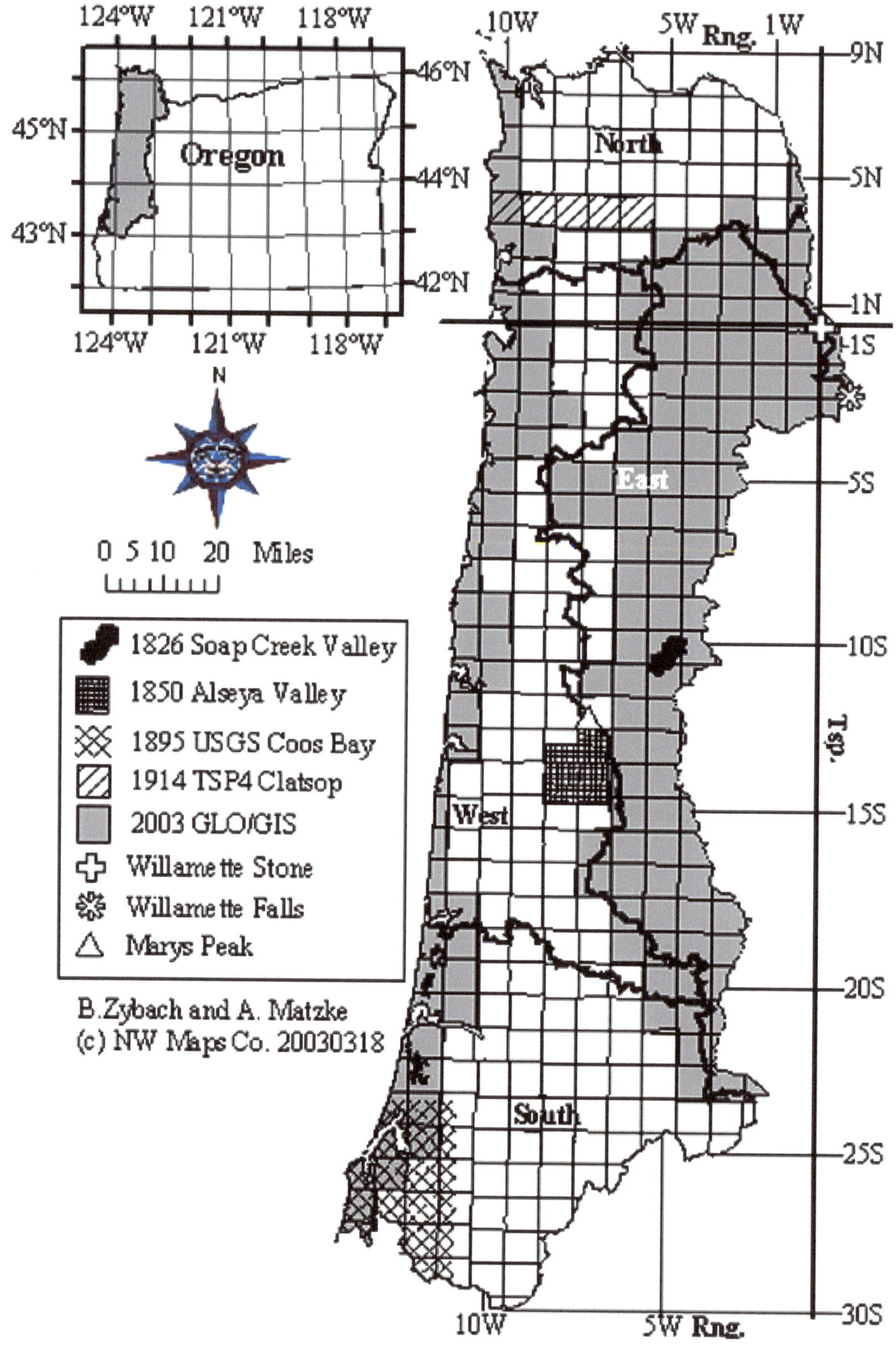


Map 2.01 Oregon Coast Range study index.

### 2.1.1 North: Tsp. 4 N., Clatsop County, 1913

The northern subregion ("North" or "northern Coast Range") follows the established Coast Range boundaries on the north (Columbia River), east (Willamette River), and west (Pacific Ocean). Its southern boundary is the southern watershed boundary of the Nehalem River and the northern watershed boundary of the Tualatin River. An important part of the fire history of the northern Coast Range is that three of its four sides--north, east and west--are bounded by wide expanses of water and, further, those borders are affected by daily tidal actions, seasonal flooding, catastrophic flood events, and moist, foggy climates; any of which can affect local fire patterns. The other three subregions have only one major water boundary each: the Pacific Ocean for West and South, and the Willamette River for East.

The focal point of research, done on a somewhat representative and much finer scale than the subregional patterns, is the entire 30-mile east-west length of Township (Tsp.) 4 N. (North of the Willamette Stone Baseline), as shown on Map 2.01. This focus was selected due entirely to the ready availability of an original 1913 timber cruise of the area (Cabe et al 1913: map). (Timber cruises are usually systematic and detailed accountings of the amount and value of principal tree products within a given area. They typically result in both a detailed map with vegetation patterns, roads, and trails drawn to a local 40 or 640-acre scale, and a table of timber volumes by diameter, species, and length, and product potential.) Both the date (pre-automobile and pre-World War I) and the detail (see Map 4.02) worked very well for comparisons with other historical maps (see Map 4.03). Larry Fick (2003) and George Martin (2003) of Forest Grove Department of Oregon Department of Forestry (ODF) made the timber cruises available for extended study. The Clatsop County portion of Tsp. 4 N. appears to be slightly smaller than five full Tsps. (180 square miles) because of the jagged Pacific Ocean western boundary, and appears to be about 110,000 acres in size.

### 2.1.2 East: Soap Creek Valley, 1826

The eastern subregion ("East" or "eastern Coast Range") includes all of the Coast Range rivers flowing into the Willamette River. Its northern boundary is the northern watershed boundary of the Tualatin River; its eastern boundary is the Willamette and Umpqua rivers; the western boundary is the north-south ridgeline boundary that separates the eastward flowing rivers tributary to the Willamette, from the westward flowing rivers that empty directly into the Pacific Ocean (see Maps 1.02 and 1.03); and its southern boundary is the northern watershed boundary of the Umpqua/southern headwaters of the Willamette. East has the least rainfall, most people, and is the flattest of the four subregions (see Table 1.03).

The focal point of research for East is Soap Creek Valley (see Map 2.01), a 15,000-acre subbasin tributary to the Luckiamute River (see Map 3.03). One reason for selecting this area was ready access to a significant amount of thematically related data assembled via similar research methods (Zybach 1999). Other reasons include its central location in the eastern Coast Range, the availability of local records, the somewhat typical nature of its general forest history as a small, east-facing, foothill valley, and the availability of GIS layers based on 1851-1910 GLO surveys (see Map 2.01; Christy & Alverson 2003; Map 3.04).

### 2.1.3 West: Alseya Valley, 1850

The western subregion ("West" or "western Coast Range") includes all of the Coast Range rivers south of the Nehalem River basin, west of the Willamette River basin, and north of the Siuslaw River basin, that flow into the ocean (see Map 1.03). Its northern boundary is the southern watershed boundary of the Nehalem River; its eastern boundary is the north-south ridgeline divide from the Willamette River; its western boundary is the Pacific Ocean; and its southern boundary is the northern watershed boundary of the Siuslaw River. This subregion has the greatest percentage of Douglas-fir forestland and the longest stretch of coastal fog belt forest of the four subregions.

The focal point of research for West is Alsea (or "Alseya") Valley, a 90,000-acre subbasin headwaters tributary to the Alsea River, to the immediate southwest of Marys Peak (see Map 2.01). This area was the focus of a paper presented to the 5th annual Coquille Cultural Preservation Conference in North Bend (Byram and Ivy 2002; Zybach 2002b), and all research data was readily available. The principal map associated with the paper is Map 2.13. Most of the precontact map projections and early historical vegetation map patterns for this area were developed with GLO survey data recorded between 1853 and 1897 (see Appendix D).

### 2.1.4 South: Coos Bay Quadrangle, 1895

The southern subregion ("South" or "southern Coast Range") is bounded on the north by the southern watershed boundaries of the Siuslaw and Willamette rivers (see Map 1.03). Its eastern boundary is the mainstem Umpqua River; its western boundary is the Pacific Ocean; and its southern boundary is the Middle Fork of the Coquille River (Orr et al 1992). South is bounded by the Cascade Range to the east and the Klamath and Siskiyou Mountains to the south. It has greater amounts of black oak, tanoak, myrtle, and Port Orford whitecedar than the other three subregions combined.

The focal point of research for South is that portion of the 1895-1896 USGS 30-minute quadrangle map (see Map 2.01; Map 2.06) to the north of the Middle Fork of the Coquille River and to the east of the Pacific Ocean. A principal reason for selecting this area is ready access to the original map in question, and shared experience with ODF in 1995 to get the map digitized into current GIS layers (see Map 2.07). Familiarity with its contents, age, quality of data, and ready access to the GIS version were other reasons this area was selected. An acreage estimate of the total land area of this focal point remains to be done, but its scale is subbasin, in common with the other three focal points.

## 2.2 Sources of Information

There were seven basic types of information used to complete this study: names on the landscape; persistent vegetation patterns (including tree rings and pollen cores); literature (principally documentary and scientific); living memory (consultants, oral histories, and oral traditions); historical maps and surveys; aerial photographs; and digitized GIS map layers. This section of the chapter provides examples and brief descriptions of each type, including representative examples of how they were used.

The value of different types of information varies significantly, depending on the time frame being researched, or the specific focus of a query. For precontact time the most useful information came from paleoecological studies of pollen and tree rings, from archaeological and anthropological studies of local artifacts and native peoples, from historical forest maps, from consultations with experts, and from persistent patterns of native vegetation that can still be identified on aerial photographs and satellite imagery. For early historical time, from 1770 to 1850, the most valuable information was obtained from journalists, correspondence, business and government reports, and documentary artists. From 1851 to 1900, General Land Office survey notes and maps, and landscape drawings and photographs, provided the best methods to document the region's forest and fire history, and also provided the best method to predict late precontact conditions. In 1880 the federal government began publishing the first of several reports on the forests of the Pacific Northwest, including those of the Oregon Coast Range, that provided maps, photographs, tabular data, and detailed written descriptions (e.g., Sargent 1880; Leiberg 1900; Gannett 1902). In 1914, following the lead of the federal government, and in response to striking changes in property tax law, Oregon began producing detailed maps and reports through its own Department of Forestry (Rowland and Elliott 1914: map) and through the Timber Tax Division of the Department of Revenue (Zybach and Maeder 1996). Much agency work was based on the maps and logs of private timber cruisers, who created their own body of work from the late 1800s until the end of the study period (e.g., Cabe et al 1913: map; Benton County Commissioners 1914; Bagley 1915: map). Other critical timeframes were the 1930s, and the practical introduction of aerial photography, and the 1980s, with the practical beginnings of GIS.

### 2.2.1 Names on the landscape

A significant number of named landscape features on the Coast Range provided strong clues and information regarding precontact cultural and ecological conditions important to this study. Names of Indian nations, native food plants, wildlife habitat patterns, and past environmental conditions were often attached to prominent rivers, peaks, creeks and other landscape features and easily located through personal knowledge and documentary records of such names. The most obvious example is the names of most Coast Range Rivers--including the Clatskanie, Coos, Yaquina, Nehalem, and many others (see Map 1.03)--which clearly identify the Indian nations and tribes that owned and occupied those areas at the beginning of historical (and probably during late precontact, at least) time. In a similar fashion, Oak Point, Onion Peak, Grass Mountain, Camas Valley, Peavine Ridge and the numerous Burnt and Bald peaks, ridges, and mountains provide additional information and insights (see Appendix B).

Coast Range landscape names related to native species, plant environments, and plant conditions are of particular interest to this study. The patterns of these names across the countryside and in correlation to prehistoric cultures proved to be good indicators of the relative importance of certain landscape features, the location of specific plant species, and the occurrence of past events. Named landscape features were also good focal points for additional research: Oak Point, for example, was named October 28, 1792 by Broughton, who noted the furthest point west that oak trees were found along the Columbia River at that time, "one of which measured thirteen feet in girth" (Barry 1926: 401-402). Lewis and Clark documented the same location on March 26, 1806 (Thwaites 1959b: 204); Thompson on January 11, 1814 (McArthur 1982: 551), and Cox on April 19, 1817 (Cox 1957: 268). The name Oak Point has since migrated across the river to Washington State, but it originally designated an oak grove located in Oregon on the northern edge of Fanny's Bottom, a flat bottomland prairie in Chinook country that took its name from nearby Fanny's Island, named by Clark for his sister, Frances (McArthur 1982: 551). Fanny's Bottom may have also been the location of a large Klaskani town site, Whill Wetz (Ross 1986: 117), allowing these Athapaskan speakers a direct access to Columbia River trade routes otherwise controlled by Chinookans (see Map 3.01). Another example is Camas Valley, on the Middle Fork Coquille. This area had previously been named "Wheat Prairie" about 1850 by a group of "explorers" said to have noted, "here and there were patches of wild wheat supposed to have been planted by Indians" (Connolly 1991: 7). The "wild wheat" was also known as "Indian oats" and "Indian wheat" and was probably tarweed, whose seeds were roasted on the stalk via broadcast burning, before being harvested for further processing. Archaeological investigations show Camas Valley to have been occupied and a source of camas and other native plant foods for thousands of years (ibid: 39-41; 189-191).

Appendix B is a listing of Coast Range landscape features named for native food plants (e.g., Salmonberry River and Fern Hill), native tree species (e.g., Cedar Creek and Yew Creek), native plant environments (e.g., Enchanted Prairie and Forest Peak), and native plant conditions (e.g., Deadwood Creek and Burnt Ridge). Names of towns and post offices are generally excluded as they are usually of more recent vintage and not always good indicators of local conditions: Forest Grove, for example, was named after a man's land claim, rather than a grove of forest trees (McArthur 1982: 283); Berry Creek was named after Thomas Berry rather than a local plant population (ibid: 57); the community of Laurel wasn't named until 1879 (ibid: 436); and the Meadow post office wasn't established until 1887 (ibid: 487). Names for Appendix B were generally obtained from historical texts and maps (e.g., Barry 1926; Ross 1986; Metsker 1929a: map; 1929b: map; 1929c: map) and from two current atlases (Pittmon 1997: maps; Benchmark Maps 1998: maps). Landscape names are listed by county and river basin to establish provenance and distribution, and key names are located on several maps included in this study. Whenever possible, names were researched for authenticity (usually McArthur 1982) before being listed or mapped.

### 2.2.2 Persistent vegetation patterns

Persistent patterns of vegetation provide some of the most detailed information regarding precontact forest and fire history. Three basic sources of this type of data are pollens, tree rings, and documented populations of perennial vascular plants. Several native species of trees, shrubs, forbs, and grasses are useful for reconstructing precontact and early historical landscape patterns of vegetation (see Appendix A). Trees (e.g., Douglas-fir, white oak, and Sitka spruce) are particularly valuable for such uses for at least four reasons: 1) they are long-lived and relatively pure stands of individual species have regularly been documented that are hundreds of years old (e.g., Weyerhaeuser Timber Company 1947; Associated Press 1997), therefore the patterns of these stands have remained consistent from precontact time well into the current period of living memory; 2) they are usually the dominant form of vegetation in a stand, are readily identifiable from a distance (therefore providing consistency in interpretation from a variety of sources, including maps, drawings, landscaped photographs, and aerial photographs), and can usually be characterized by only one or two principal species (e.g., an oak grove, a stand of Douglas-fir, or a spruce-hemlock forest); 3) following death, their remains, including snags, logs, and stumps, can persist dozens or hundreds of additional years, thereby providing opportunities for additional interpretation (Franklin et al 2002); and 4) annual and seasonal growth is recorded in "rings" (Drew 1975), which can be used to age stands (e.g.,; Impara 1997), measure relative variations in growth over time (e.g., Shea 1963), and interpret past climatic conditions (e.g., Fritts and Shao 1995). While shrubs (such as salmonberry and huckleberry), forbs (such as camas and bracken fern), and perennial grasses are not so reliable or versatile for interpretation as trees, they do form identifiable patterns across the landscape and can also persist in the same locations for decades or centuries (see Appendix D and Appendix H). Many examples of this type of persistence are provided by archaeological evidence throughout the Oregon Coast Range. One good example is provided in the southern Coast Range:

> Camas, filberts, and cherry pits were identified among the charred remains of excavated ovens in Camas Valley, near the headwaters of the Coquille River in Douglas County, that were radiocarbon dated from 310 to 2430 years of age (Connolly 1991: 39-41; 189-191).

Pollen. A temporal test of long-term persistence is provided by fossil plant pollens, which can track change and migration of local plant species in an area over periods of thousands of years' time. They provide a far more general and less reliable form of information than provided by tree rings or documented vegetation patterns, but extends the record of fire and species composition much further into the past. Hansen (1947) conducted the first series of pollen studies in the Pacific Northwest, beginning in the late 1930s. Core drillings were made of peat deposits that had accumulated in relict bogs throughout the region. Many of these bogs dated to the last deglaciation, more than 12,000 years ago. Tree pollens preserved in the peat were counted and identified at specific intervals along the cores, showing the changes in local tree species assemblages through time. Although pollens are deposited and preserved chronologically, such factors as wind, gravitational pressure, floods, disease, fire, and long-term climate changes can create problems with dating (Hansen 1947: 6-8; 1949: 1). Hansen helped resolve some of these problems by cross-referencing his findings with a geologist (Allison (1946: 63-65) and an anthropologist (Cressman 1946: 43-51). These results were subsequently verified by radiocarbon-dating of volcanic ash deposits from Mount Mazama, Mount St. Helens, and the Three Sisters that appeared as "prisms" in his western Oregon pollen core samples (Hansen 1966).

According to Hansen (1947: 84-86), lodgepole pine (shorepine) was the pioneer invader species in the northern part of the Willamette Valley after the most recent subsidence of Lake Allison, about 12,800 BP (years before the present time) (Allison 1953; Allen 1984). Lodgepole had been entirely eliminated from the valley well before the beginning of historical time, its decline and extirpation paralleled by the advent and persistence of white oak and increases in local Douglas-fir populations. Hemlock seems to never to have been prevalent in the valley since postglacial times, possibly because of the dry summers that characterized the region's climate during this period (Ruth and Harris 1979: 3-4). The Willamette Valley pattern for lodgepole is the reverse of his analysis of coastal bogs covering "most or all of postglacial time" (Hansen 1944: 632), in which deposits of predominantly hemlock and Sitka spruce pollens are replaced over a 10,000 year or so period primarily with deposits from lodgepole. A radiocarbon-dated study of the ash and pumice profiles noted by Hansen was used to create a more specific timeline of his findings in 1967 (Hansen 1961; 1967: 102). The following summary of general climate and tree migration patterns for the Oregon Coast Range during postglacial time is based on a discussion of the 1967 timeline with Hansen (1989: personal communication), after he'd had nearly 25 years to reflect on his findings:

<u>12,800 BP to 8000 BP</u>. A period of cooler, wetter time than the present, resulting in a coastal forest dominated by hemlock and spruce. People began to arrive in western Oregon from Asia during this period, although it may be that other people were already here. Fire would have arrived with the first people and has been a constant presence since.

<u>8000 BP to 4000 BP</u>. Hansen termed this period the "thermal maximum," because it was characterized by warmer and drier conditions than for any other period in the last 20,000 years, and included the apparent introduction of oak and the expansion of ponderosa pine populations in the Willamette Valley near the beginning of the period. Several ice age mega-faunas became extinct during this interval, a probable result of climate changes and/or human hunters (Kay 2002: 238-246). Coastal forest populations of Douglas-fir, lodgepole, and grand fir increased dramatically, corresponding to an apparent increase in the number and frequency of "holocaustic" fires and decreases in spruce and hemlock. Heusser (1960: 184-185) referred to this period as the "Hypsithermal," which he defines as beginning and ending with a "mean annual temperature higher than at present," and which he dates at 8500 to 2500 BP. Hansen was uncertain whether Pyne's (1984: 232-235) observation that people bring "regularity" and "predictability" to a fire history had any bearing on the proliferation of Douglas-fir during this period (see Chapter 1.5.2), but did not entirely discount the possibility. He also acknowledged the possibility that oak had been introduced into the Willamette Valley by people, but thought the same process may also have been achieved by way of climate change and wildlife transport (Hansen: personal communications).

<u>4000 BP to now</u> (Heusser says 2500 BP to now). As the climate cooled to current temperatures, the end of the thermal maximum interval was followed by a fluctuating succession of Douglas-fir, spruce, lodgepole, grand fir, and hemlock forests to the general historical mix of shorepine, "fog belt" spruce, hemlock, and cedar combinations, and inland even-aged stands of Douglas-fir and alder. Worona and Whitlock (1995: 873) mark the beginning of this cooling period at 5600 BP

(Hansen's "oak maximum"; the warmest point of the thermal maximum period), with a possible reduction in seasonal rainfall at 2800 BP (Heusser's beginning of the current cooler period).

Although Hansen's research gives us information and insights into climate changes and basic forest migration patterns, it provides few clues as to specific forest locations, plant associations, populations, or structures, and very little specific to Indian burning or wildfire events. Additional inferences for developing that type of information must be derived from geological (slope, aspect, elevation), ecological (plant assemblages, soil types, landscape patterns), and anthropological (ethnobotany, cultural fire uses, archaeological findings) sources. Since 1989, Whitlock has done pollen studies (Worona and Whitlock 1995; Grigg and Whitlock 1997) and analyzed sediment cores (Long et al 1998) from Little Lake, a small subbasin of the Coast Range in Lane County. Tree ring studies (Impara 1997; Tappeiner et al 1997) have also been performed in the same area. These recent studies constitute the majority of paleoecological work that has been performed in the Coast Range during the past 50 years, and most of it is within a few miles of Hansen's 1941 Swamp Creek study (Hansen 1941). Otherwise, there remains little paleoecological research performed on the Oregon Coast Range and all conclusions must be very general, with the possible exception of the Little Lake subbasin in Lane County.

The value of isolated and localized pollen analysis and lake sediment interpretations for this study is marginal, except to provide broad climate, fuel, and fire backgrounds. However, tree rings and historical patterns of vegetation are key sources of information, particularly for late precontact and early historical time, from ca. 1600 to ca. 1850. Very little information of this nature exists in the Coast Range for the 1491 to 1600 time period (Weisberg and Swanson 2003: 21), and even less has been discovered from before then (see Appendix G).

Tree rings. Tree-ring analysis dates back at least 400 years, when Da Vinci observed the relation of ring width to wet and dry years (Bennett 1948: 2). In the US, Keuchler completed a study in 1859 that related oak tree ring widths to annual precipitation; however, the science of dendrochronology is said to have begun with the 1904 studies commenced by Douglass, of the University of Arizona (ibid: 2). This methodology was initiated in the late 1930s in the Pacific Northwest (about the same time as Hansen's pollen studies), largely in an attempt to uncover evidence of cyclic precipitation patterns to predict drought (Graumlich 1987: 19). Shinn (1977: 92), for example, reproduces a 1937 "tree ring calendar" by Keen that charts tree-ring variations in eastern Oregon from 1268 to 1937. Graumlich (1987: 25-26) also uses tree rings to analyze occurrences of droughts in the Pacific Northwest. Her findings of significant episodes during the late 1700s, the 1840s, the late 1860s, the 1920s, and the 1930s correspond to catastrophic forest fire events in the Coast Range during the same periods (see Chapter 4). In addition, a drought episode recorded for the time around 1680 could have contributed to wildfires that helped define the parameters of later events (see Chapter 4.1.1; Chapter 5.3). More recently, tree ring research has developed its own specialized terminology (Pyne 1984: 103-106, 232-235) in an attempt to develop statistical "forest histories" and "fire histories" (e.g., Agee 1993: 75-112) from this data source.

Current vegetation patterns. Patterns of tree species important to this study--both for their roles as fuels and as forming the principal landscape patterns in the study area--can be characterized as conifer or hardwood. Most Coast Range forestland is dominated by conifer trees, which are, for the most part, nearly pure stands of even-aged Douglas-fir (see Map 1.09). Other important species include a thin strip of shorepine that occurs along the western-most forested area of the coast; a belt of spruce-hemlock stands that dominate the coastal fog belt to the immediate east of the shorepine strip; and stands of true fir, mostly grand and noble, that exist at higher elevations, such as Euchre Mountain and Marys Peak. Other conifer species, such as hemlock and Ponderosa pine, also exist in fairly pure stands, but they are scattered throughout the Range and typically constitute a minor part of the environment. Native species such as yew, redcedar, and chinquapin also occur throughout most of the Coast, but rarely create pure stands of any magnitude. These patterns of trees and tree species--large and small, living and dead--form the primary patterns of fuels that exist at the time of a forest wildfire (see Chapter 1.5). These same patterns reveal effects of past wildfire and prescribed fire events, and their depiction and comparison in maps and figures forms the basis for much of this dissertation.

### 2.2.3 Literature

Scientific and gray literatures were researched for this study. Scientific literature, including studies of tree rings and pollens described in the previous section of this chapter, focused on disciplines related to studies of people, fire, and forest history specific to the Oregon Coast Range. Gray literature included newspaper and magazine articles, local histories, memoirs, government reports, and other published accounts of specific interest to this study.

Scientific studies of precontact and early historical people in the Oregon Coast Range usually fall under the anthropological headings of ethnography and archaeology. Archaeology is the science of reconstructing the history of people by studying the physical evidence they have left behind. Three basic methods are used in western Oregon: stratigraphic excavation, radiocarbon dating, and topological cross-dating (Aikens 1975: 1). Analysis of artifacts from a site can show the types of animals eaten, species of wood used for fuels, approximate size of the local human population, and so on, for a given time and area (e.g., Reckendorf and Parsons 1966; Freidel et al 1989). The location of specific sites is generally withheld from the public to discourage looting or other damage. Environmental changes can also be inferred from the analysis of such sites. Evidence from the Yaquina Head site, for instance, was used to support the possibility of a major coastal earthquake during the 1600s (Minor 1989: 77). The primary problems with coastal archaeological sites in this study is that these populations largely depended on shellfish and marine mammals for much of their subsistence--commodities that were locally unavailable in inland valleys such the Willamette or Umpqua--and that many sites in excess of 4000 years of age may have been covered by rising seas caused by deglaciation (Aikens 1993: 143-144) or by soil layers produced through erosion (Minor et al, 1987: 19). The principal value of ethnographic studies was recorded interviews with representatives of precontact tribes (e.g., see Appendix C). These are covered in more detail in the Living Memory section (2.2.4) of this chapter. Other types of relevant scientific literature include forest ecology, fire ecology, and wildlife sciences. Scientific literature related to Indian burning has increased in the past ten years (Williams 1992: 43-49; Williams 2003), but--other than the Willamette Valley--little of it is directly applicable to the remainder of the Coast Range (e.g., Boyd 1999b; Vale 2002). The same pattern holds true for Coast Range catastrophic fire history. With the exception of the Tillamook fires of 1933-1951, little research attention has been paid to the other Great Fires of the Coast Range. Morris (1934b) and Pyne (1982) have described these fires, but Morris' article was written nearly 70 years ago, and Pyne's book does little more than paraphrase Morris. There is no literature that relates patterns of Indian burning in the Oregon Coast Range to subsequent patterns of forest wildfires.

Scientific literature was also reviewed to compare research methods and findings used to conduct similar research in other areas of North America. Recognized experts in fire history and research were consulted to develop a list of 25 studies that included "continental" North America or US (e.g., Pyne 1982; Williams 1992; Bonnicksen 2000), "regional" portions of a state, or larger (e.g.,

Braun 1950; Heinselman 1981; Stewart 2002), and "local" subbasin, forest, or field trial (e.g., Raup 1966; Gottesfeld 1994; Cromack et al. 2000), scale studies. The sampling was purposefully biased toward research on fire history (e.g., Pyne 1982; Heinselman 1993), Indian burning (e.g., Boyd 1999; Stewart 2002), and forest wildfires (e.g., Botkin 1992; Vale 2002), with particular attention paid to the methodologies used by each author. In addition to forest history and fire studies, recognized texts on North American forest ecology (e.g., Spurr 1964; Kimmins 1987; Agee 1993) and anthropological literature regarding Indian burning (e.g., Gottesfeld 1994; Anderson 1996) were also considered in the survey.

Table 2.01 is a select listing of the types of studies just described. The geographic basis of the selection was intended to represent the principal temperate forests of North America, with a focus on the forests of the United States. Selection was made following recommendations by Buckman (2003: personal communication), Cromack (2003: personal communication), Lake (2003: personal communication), and Stout (2003: personal communication). This list, much like the lists of early journalists (Table 2.02) and early land surveyors (Table 2.03), is comprised almost entirely of white males; although Braun, Anderson, Russell, and perhaps one or two others, are notable exceptions. Because this dissertation is focused on forest fires and Indian burning, and because archival research, anthropological research, and GIS technology are key elements of the methodology used to acquire and display their findings, Table 2.02 is principally constructed to display similarities and differences in those topics and methods. For example, writers that have relied primarily on literature reviews or pollen analysis for their findings will have mostly "No" entries on their rows, because this research is not heavily dependent on those methods. Likewise, researchers with little focus on, or scant mention of, forest fire or Indian burning histories will show "No" entries in the "Indian" or "Fire" columns. This is generally no reflection on the quality of research methods, findings, or work of these individuals but, rather, indicates similarities and differences of related studies with the focus and methods of this dissertation. In instances where an author had written several books or articles on a chosen topic, a representative work was selected and assigned that date to the "Year" column. If the author paid fair attention to the details of Indian burning in the geographical area of their study, the column was marked Yes (e.g., Whitney 1994; Boyd 1999); otherwise, No (e.g., Spurr 1964: 300; Botkin 1992: 53). If the author mentioned the roles of wildfire or prescribed (non-Indian) fire in forest history or management, a Yes was entered into the "Fire" column (e.g., Pyne 1982); otherwise, No (e.g., Kimmins 1982).

Table 2.01 Comparative North American forest research methods.

| Author | Year | Scale | Indian | Fire | GLO | GIS | Oral | Trial |
|---|---|---|---|---|---|---|---|---|
| Agee | 1993 | Region | Yes | Yes | No | No | No | Yes |
| Alban | 1978 | Local | No | Yes | No | No | No | Yes |
| Anderson | 1996 | Region | Yes | No | No | No | Yes | No |
| Bonnicksen | 2000 | Continent | Yes | Yes | No | No | No | No |
| Botkin | 1992 | Local | No | Yes | No | No | No | No |
| Bourdo | 1956 | Local | No | No | Yes | No | No | Yes |
| Boyd | 1999b | Region | Yes | No | No | No | Yes | No |
| Braun | 1950 | Region | No | No | No | No | No | No |
| Buckman | 1964 | Local | No | Yes | Yes | No | No | Yes |
| Croker | 1987 | Region | Yes | Yes | No | No | Yes | No |
| Cromack | 2000 | Local | No | Yes | No | Yes | No | No |
| Gottesfeld | 1994 | Local | Yes | No | No | No | Yes | No |
| Heinselman | 1996 | Local | No | Yes | No | Yes | No | No |
| Keeley | 2002 | Region | Yes | No | No | No | No | No |
| Kimmins | 1987 | Continent | No | No | No | No | No | No |
| MacCleery | 1992 | Continent | Yes | Yes | Yes | No | No | No |
| Pyne | 1982 | Continent | Yes | Yes | No | No | Yes | No |
| Raup | 1941 | Local | No | No | Yes | No | Yes | No |
| Rouse | 1986 | Region | No | Yes | No | No | No | No |
| Russell | 1980 | Region | Yes | No | No | No | No | No |
| Spurr | 1964 | Continent | No | No | Yes | No | No | No |
| Stewart | 2002 | Continent | Yes | Yes | No | No | Yes | No |
| Vale | 2002 | Region | No | Yes | No | No | Yes | No |
| Whitney | 1994 | Continent | Yes | Yes | Yes | No | No | No |
| Williams | 1992 | Continent | Yes | Yes | Yes | No | No | No |
| **Totals** | **1941-2002** | **8C: 8L: 9R** | **12N: 13Y** | **10N: 15Y** | **18N:7Y** | **23N: 2Y** | **17N: 8Y** | **23N: 4Y** |

If land survey methods--or other, similar, detailed archival research methods--were used as a significant part of research, a Yes was entered in the "GLO" column (e.g., Raup and Carlson 1941: 14; Bourdo 1956: 754); otherwise, No (e.g., Alban 1978, Agee 1993). If GIS methodology were used to depict forest history patterns and conditions, a Yes was entered in the "GIS" column--(note that GIS technology is new and was not generally available to researchers before the late 1980s)--(e.g., Heinselman 1996; Cromack et al: 2000); otherwise No (e.g., Williams 1992; Stewart 2002). If living memory, as acquired through oral traditions, oral histories, formal interviews, or consultations, was used as a principal research method, a Yes was marked in the

"Oral" column (e.g., Anderson 1996); otherwise, No (e.g., Alban 1978; Croker 1986). Finally, if field trials or other statistical methods of gathering and analyzing data were used, a Yes was marked in the "Trial" column (e.g., Buckman 1964; Agee 1993); otherwise, No (Braun 1950; Bonnicksen 2000).

Table 2.01 has a total of 25 entries, selected and codified as described. These texts and studies were published over a sixty-year period, representing eight works performed on a local scale, nine on a regional scale, and eight on a continental scale; 13 of the 25 discussed Indian burning and 15 discussed forest wildfires, but only eight discussed both topics, and five discussed neither topic. Two of the latter five texts (Spurr 1964; Kimmins 1987) were principally concerned with forest ecology in general, rather than fires or Indian history in particular, and the other three (Bourdo 1956; Braun 1950; Raup 1941) regarded a time frame (largely post-contact) and forest region (largely eastern hardwood) in which Indian burning and catastrophic wildfires are not critical factors. Of the eight texts addressing both Indian burning and forest fires, six were continental in scale (e.g., Pyne 1982; Whitney 1994), and the other two focused on regional conifer species (Agee 1993; Croker 1987); of which Indian burning and wildfire history are important factors. This research, then, varies principally in scale from six of the eight other studies that include Indian burning and forest wildfires in their focus, and shares basic themes and scales with only two of the selected readings.

Comparison of methods. Four different research methods were considered for each of the 25 studies listed in Table 2.01, to compare with the methods used in this dissertation. Most of the authors relied heavily on literature reviews: only seven researchers used GLO survey, or similar, historical documentation (see Chapter 2.2.6; Appendix D); only two studies (Heinselman 1996; Cromack et al 2000) used GIS technology or techniques to research, analyze, or display findings (see Chapter 2.2.7); only four studies used field research trials and statistics to develop the principal portion of their fire history data (e.g., Alban 1978; Buckman 1964); and only eight researchers--including most of the anthropologists--used oral interviews and consultations to obtain a significant portion of their findings (see Chapter 2.2.4). The two GIS studies were used to analyze and display wildfire histories and were not used in conjunction with GLO survey records, or to consider Indian burning patterns. None of the studies compared Indian burning patterns--whether spatial or temporal--with forest wildfire patterns. Of the two studies with shared themes and scale, Agee (1993) relied almost entirely on statistical tree ring studies for his data, and Croker's (1987) work regarded another geographic region in which forest fires are a relatively minor occurrence. In sum, although most of the reviewed research addressed topics in common with this dissertation--and most employed similar methods--none used the same (or even similar) combination of methodologies, and none addressed the specific focus of Indian burning patterns and catastrophic fire history. A good share of this result, though, is tied to recent advances in computer technology and GIS.

Gray literature review involved historical studies, newspaper and magazine articles of current events, and primary documentation, such as memoirs, journals, diaries, and letters. For certain time periods and/or events, these forms of literature are the principal data used. Individual journals kept from 1788 to 1849, for example, are the most complete and detailed records and eyewitness accounts that exist for early historical Indian burning practices in western Oregon. Table 2.02 shows the disparity in historical record-keeping chronology that exists in the Coast Range. Spanish, British, Russian, African, and American sea traders traveled up and down the Oregon Coast for several decades in the late 1700s, but had little information regarding inland areas. In the early 1800s, white exploration of the Columbia River (northern Coast Range) increased dramatically, resulting in the establishment of Fort Astoria in 1810, and of Fort Vancouver in 1825. At the latter time, overland explorations of the Oregon coast, and the Willamette, and Umpqua valleys began in earnest. By 1850, most of the Coast Range was fairly well known to informed residents and certain US government agencies. In all, it took about 75 years before most of the coastal rivers had been documented and mapped, resulting in a 70+ year history for the western Coast Range, a 50+ year history for the North, and about 25 years for the East and inland South. All of the journalists were white males, so a certain amount of demographic bias needs to be taken into account for these narratives.

In addition to eyewitness journal descriptions of Coast Range peoples and forests, early cartographers, artists and draftsmen also made and kept records. Vancouver left a detailed set of maps behind his Northwest explorations (Speck 1970: 128-150), and Wilkes (1845a; 1845b) had an artist among the ranks of his 1841 expedition (see figures 3.01, 3.02, and 4.01). Figure 2.01 (Fagan 1885: 80) shows an 1885 drawing of Marys Peak in the background of a house that still exists in the same location. Notice the smooth profile and lack of timber on the peak--a near certain indicator of Coast Range Indian burning, and a landscape condition that exists largely to

Table 2.02 Early journalists of the Oregon Coast Range, 1788-1849.

| Name | Year(s) | Reference | N | E | W | S | Sp | Su | F | W |
|---|---|---|---|---|---|---|---|---|---|---|
| James Cook | 1778 | Elliot 1928b: 269-272 | | | W | | | | | W |
| Robert Haswell | 1788 | Elliott 1928a: 167-177 | | | W | | | Su | | |
| John Meares | 1788 | Elliott 1928c: 280-287 | | | W | | | Su | | |
| Wm. R. Broughton | 1792 | Elliott 1917: 231-243 | N | | | | | | F | |
| | | | | | | | | | | |
| Lewis & Clark | 1805-06 | DeVoto 1953: 273-346 | N | | | | Sp | | | W |
| Alexander Ross | 1810-13 | Ross 1923: 59-117 | N | E | | | Sp | Su | F | W |
| Gabriel Franchere | 1811-14 | Franchere 1967: 37-125 | N | | | | Sp | Su | F | W |
| Ross Cox | 1812-16 | Cox 1957: 69-269 | N | | | | Sp | Su | F | W |
| | | | | | | | | | | |
| David Douglas | 1824-27 | Douglas 1904; 1905 | N | E | | S | Sp | Su | F | W |
| Alexander R. McLeod | 1826-28 | Davies 1961: 148-219 | N | E | W | S | Sp | Su | F | W |
| Harrison Rogers | 1828 | Dale 1918: 260-271 | | | | S | | Su | | |
| Jedediah Smith | 1828 | Morgan 1964: 256-438 | | E | | S | | Su | F | |
| | | | | | | | | | | |
| John Work | 1832-34 | Scott 1923: 238-268 | | E | | S | Sp | Su | F | |
| David Lee | 1835-43 | Lee & Frost 1968 | | E | W | S | Sp | Su | F | W |
| | | | | | | | | | | |
| Gustavus Hines | 1840 | Hines 1973: 92-119 | | E | W | | | | | |
| John Frost | 1840-43 | Pipes 1934: No. 1-4 | N | E | W | | | Su | F | |
| Charles Wilkes | 1841 | Wilkes 1845a; 1845b | N | | | | | | | |
| William Brackenridge | 1841 | Sperlin 1931: 9-64 | | E | | S | Sp | Su | | |
| Joel Palmer | 1846 | Palmer 1983: 38-107 | | E | | S | | Su | F | W |
| James Clyman | 1846 | Clyman 1984: 83-117 | | | W | | | | | |
| Theodore Talbot | 1849 | Haskin 1948: 1-14 | | E | W | | | Su | | |
| "Albert" | 1849 | Oregon Spectator: Sept. 9 | | | W | | | Su | | |

this time. Marys Peak is a well-known landmark clearly visible for dozens of miles from the surrounding countryside. It marked the entry to precontact and early historical "Alseyah Valley" by established foot and horse trails (Zybach 2002b) from the north and east (see Map 2.01). This drawing is of Felger's Mill (ibid: 453, 513; Phinney 2000: 214), near present-day Philomath, to the northeast of Marys Peak. It clearly shows the grassy bald that characterized a significant number of Coast Range peaks at that time (Aldrich 1973). The Indian name for Marys Peak "is said to have been" *Chintimini* (McArthur 1982: 474). Nearby Grass Mountain was called

Figure 2.01 Grassy bald on Marys Peak, 1885.

Holgates Peak (Hathorn 1856a: 235), Prairie Peak called Blue Mountain (Gesner 1891b: 418), and Old Blue Mountain called Green Mountain (Collier 1891: 388) in early land surveys. The Indian names for the latter peaks are unknown.

### 2.2.4 Living memory

Because the practices and events that are the focus of this study took place so long ago, living eyewitness informants are limited, for the most part, to a decreasing number of older citizens who can recall or interpret only certain things about the most recent Tillamook fires (see Table 4.01). There are a few exceptions--Larry Fick (1988-2003: personal communications) and George Martin (1988-2003: personal communications), for example, spent their careers studying and reforesting the Tillamook Burn and are intimately familiar with its history (Fick and Martin 1992). Likewise, Jerry Phillips (1988-2003: personal communications) spent his career managing the Elliott State Forest that was carved out of the 1868 Coos Fire, and has mapped the extent of historic fires in Coos County (Phillips 1988: map). All three were career employees of the Oregon Department of Forestry. Too, until his later years, Rex Wakefield (1976-1994: personal communications) enjoyed a similar knowledge of the Yaquina Burn, both as a youngster performing "fern burning" activities to provide pasturage for his family's goat herd in Eddyville, and as Supervisor for the Siuslaw National Forest in the 1950s and 1960s (Kirkpatrick ca. 1940: appendix). Formal and informal communications with these four individuals over the past 15 to 30 years have helped to shape much of the documented history of the Great Fires that appears in this dissertation. Because Indian burning effectively ended in the Coast Range more than 150 years ago, there are no living people with first-hand experience with these practices. What few accounts remain are in the forms of ethnographic interviews and brief observations by early historical journalists and residents (see appendices C and D).

Methods of obtaining living memory data include personal interviews, oral history research, and archival research. Personal interviews regarding the Great Fires were limited primarily to the four people mentioned in the previous paragraph, and for the reasons mentioned. Focused oral histories that can be cross-referenced provide the most accurate and complete form of interview (Baum 1985; Dunaway and Baum 1996), but they require that the interviewee(s) be alive and willing. Additional value is gained by willingness to help, acuteness of memory, good verbal skills, and a skilled and prepared interviewer. Local informants are the best source of contemporary local history. The problem is that the first oral history recordings were not made until 1948 (Dunaway 1996; Nevins 1996), and none could be located that focused on the topics of this research. Further, there seemed to be little purpose in initiating such a process, given the limited number of possible participants. However, published interviews that appeared in local histories, newspapers, magazines, radio scripts, and other formats were useful sources of information, as were ethnographic interviews performed with Coast Range Indians in the early 1900s (see appendices C and F). The primary problem is identifying the sources and locations of these interviews. County historical societies maintain 1930s WPA (Works Progress Administration) files, which contain numerous interview summaries with Coast Range pioneers, as do the University of Oregon (UO) Knight Library, the OSU Valley Library, and the Oregon Historical Society (OHS) Museum. A familiarity with local family names from pioneer times through World War II is critical to efficiently locating many of these documents. Tape recorded interviews are also stored at the repositories listed above, as well as by the Forest History Society and OSU Archives, and they are occasionally maintained by individual families as well.

Figure 2.02 Sources of living memory, Siletz Tribe, 1924.

Ethnographies are the formal studies of human cultures. Interviews made during the late 1800s and early 1900s by anthropologists with members and descendants of local families who inhabited the coast and Willamette Valley during late precontact and early historical time are of particular value (Seaburg 1982). Individual members of the Alsi, Siuslaw, Coos, and Kalapuyan tribes were interviewed in this manner between 1910 and 1936 (e.g., Frachtenberg 1913; 1920; Jacobs 1945). One or more Yakonans may also have been recorded in 1884 (Minor 1989: 3). Although concerned with lands east of the Cascade Range in Oregon, Shinn's (1977: 23-28) analysis of Native American myths about fire use demonstrates several methods by which ethnographic data can be interpreted. Fagan's (1885: 320-323) legend of romance involving Alsea, Klickitat, and "Drift Creek Indians," although not an ethnography, contains information of this sort that might be worth further study. Figure 2.02 shows members of the

Siletz tribe in a mixture of local, European, and Plains Indian clothing and decorative styles 80 years ago, in 1924. They are shown attending the "Last Official Indian Ceremonial Ever Held in North Lincoln" (Nabut 1951: 105), a few miles from the present location of Chinook Winds Casino in Lincoln City. These people were descendants of Salish and Yakonan tribes that had been assimilated into the Siletz Indian Reservation, beginning in 1856 (Kent 1977; Downey et al 1993). It is unlikely any of them had clear recollections of precontact time, even in 1924. Most ethnographic interviews with Coast Range Indians were made between 1910 and 1936 and involved members of the Siletz or Grand Ronde reservations, perhaps involving individuals shown in this photograph.

Wasson (2000: 24-30) hypothesizes that oral traditions describing geological and ecological events that occurred 5,000 to 10,000 and more years ago, may have been based on eyewitness accounts. He acknowledges Jacobs' 1972 observation:

> Myths of most variable merit that have been collected over the region total less than a thousand, and will never exceed that number. Tales amount to a few hundred, forever so. The bleak harvest . . . maybe one percent of what could have been obtained if the culture-bound, condescending, and racist invaders had had the slightest capacity to perceive merit in the heritages of non-Europeans (Jacobs cited by Wasson 2000: 23).

Wasson (2000), however, fails to consider alternate hypotheses to account for stories detailing a time "before salmon," the coming of the "first people," or even possible accounts of the Bretz Floods that occurred more than 12,000 years ago. Losey (2001) ties ethnographic interviews of Coast Range Indian "traditional tales" with accounts of precontact earthquakes and tsunamis, but specific dates and details are lacking. Current studies (e.g., Finnegan 1996; Vansina 1996) suggest that oral traditions routinely change from generation to generation (about every 20 years), and that it is difficult to recover factual historical data in excess of 100 years' time (Zybach 1999: 53-56).

### 2.2.5 Aerial photographs

Aerial photography was first implemented in Oregon in the 1920s by such organizations as the USDA Forest Service, the Oregon State Highway Commission, and the US Soil Conservation Service. The WPA assembled a good inventory of aerial photographs taken in Oregon before the summer of 1937 (Bennett and Stanbery 1937). After World War II, the availability of military aircraft and sophisticated photographic equipment and techniques developed during the war led to significant improvements in image quality and flight frequency (Spurr 1948). The University of Oregon has the most complete collection of aerial photographs available in Oregon, but businesses such as W.A.C. (Western Aerial Contractors) in Eugene and organizations such as the Forest Service maintain extensive collections that allow for the purchase and/or use of individual "site-and time-specific" sequences.

Aerial photographs are, in effect, highly detailed and time-specific maps of landscape patterns. Before aerial photography came into widespread use in the 1930s, people had to use maps or visit scenic vistas to consider landscape scale patterns of vegetation and human constructs. Because aerial photography didn't exist during the time of Indian burning or, in a practical sense, prior to the 1933 Tillamook Fire, its use in this dissertation is limited principally to interpretive--rather than documentary--purposes. The exception is the Tillamook fires of 1933-1951, which have been extensively photographed, otherwise documented, mapped, and analyzed over the course of the last 70 years (Fick and Martin 1992).

Despite stated limitations, aerial photographs were used extensively over the course of this research, particularly photos taken between 1939 (e.g., Zybach and Badzinski 1998) and 1966 (e.g., Figure D.01). Figure 2.03 (courtesy of the University of Oregon Knight Library collection) is an aerial photograph of the town of Alsea, Oregon, taken on June 29, 1948. The juncture of the North Fork and South Fork of the Alsea River can be seen near the bottom of the photo. General Land Office (GLO) subdivision and Donation Land Claim (DLC) boundary lines dating to the early 1850s are clearly apparent a century later as the straight courses of roads, fence lines, crop borders, and logging boundaries. Fenced and plowed bottomland prairies, grassy meadows, ridgeline trails, hillside scatterings, and other pre-contact landscape patterns can also be easily discerned. A principal difference between Indian and white cultural patterns, including firewood gathering and broadcast burning patterns, and--as evidenced by this photo--is the imposition of straight lines on the landscape, indicating both cultural and land ownership changes in the environment (see Figure 3.04).

Today (2003), the Alsea River bottomlands are still dominated by the town of Alsea; a rural, white community descended from the "Alseya Settlement" of 1855 (McArthur 1982: 12-13), in turn descended from the Alsi Indian community of *Ts! Iphaha* (Drucker 1933: 82), at the juncture of the North and South forks of the Alsea River. Surrounding bottomlands consists largely of family-owned farms and ranches. Adjacent uplands are heavily forested with a variety of hardwoods and conifers; particularly Douglas-fir, which forms about 90% of the population and volume of trees in the area (Bagley 1915: map; Munger 1940: 451).

Figure 2.03 Aerial photograph of Alsea, Benton County, June 29, 1948.

2.2.6 Historical maps and surveys

Three types of historical maps proved particularly useful to this research: original land surveys (see Appendix D), local and county timber cruises (e.g., Cabe et al 1913: map; Bagley 1915: map), and statewide vegetation types (e.g., Thompson and Johnson 1902: map; Rowland and Elliot 1914: map). In addition, all of the GLO data (1851-1940), a good number of timber cruises (1913-1928), and three pre-aerial photograph/vegetation type maps (1900-1936) fell within the range of the 1849 to 1951 catastrophic fire history period that is the focus of this study. Other types of maps--including USGS quadrangle maps, property tax lot and land ownership (cadastral) maps, and highway maps and atlases--also proved helpful, but did not contain the types of data that could be readily used to represent burning and wildfire patterns.

GLO surveys completed throughout western Oregon between 1851 and 1910 provide an exceptional source of information for documenting the transition of environmental conditions from traditional American Indian land management practices to introduced European American practices (see Appendix D). Although these surveys were performed during the early 1850s and after, they can also be used to make reasonable estimates of environmental conditions at the time of contact, or even before (Zybach 2002b). GLO land survey records have been long accepted as useful measures of past environmental conditions (Whitney and DeCant 2001: 147). These records have specific advantages over other early historical sources of information in being a definite sampling of vegetation, written on location, and in accordance to a predetermined plan (Bourdo 1956: 755).

GLO survey notes and maps document environmental conditions systematically measured for the entire western US and a large portion of the eastern US from the late 1700s to the present. Original land surveys for western Oregon, including the Coast Range, were completed between 1850 and 1910 (Stewart 1935: 72-73). This voluminous source of data includes detailed information that is both quantitative and qualitative; is arranged in precise spatial components at a relatively fine scale; reflects a critical period of time in which American Indian resource management practices were dramatically altered by European American immigration and land appropriation; and was--significantly--gathered by a large number of surveyors operating under a single set of specific instructions (Bourdo 1956; Schulte and Mladenoff 2001). Figure 2.04 (Hathorn 1856c: 154) is a copy of a page from the Benton County Surveyor's Office that was typewritten, possibly in the 1930s, from handwritten field notes. The subsequent colored notations were used to help construct maps and GIS layers for this study. Several such excerpts have been indexed in Appendix D. The blue marks show the exact distance and compass bearing from an established survey point that the surveyor enters a distinct vegetation type, such as a "bottom" or "thicket" (see Table 3.03). The pink highlights show precontact cultural artifacts,

156 County Surveyor's Record, Benton County, Oregon

West Boundary of T. 14 S., R. 7 W.

| CHAINS | | |
|---|---|---|
| 8.00 | Leave bottom, course E. and SW. | level |
| 16.75 | Summit of a ridge, course NW. and SE. | +50 |
| 16.75 | Var. 21°30'E. | |
| 25.00 | Enter bottom, course NW. and SE. | -50 |
| 31.50 | Enter thicket, course NW. and SE. | level |
| 35.00 | Leave same and enter prairie, course E. and W. | level |
| 40.00 | Set ½ sec. post; from which | level |
| | An oak, 20 ins. diam., bears N.8°W., 324 lks. | |
| | An oak, 20 ins. diam., bears N.82°E., 103 lks. | |
| 44.25 | Indian trail, course N.25°W. and SE. | level |
| 44.25 | Same house, bears N.28°W. | |
| 63.00 | Enter thicket, course NW. and SE. | level |
| 63.10 | To right bank of South Fork of Alseya. Set flag; from a pt. W. on opposite bank offset South 130 lks., thence E. to line; river is about 60 lks. wide, course NW. | |
| 75.00 | Enter prairie, course NW. and SE. | level |
| 80.00 | Set post, cor. secs. 12, 13, 7 and 18; from which | level |
| | A fir, 36 ins. diam., bears N.2°W., 675 lks. | |
| | A fir, 36 ins. diam., bears N.10°E., 590 lks. | |
| | An oak, 24 ins. diam., bears S.60°E., 578 lks. | |
| | An oak, 20 ins. diam., bears S.72°E., 615 lks. | |
| | No tree convenient SW. | |
| 80.00 | David Fudge's house, bears S.52°E., 5.50 chs. | |
| | Land, principally level prairie. | |
| | First rate soil. | |
| | June 9, 1856. | |

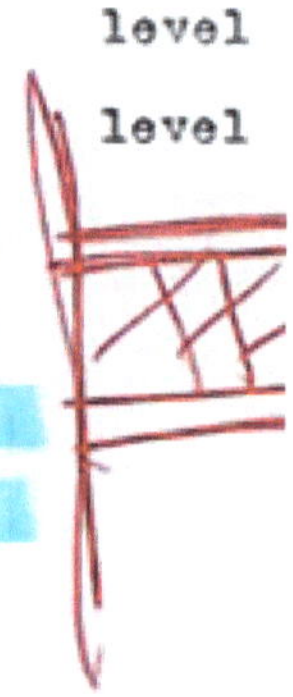

Figure 2.04 T. 14 S., R. 7 W. annotated GLO field notes, June 9, 1856.

such as prairies, Indian trails, and oak groves, and early historical artifacts, including Indian trails and David Fudge's house. The red ink and cross-hatchings mark the great distance--600 links ("lks") equal about forty feet--between bearing trees. This corner must have been within a "scattering" or savannah environment for oak and open canopy to be present. Further, the (probably) Douglas-fir trees were only 36 inches in diameter, making them fairly young second-growth trees that had probably been seedlings sometime after 1790, while the slower-growing oak were probably around 120 to 240 years of age (Thilenius 1968). None of these trees were likely living before 1600. It is possible that first the oak, and then the Douglas-fir, encroached on what had once been an open field or prairie.

GLO surveys began with the establishment of initial points ("meridians"), located for practical geographic reasons and recorded by astronomical observations. Thirty-five meridians were established in the continental US (Stewart 1935: 74). The initial point for Oregon Territory (then including Washington and Idaho), the "Willamette Meridian," was determined on June 4, 1851 (McArthur 1982: 798), in the West Hills above Portland. The location, later marked with the "Willamette Stone" (see Map 2.01), was selected on the basis of visibility, "that the Indians were friendly on either side of the line for some distance north and south" (Burnham 1952: 229), and because the east-west Base Line of the survey intersected the Columbia River in only one place. Map 2.01 locates and names each of the six-mile square (36 square mile) townships in this study (Moore 1851: 227-230), relative to their position in the Coast Range. Beginning at the Willamette Stone, a north-south "meridian" and an east-west "baseline" serve as the beginning to a succession of six-mile wide tiers. Each east-west tier (also called "Township" and written as "Tsp." or "T.") is numbered consecutively north ("N.") or south ("S."), and each north-south tier (called "Range" and written as "Rng." or "R.") is numbered consecutively east ("E.") or west ("W."). The intersection of numbered township and range tiers form square townships designated by this numbering system. For example, the township containing Marys Peak (Figure 2.01) is named Tsp. 12 S., Rng. 7 W. (or, T. 12 S., R. 7 W.) and can be located by counting 12 tiers south and seven tiers west of the Willamette Stone. Every surveyed township in Oregon and Washington is located and named in the same manner and commences from the same starting point (Stewart 1935: 74). Most historical maps, including almost all timber cruises and vegetation type maps, and all current landownership boundaries and tax lot maps, are based on these lines. Before 1851, these lines were absent from the environment. Since that time they have served as an exacting index to specific trees, points, locations and cultural artifacts on specific days in time.

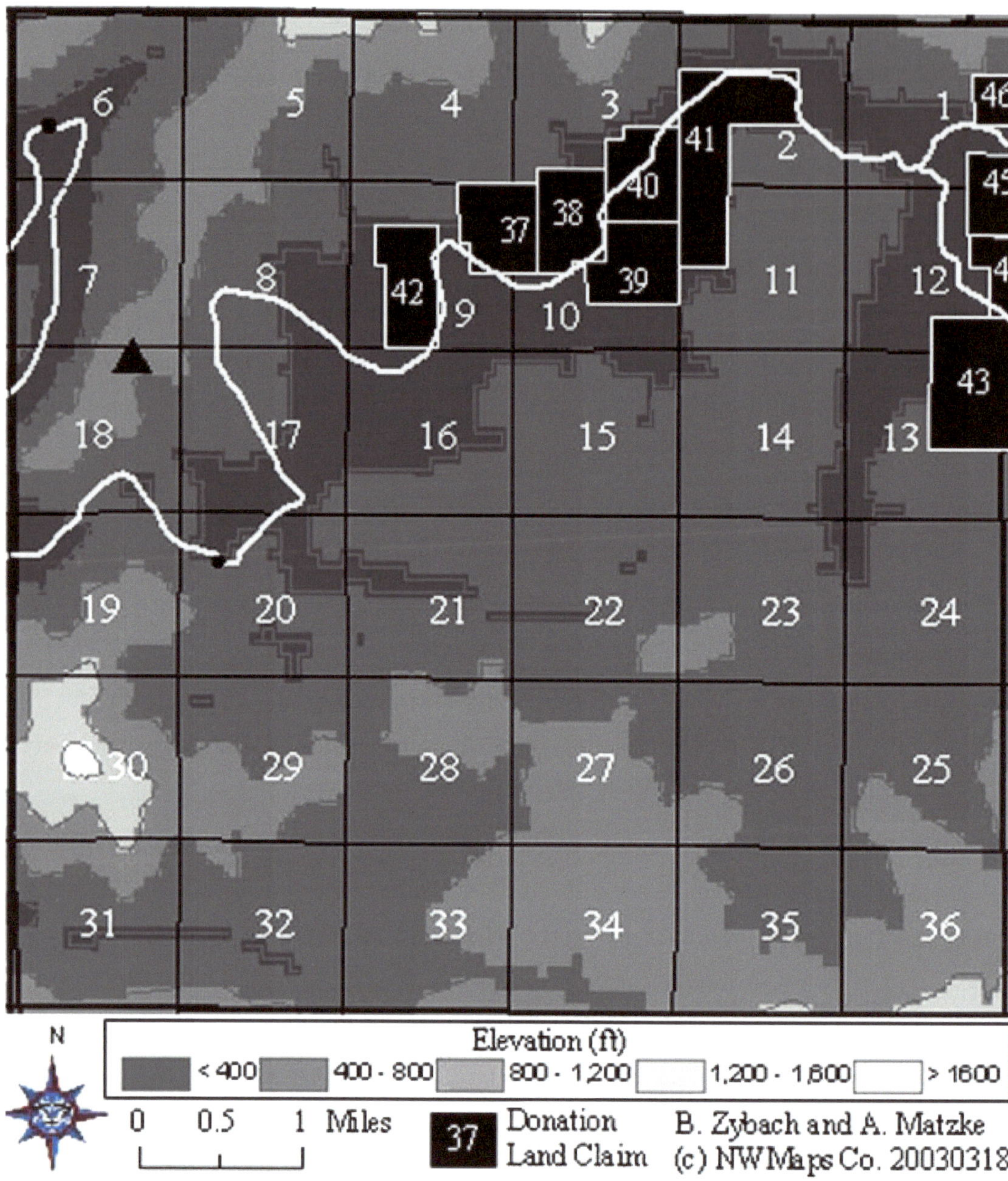


Map 2.02 T. 14 S., R. 8 W. GLO subdivision and DLC survey index.

Map 2.02 shows the pattern of 36 square mile "sections" ("Sec." or "S.") created when surveyors subdivided a township. Table 2.03 lists the names of the surveyors that completed this map and others in the "Alseya Valley" area (see Map 2.13). The numbering is always the same--beginning in the NE corner with Sec. 1, continuing west to Sec. 6, then south one mile to Sec. 7, proceeding east through Sec. 12, and then repeating the sequence two more times, continuing to number consecutively, until Sec. 36 is reached in the SE corner of the township (Moore 1851: 229). If original land claims (called a "Donation Land Claim" or "DLC") or portions of DLCs are contained within the township, they are also numbered consecutively, beginning with the number 37, so as to not confuse them with section numbers. For example, the DLC of Basil Longworth, (Longworth 1972), No. 37, is contained in T. 14 S., R. 8 S., Secs. 9 and 10 (Hathorn 1856c: 491-493). When a land claim straddled more than one township, it was given separate numbers for each township. Squire Rycraft (Fagan 1885: 525-526), for example, was given DLC No. 45 for the 94 acres he claimed in T. 14 S., R. 8 W., and No. 40 for the remaining 66 acres of his claim to the east, in T. 14 S., R. 7 W (Hathorn 1856c: 496-497). Boundaries of Rycraft's 1856 survey are still readily apparent today, as evidenced by aerial photography (Figure 2.03).

Table 2.03 lists the names of the men who conducted the original GLO land surveys in "Alseya Valley," the years and times of year they conducted their surveys, and the locations they described and mapped. It also indicates whether Indian foot or horse trails were noted ("Trails"), and when and where forest fires had occurred ("Burns"). In common with the earliest Coast Range journalists (Table 2.02) and the Valley's original land claimants (see Map 2.02), all of the surveyors were white males.

Table 2.03 Original GLO surveyors of "Alseya Valley," 1853-1897.

| **Surveyor** | **Dates** | **Townships** | **Burns** | **Trails** |
|---|---|---|---|---|
| Webster, Kimball | Oct.-Dec., 1853 | 14-7 | X | X |
| Gordon & Preston | Sept., 1854 | 13-7 | X | |
| Hathorn, Dennis | June-July, 1856 | 13-7; 13-8; 14-7; 14-8 | X | X |
| | | | | |
| Mercer, George | Aug., 1865 | 13-7; | X | |
| Dick, J. M. | Aug., 1873 | 13-8; 14-8 | X | |
| Mercer, George | Jan., 1878 | 14-7; 15-7 | X | X |
| | | | | |
| Gesner, Alonzo | May-July, 1891 | 14-7; 14-8; 15-7; 15-8 | X | X |
| Collier, Charles | Sept.-Oct., 1891 | 13-7; 13-8 | X | X |
| Collier, Charles | Feb., 1892 | 13-7 | X | |
| Collier, Charles | Apr.-May, 1893 | 14-7; 15-7 | X | X |
| Sharp, Edward | July, 1896 | 15-7 | X | |
| Sharp, Edward | Jun.-Sept., 1897 | 13-8; 15-7 | X | X |

In July 1856, Dennis Hathorn surveyed each of the 15 Donation Land Claims (DLCs) that had been established in the Valley during the previous four years (Hathorn 1856c). Eight of the claims were each for 160 acres (1/4 section), and seven of the claims were for 320 acres (1/2 section). The larger claims indicate married couples. All of the claims were established along the mainstem Alsea River, along the main foot and horse trails from the Willamette Valley to tideland, on the best 2,500 acres of bottomland prairie in Alseya Valley (see Figure 2.04; Map 2.12). By law, no Indians, blacks, or Chinese were allowed to make or purchase land claims (Carey 1971: 253).

Surveyors described native plants in terms of species, general locations, and plant mosaic types. Vegetation types were also described in reference to land and property developments and often noted as they were encountered along GLO survey lines, DLC property lines, and riparian intersections. Bearing trees (Moore 1855: 8-9) were precisely located and listed by species and diameter in relation to the mosaics (e.g., Zybach 1999: 275-292); typically six or more trees per running mile, with a minimum of eight trees per forested section, and often more than eight for sections adjacent to double corners (Moore 1855: 12) or containing DLCs or riverine meanders (ibid: 14). This process allows us to compare the measures and observations of individual surveyors and better evaluate our estimates of past environmental conditions. Surveyors regularly noted specific types and species of plants and whether forest fires had entered the area, or not. Collier gave this general description of Tsp. 14 S., Rng. 7 W. (approximately 36 square miles; or about 23,000 acres) in May 1893, for example:

> The usual vegetation of the Coast Range grows strongly everywhere. Wild grass and pea vine grows thick where other growths have been killed. Most of the openings are taken by a rank growth of fern. The whole has been covered by heavy fir, cedar and hemlock forests which has been deadened by fire on about half the area and the slopes exposed afford excellent range for cattle and sheep (Collier 1893: 228).

In addition to property lines, bearing trees, and vegetation patterns, GLO maps and survey notes contain a significant amount of information about cultural artifacts and developments. Map 2.03, (a segment of the 1852 Tsp. 2 S., Rng. 2 E. plat; Corning 1973: 28) for example, clearly shows the location of an Indian village and burial ground a few miles northeast of Willamette Falls. Map 2.03 is a good representation of the type of subdivision map that was developed from GLO field surveys. A second type of map for property owners, the DLC plat map (see Map D.02), was

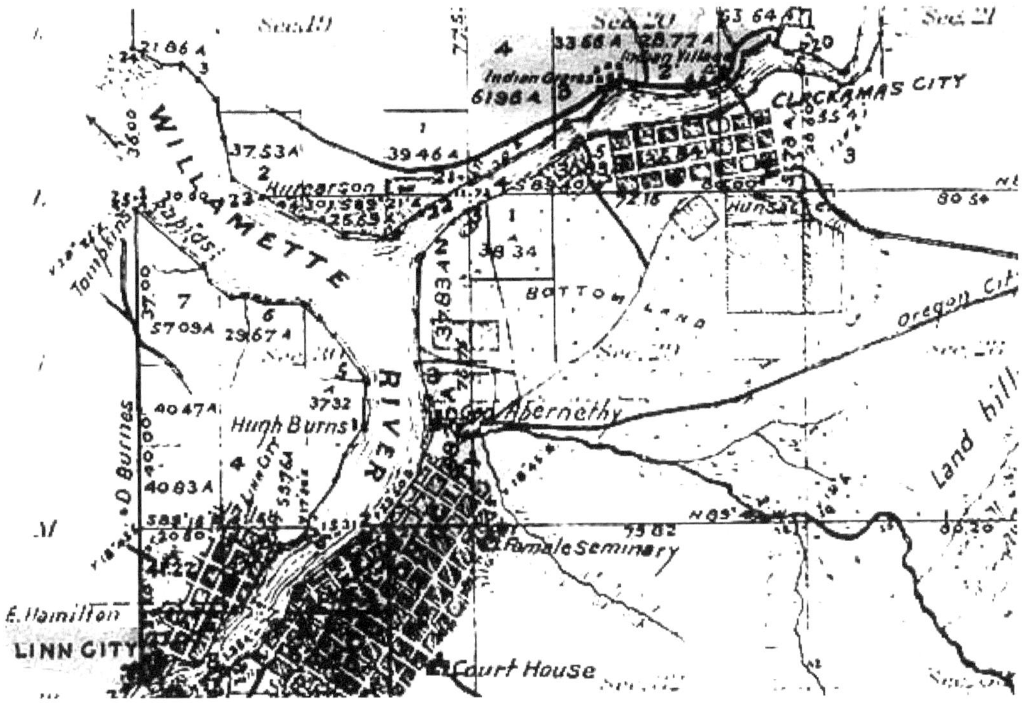


Map 2.03 Willamette Falls, Clackamas Indian burial ground, 1856.

also produced at the same scale as the subdivision maps. GLO surveyors were specifically instructed to note cultural "improvements" to the environment, including "Indian towns and wigwams," and "natural curiosities," including "all ancient works of art, such as mounds, fortifications, embankments, ditches, or objects of like nature" (Moore 1855: 17-18). They were probably not trained to recognize such localized features as elk pits, peeled trees, or rock ovens. Figure 2.03 illustrates lasting changes brought to Alseya Valley landscape patterns by white settlement and GLO surveys. Beginning with the establishment of DLC boundaries, native plant populations were soon limited and infiltrated by the plantations and structures of the new settlers. Surveyors duly noted and mapped fields, orchards, gardens, wagon roads, ditches, a bridge, a walk log, fences, barns, houses, an outhouse, a school house, a post office, and an old flume to "Inman's sawmill" in Tsp. 14 S., Rng. 7 W. (Fagan 1885: 500; Collier 1893: 178). As might be expected, more of these structural and vegetational changes are indicated in the GLO records from the 1890s than the 1850s, as the Valley became more heavily populated with immigrant farmers, ranchers, loggers, sawmill workers and their families.

This study demonstrates the potential of GLO maps and survey notes to make good estimates of precontact land use patterns and resource management practices. The detailed mapping of Indian canoe routes, foot and horse trail networks, homesites, and campsites in relation to managed fruit and nut orchards, berry patches, and extensive fields of root and grain crops cannot be duplicated by any other source of information. The location and timing of daily, seasonal, and annual practices of burning, tilling, and harvesting that took place in pre-contact times can be easily and reasonably inferred from these records. The precise location of individual trees and possible elk-fall pits provides exacting details of how and where people hunted, where they shaded themselves from the sun, and when they met to gather nuts or firewood. The great diversity and ready access of managed environments, including grassy prairies, oak groves, meadows, brakes, balds and berry patches provide strong insights into the lives of people and cultures that no longer exist. This combination of detailed records and reasoned insights are just as significant for considering pre-contact wildlife populations, habitat patterns, and food sources.

The other types of historical maps used extensively in this research are timber cruises and vegetation type maps, typified by Map 2.04 (Bagley 1915: map) and Map 2.05 (Peterson 1994: 12; photo by B. Zybach 1988), both produced within a year or two of each other. The timber cruise map covers a single, 640-acre section of property. Each section is divided into four square "quarter secs" of about 160-acres each, called NE, NW, SW, and SE. Each quarter sec is also quartered, and also designated NE, NW, SW, and SE, but these "1/4 of a 1/4 secs" are 40-acres in size. Therefore, the NW 1/4 of the NW 1/4 of this section would be a square 40-acre area of land on the extreme northwest corner of the map. The tally sheet can then be used to find the same location, where precise tree species, general diameters, and reasonably estimated volumes of each species are listed. The maps provide polygon shapes that couldn't be shown on GLO maps. Typed and signed notes regarding timber markets, fire hazards, etc., also add details. These are the most detailed and exacting patterns of forest fires that predate aerial photography. Unfortunately, timber cruises of this quality do not cover all of the forested lands in western Oregon, and existing cruises are of a highly uneven quality.

Oregon's entire pattern of wildfires was mapped for Oregon State Forester F. A. Elliott in 1914 (Map 2.05; Peterson 1994: 12; Botkin 1996: 52-53). This map was prepared in advance of the 1918, 1933, 1939, and 1945 Tillamook Fires and clearly depicts over half of the Coast Range (south to the middle fork of the Coquille River) as "Burned & Restocked," "Burned & Not

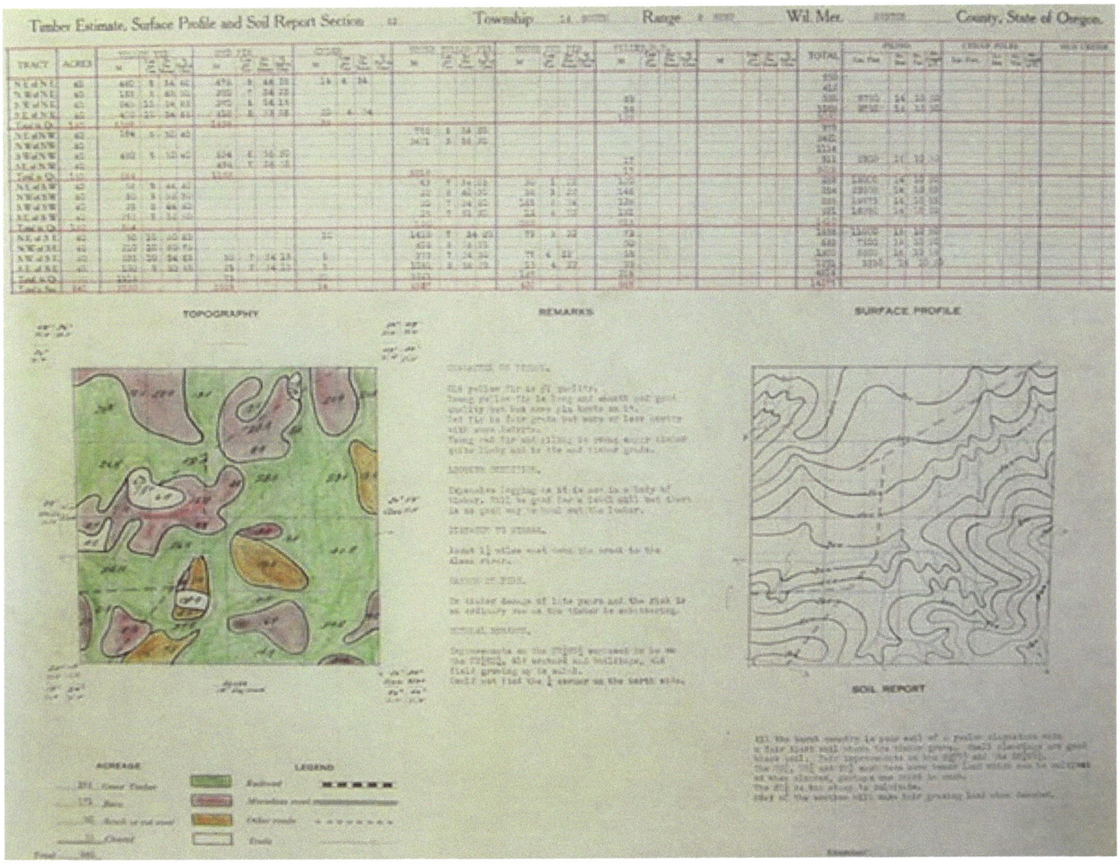


Map 2.04 Tsp. 14 S., Rng. 8 W., Sec. 22 timber cruise, 1915.

Restocked," "Brush," and "Prairie." There is also a large burned area around Tillamook Bay, possibly a result of the 1902 fire, which may have served as a partial buffer for the Tillamook Bay community against later events in 1910 to 1951. The boundaries of the Yaquina, Coos, and Nestucca fires of ca. 1849 to 1868 are clearly depicted, although of varying quality of detail. This statewide map has several advantages over the timber cruise map: it covers the entire region at a similar point in time (probably a six-month to six-year range of time the information was gathered before the map was printed); it used the same uniform vegetation type codes for the entire State , and it was available to the public. Disadvantages are that it had little practical value for land managers used to working in 40-acre increments, it contained far less information, and of far more general scale than standard cruise maps, and it was generally too bulky for field work. For purposes of this study, the 1914 ODF Map (Rowland and Elliott 1914: map) also had value because it had been converted to a GIS coverage (see Map 2.09).

Map 2.05 Oregon State Forester's Map, 1914.

### 2.2.7 GIS layers

This study used existing computerized GIS map files to help construct new maps for analysis and display (Goodchild 1987). GIS map files are called layers or coverages and can be in the form of points, lines, text, or polygons. Map information is stored in relational databases that are georeferenced: that is, they are created in precise relationship to their relative position on the surface of the earth in terms of longitude, latitude, and elevation. In this manner, individual coverages can be "layered" on one another with assurance that the data is displayed in accurate spatial relationships. For example, a point representing the center of the City of Corvallis might be used to locate the city's name in text; a polygon is layered over the point and text that represents the city's corporate boundaries; a line is layered over the polygon that represents Highway 99 W; viewed together, a labeled map of Corvallis' city limits has been created, showing where Highway 99 enters, passes through, and leaves town. Other points, lines, polygons and labels could be added to show and identify streets, blocks, creeks, property lines, etc., as desired.

Nearly 60 existing GIS files were identified, inventoried, and used during the course of this research (see Appendix I). They were obtained from twelve different sources, mostly federal and state agencies. These layers were then modified and combined with newer layers to create the maps used in this text. In this manner--by mixing and matching GIS layers from a variety of existing sources--new GIS layers can be created from GLO surveys made in the 1850s (Christy and Alverson 2003) and other historical maps. Map 2.06, for example, is an historical 30 minute quadrangle map assembled along latitude and longitude lines, rather than the rectangular survey lines used by the GLO (see Appendix D). Map 2.07 is the GIS version of the same map, as digitized under the direction of ODF in the early 1990s. As with the digitized version of the 1914 ODF map (Rowland and Elliott 1914: map), this area was selected for further study in part because the GIS layer already existed. One advantage of the 1895 GIS map is that it can be colored and shaded in distinctive patterns. In this case, the intent was to emphasize burn patterns (red and pink), old-growth (dark green), logged (black), and non-timbered (yellow). A disadvantage of simplified patterns is that valuable text and graphic information, including the locations of roads, trails, springs, and houses, is lost. The precontact Millicoma (pink) and historical Coos (red) catastrophic fire patterns on the GIS map bring some good questions to mind. Logging started in the early 1850s in Coos Bay, but the Great Fire didn't occur until 1868, and Weyerhaeuser didn't start logging the second-growth Millicoma Tract until after World War II (Smyth 2001). Note the exacting boundaries between early prairie lands (yellow), floodplains (yellow), logging (black), and the 1868 fire (red): Did logging blur the fire boundaries, by salvaging burned timber in addition to logging green timber? Or, did the Coos Fire simply stop burning as it entered logged land and prairies? Or something of both? And, did the timber owners log or salvage mostly tanoak, old-growth, second-growth, spruce, or cedar? Or some of each? How did those harvests affect future fuel patterns? What was the impact on subsequent wildfire events?

The assignment of GIS map legend colors and descriptions can emphasize certain patterns and relationships, as shown by Maps 2.06 and 2.07. One of the interesting problems to resolve with this research: What is the minimum number of vegetation type codes (shades, colors, or patterns) needed to display basic patterns of Indian burning and catastrophic wildfires? Next: how to take

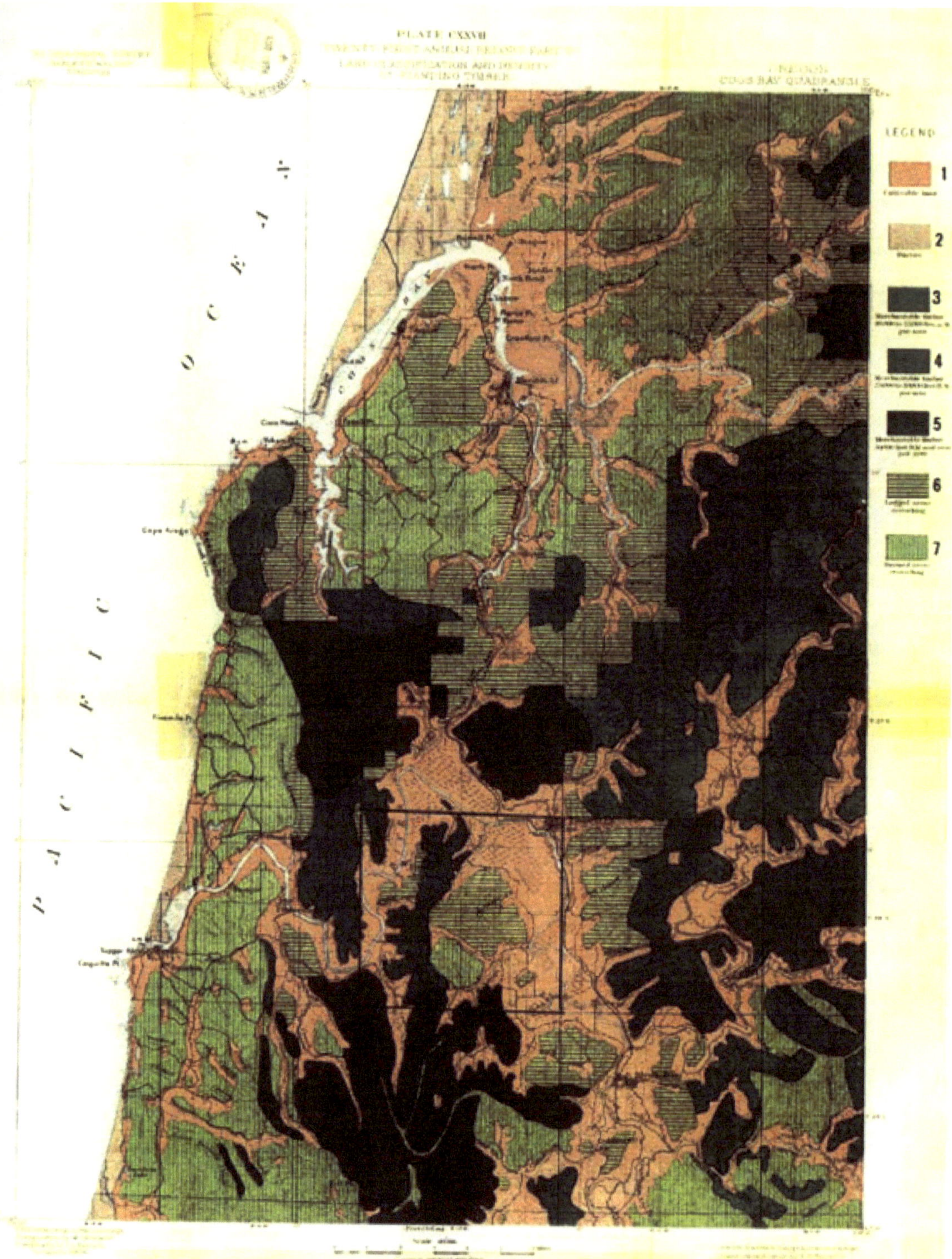


**Fig. III. 1895-96 Land Use and Timber Volume Patterns, Coos Bay. (US Geological Survey 1900: Atlas Plate CXXVII). 1:250,000**

Map 2.06 USGS Coos Bay 30 min. quadrangle map, 1895-96.

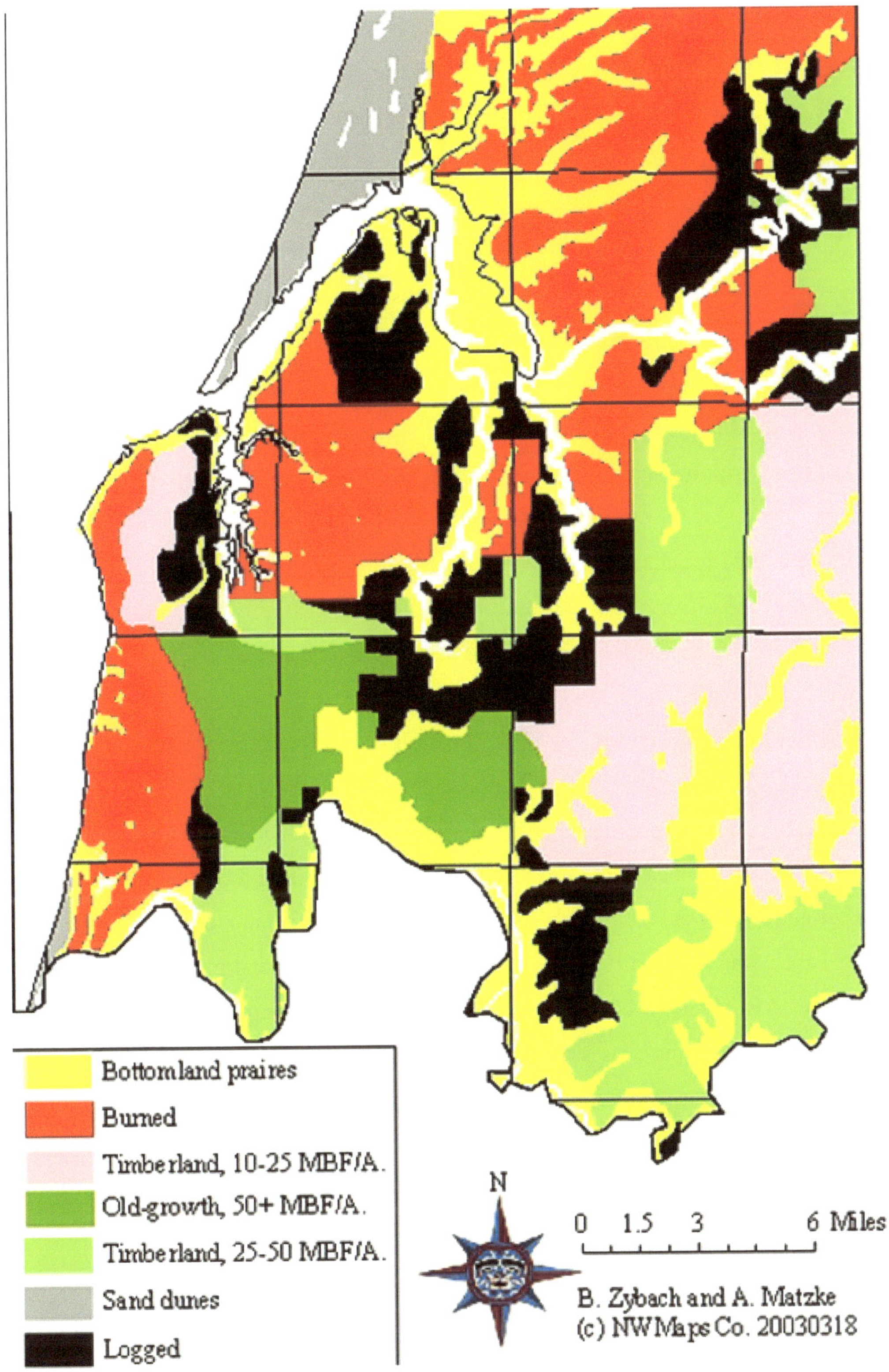


Map 2.07 1895-96 USGS Coos Bay Quad GIS map.

a lesser or greater number of vegetation types (the 1895 map had seven types, while the Christy and Alverson (2003) GLO layers had more than 50) and to translate differing measures to a common number of types and a common series of colors and patterns to display each type? This problem becomes more complex when a wide range of colors is needed to display a wide range of vegetation types, or when a series of maps using differing measures is being compared. At this point, color and shading schemes became a process of trial and error, constrained to a large degree by software options and conflicts. Table 2.04 demonstrates the end result of that process. The left hand column is the series of 27 vegetation types assigned to Oregon forests by Andrews and Cowlin (1936: map), the right hand column is the new color scheme, intended to conform (and be comparable) with Map 2.07 by adopting the same legend colors, and by combining smaller polygons into larger polygons with comparable boundaries to the older map's more general counterparts. The result of re-coloring and combining polygons can be seen in Map 2.10 and compared with Map 2.07 for changes between 1895 and 1936. Note the red and pink, dark blue, dark green, yellow, and black legend definitions that were discussed earlier, and compare changes in environment and mapping focus in the intervening 40-year period.

A total of three historical vegetation maps, covering the 1900 to 1936 time period, were selected to represent precontact vegetation types and the boundaries of the Great Fires of 1849 to 1936. They were chosen for five primary reasons: 1) their spatial extent included the entire Coast Range study area (see Appendix E); 2) each map represented the best available information from its time; 3) each map used representative precontact forest, savannah, woodland, and grassland vegetation types; 4) each map had relatively detailed depictions of the catastrophic fires of their time (and earlier); and 5) each map was available in GIS format. GLO maps, converted to GIS layers, are used in this thesis primarily to reconstruct precontact and early historical Indian burning patterns across the landscape (see earlier discussion, following Alseya Valley example, and Chapter 3); historical vegetation maps, converted to GIS layers, are used primarily to document subsequent catastrophic forest fire patterns. Three maps in particular were used for this purpose: the 1900 USDI Forests of Oregon Map (Gannett 1902; Thompson and Johnson 1902: map; Map 2.08), the 1914 ODF Oregon State Foresters' Fire Map (Rowland and Elliott 1914: map; Map 2.09), and the 1936 PNW Research Station Forest Survey Map (Andrews and Cowlin 1936: map; Andrews and Cowlin 1940; Map 2.10).

B. Zybach and A. Matzke
(c) NWMaps Co. 20030314

Table 2.04 Conversion of 1936 vegetation legend to 2003 GIS pattern.

Map 2.08, the 1900 USDI GIS map, is the earliest GIS map in the series. The original map accompanied a report by Gannett (1902), the same individual responsible for the 1895-96 USGS Coos Bay 30-minute quadrangle map (Map 2.06). In describing the 1900 USDI map, he states:

> The most startling feature shown by the land-classification map of this State is the extent of the burned areas, especially in the Coast Range and in the northern half of the cascades. It must be understood that the areas represented here as burned are only those in which the destruction of timber was nearly or quite complete. Areas which have been burned over with only a partial destruction of the timber are not here represented (Gannett 1902: 11).

This version of the 1900 USDI pattern includes the township and range index of Map 2.01.

The second historical GIS map is based on the 1914 ODF Map (Map 2.05). Map 2.09 is the GIS version of the 1914 ODF "Fire Map" pattern. County boundaries (see Map 1.03) are used as an index, as well as important political boundaries that designate the locations of important research materials (e.g., county surveyor departments, county historical societies, and county libraries). Note the greater detail of patterning that is shown, partly as a result of GLO survey projects completed by 1910 (see Appendix D). Also note that Map 2.08 has a more detailed, and likely more accurate, mapping of the woodlands and logging in the Willamette Valley and northern range than Map 2.09, but 2.09 has greater temporal and spatial detail regarding pre-1900 Great Fire events.

The 1936 PNW GIS Map pattern (Map 2.10) is based on the Forest Survey map of Andrews and Cowlin (1936: map), represented in Table 2.04. Congress authorized the Forest Survey of the United States in 1928, and the Pacific Northwest was selected as the beginning point for the inventory phase of the process. The Pacific Northwest Forest Experiment Station (PNW) was designated to conduct the survey of Washington and Oregon, and work was started in 1930. Hand-colored timber inventory maps were drawn for each county, were accompanied by tables, and have been periodically updated from that time to now (October 2003). Map 2.10 is the 1936 PNW GIS pattern, with the N/E/W/S study pattern, shown on Map 1.01 (see Table 1.03). Despite the generalizing process demonstrated by Figure 2.04, this pattern appears more detailed than the 1900 USDI or 1914 ODF patterns. Again, note the changes in the dark blue, black, red, yellow, and dark polygons from the earlier patterns.

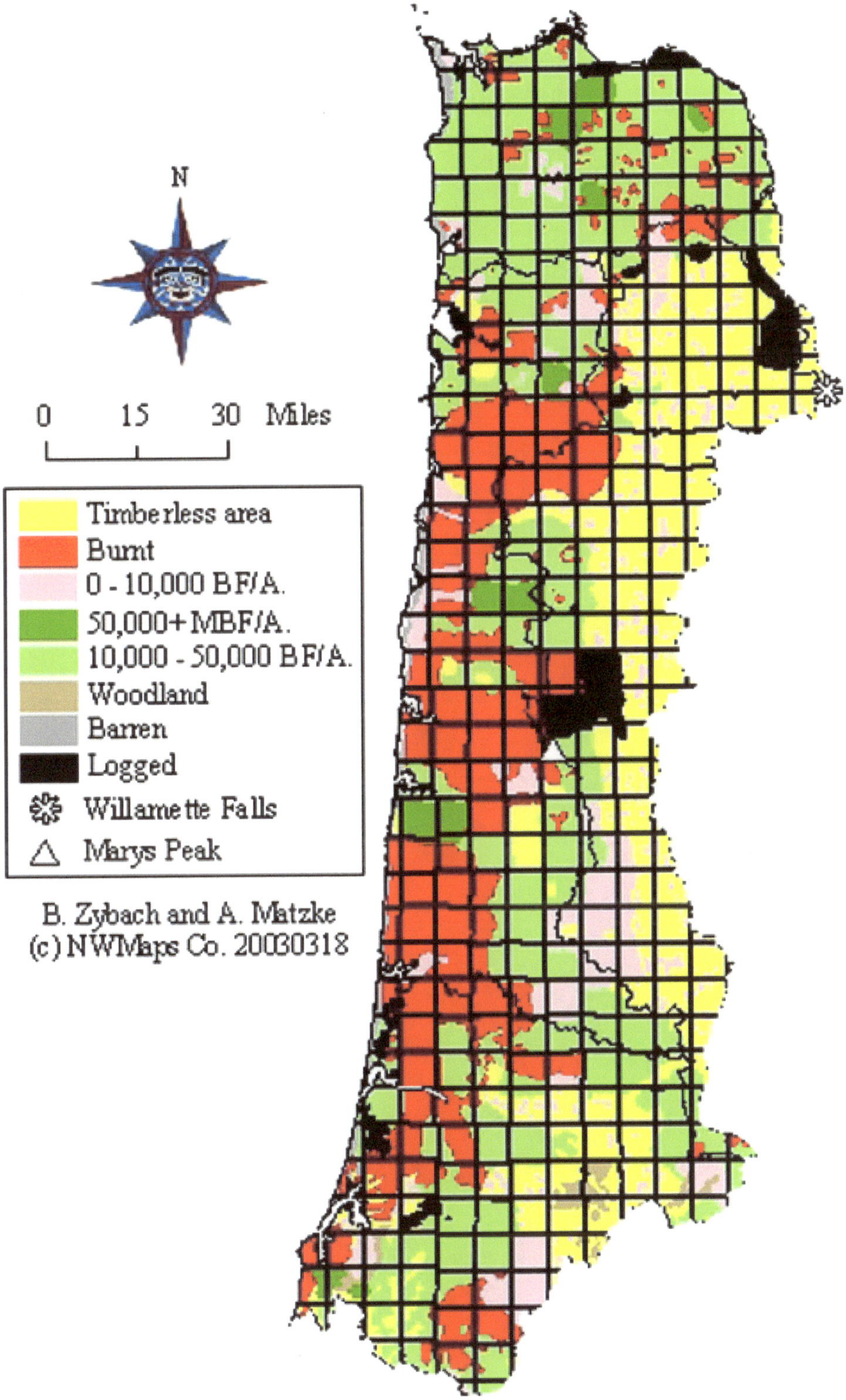


Map 2.08 1900 USDI Forests of Oregon GIS map.

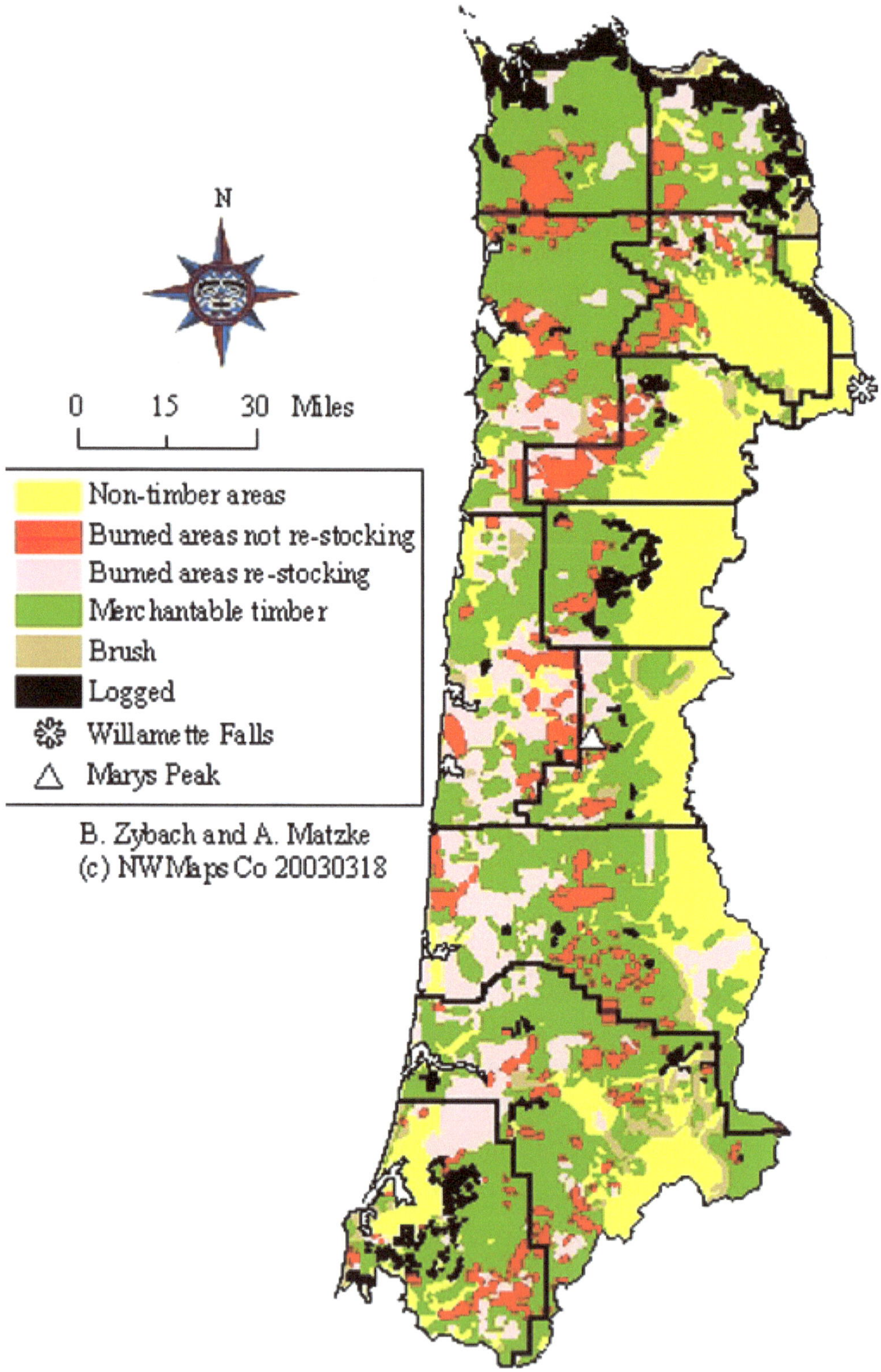


Map 2.09 1914 ODF Forest Fire GIS map.

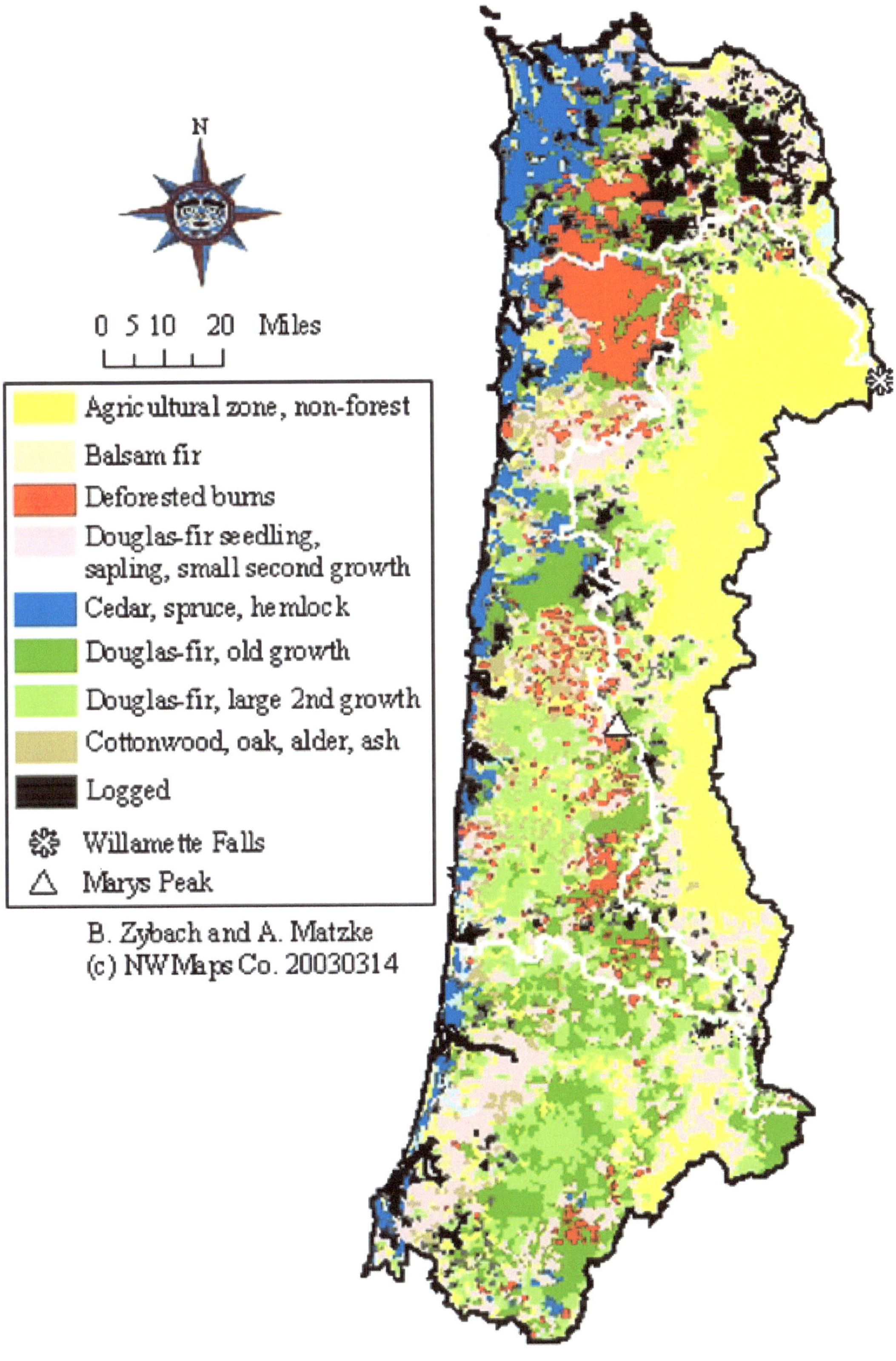


Map 2.10 1936 PNW Forest Type GIS map.

New GIS maps and map layers were produced for this research in one of two ways: the first method of inventory and generalization of existing layers available via the internet or public agency, as just described and illustrated with the historical map sequence; and the second method was to produce new maps and layers directly from GLO notes in areas where Christy and Alverson (2003) had not mapped yet, or in areas where more detail was available. Map 2.11 is a segment of the 1856 GLO subdivision map for Tsp. 14 S., Rng. 8 W. in the "Alseya Valley" study area of Benton County (See Appendix D). The "Indian Trail to Tidewater" segment noted in 1856 exists today in the form of a pre-1900 abandoned wagon road from Alsea to Tidewater. Additional information is provided in the field notes that accompany the map (e.g., Figure 2.04):

> There is a plain Indian trail leading from this township to tide water on the Alseya, which is said to be quite passable for horses. The Indians however generally travel it in their canoes from a point near the west line of the township [T. 14 S., R. 8 W.] and frequently from near the east line of the same [near present town of Alsea] (Hathorn 1856b: 278-279).

This map shows a portion of Tsp. 14, Rng. 8 W. surveyed in July 1856 (Hathorn 1856b: 265-278). Note the location of the Basil Longworth cabin (ibid: 270; Hathorn 1856c: 491-493), established late November 1853 (Longworth 1972: 38-39), on a "Level Prairie." Longworth built his home at the strategic juncture of the trail between: 1) Alseya Valley and *Yaqais*; an important Alsi Indian community at Tidewater (Drucker 1965: 82), and 2) the trail to a Siuslawan eeling spot at the mouth of Deadwood Creek (Rust 1984: 3) -- a tributary of the Siuslaw River--via Lobster River (see Map 2.12). "Old Billy's" cabin is shown on another portion of this map, in Sec. 1, near the juncture of the North and South forks of the Alsea River (see Map 2.02); the termination point of most coastal canoe traffic and a major intersection to several foot and horse trails leading to and from Alseya Valley.

> Alsea Valley (*ts !iphaha*), back in the coast ranges, was a place to which many people went in summer to harvest camas and other wild crops. (Drucker 1965: 82)

Alseya Valley was surveyed and mapped from 1853 until 1897 by at least nine different men under contract to the GLO (see Table 2.03). Note that the surveys were initiated in the 1850s, but not completed until the 1890s. This time lag of more than 40 years followed the pattern of much of western Oregon, including the Coast Range. Township and range lines were surveyed first

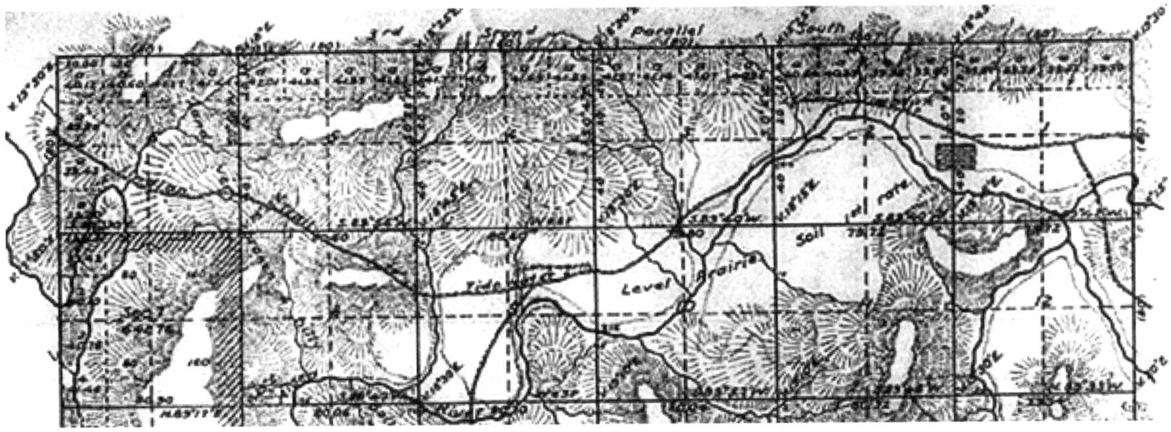


Map 2.11 T. 14 S., R. 8 W. "Indian Trail to Tidewater" segment, 1856.

(see Map 2.01), to establish the boundaries of the 36 square-mile townships, followed almost immediately by subdivisions of square-mile sections containing DLCs (see Map 2.02). Most of this work was completed in the 1850s, before the advent of Oregon statehood in 1859, and the beginning of the Civil War in 1860. Subdivision surveys of unclaimed lands--typically steep, forested, and/or burned tracts that remained in the public domain because they were unsuitable for most agriculture or grazing purposes--were often delayed for a period of twenty to fifty years after initial township boundaries had been established.

As a landform, Alsea Valley consists of bottomlands near the river and its tributaries, surrounded by gentle slopes and low hills, extending upward toward steep mountainsides and waterfalls, surrounded by several high ridges and peaks (see Map D.01). With a base elevation of about 200 feet, the valley is dominated by Marys Peak (McArthur 1982: 474) to the northeast (see Map 2.01; Figure 2.01); at 4,097 feet elevation, it is the highest point in the entire Oregon Coast Range. The median elevation for the valley is about 1,000 feet (Daly and Taylor 2002).

Map 2.12 is a new GIS Map created from GLO survey data assembled over a 40-year period, beginning in 1853. So far as known, this is the first mapping of specific Alsi, Klickitat, Kalapuyan, Yakonan, and Siuslawan canoe, foot-trail, and horse-trail trade routes connecting the Willamette Valley to the ocean (see Map 4.12). The early GLO notes, combined with knowledge of several local precontact archaeological sites, were sufficiently detailed to construct Map 2.12.

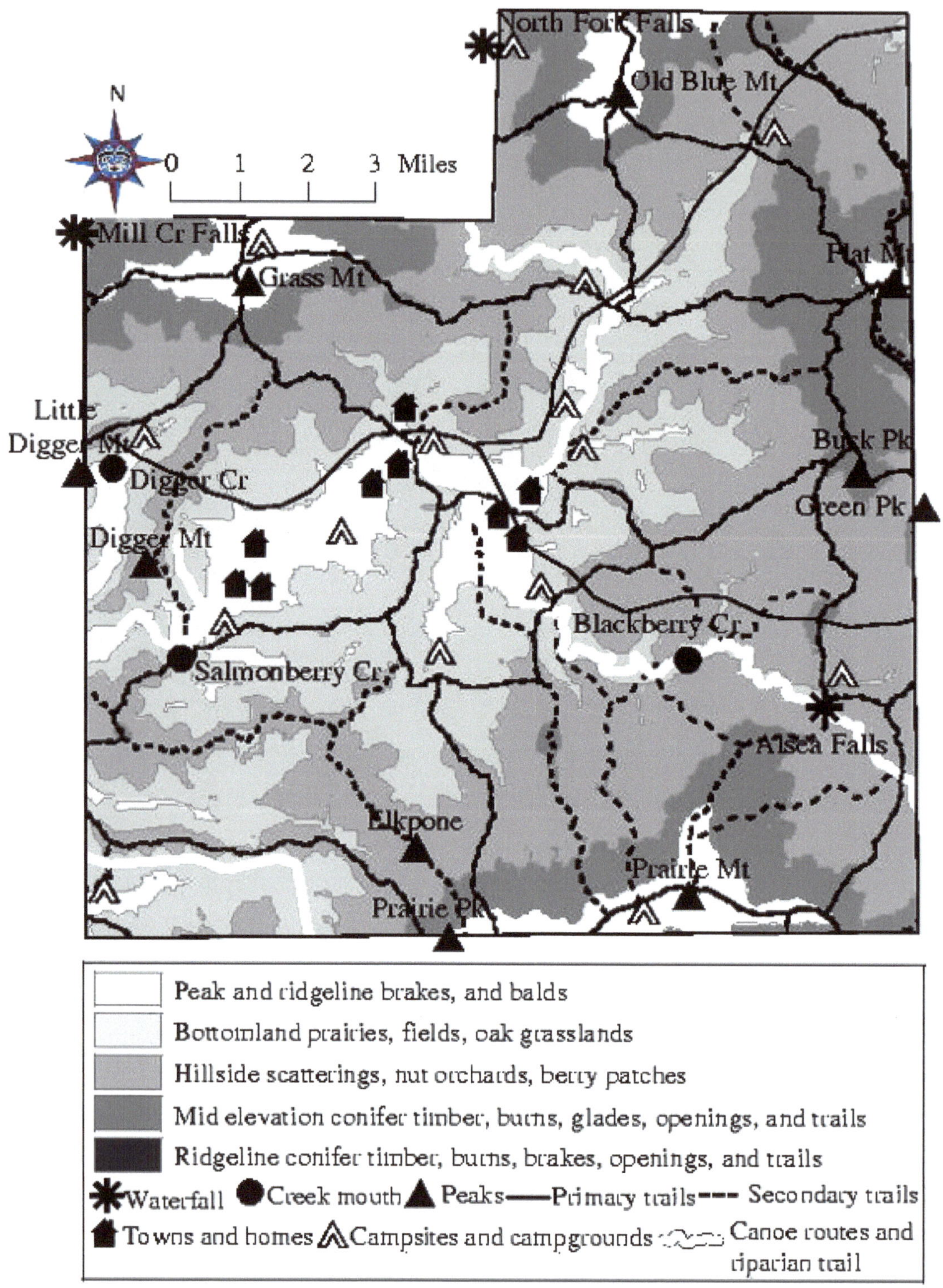


F.K. Lake, B. Zybach, and A. Matzke (c) NWMaps Co. 20030422

Map 2.12 "Alseya Valley" towns, trails, and campgrounds, ca. 1800.

For example, on July 5, 1856, as he was surveying the adjacent land claims of brothers E. Thomas and Asbury E. Ellis (Hathorn 1856c: 485-489; Map 2.02: DLC No. 38 and No. 39), Hathorn made the following series of notes:

> Enter Alseya River, course SW. . . . Leave same . . . Enter another channel of same . . . Leave same, course SW. . . .
> Enter prairie, course NE. and SW. . . . An Indian camp bears West about 25 lks. [16-17 feet] . . . Enter [Ellis] yard, course E. and W. . . . Enter [Ellis] house (Hathorn 1856c: 487)

It is difficult to establish the nationality of the Indian camp, although they were likely Alsi and likely some of the same individuals Hathorn observed traveling by canoe. Rycraft, for example, noted a "traditional" gathering of more than a hundred Kalapuyans, "saltchucks" (Alsi people living near tidewater), and Klickitats that arrived by horseback in the Valley during the mid-1850s (Rycraft ca.1922).

The trail from Wells Creek continued to the North Fork of the Alsea, from there following the stream to its juncture with the South Fork, then splitting westward toward Tidewater and south on the "South trail to the Willamette" (Hathorn 1856a: 159; see Map 2.12; Map 2.14). In addition to named trails that ran parallel to the river and major tributaries, well established trails were also noted along most ridgelines:

> Enter timber . . . Leave timber and enter prairie on summit of ridge, course NW. and SE. Trail, course NW. and SE. . . . An Indian Trail . . . Enter thicket . . . Enter prairie . . . Leave same and enter timber, course NW and SE . . . Indian trail, course E. and W. (Hathorn 1856a: 155).

> Indian trail on summit of ridge, course E. and W. (Hathorn 1856b: 278).

> Top of ridge and trail, bears NE. and SW. . . . Top of ridge and trail, bears NE. and SW. . . Summit of ridge and trail (Gesner 1891a: 285).

> Top of ridge, bears East and West . . . Trail from Alsea to Lobster, bears East and West (Gesner 1891a: 300).

> Trail to Prairie Mountain, course 750 E., and S. 750 W., on top of ridge of same (Collier 1893: 197).

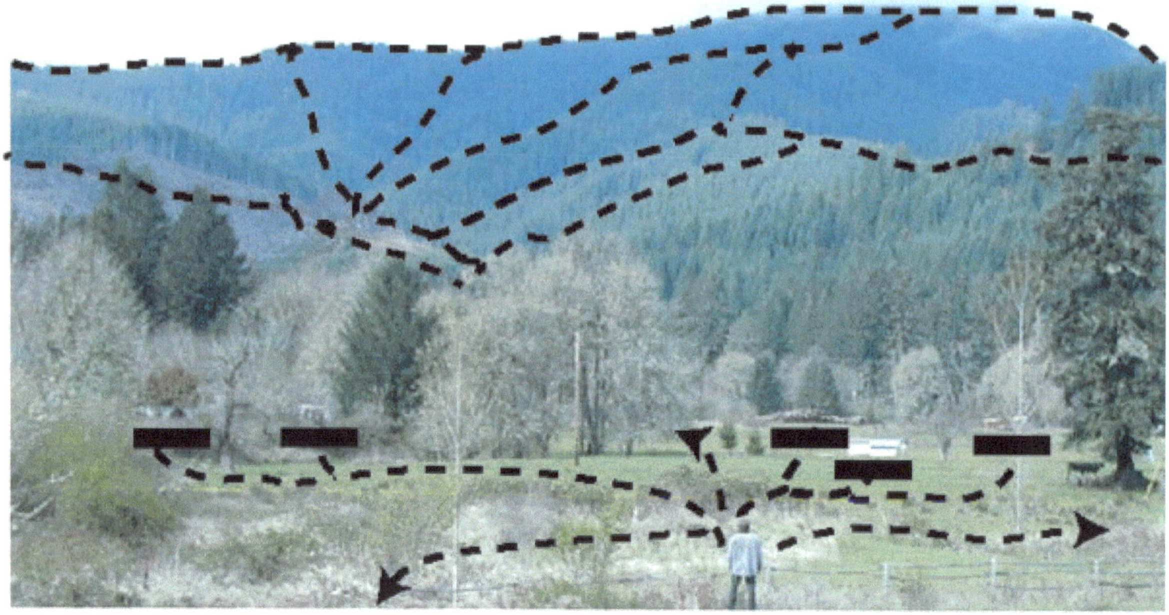

Figure 2.05 Alsea North Fork ridgeline trail network, ca. 1800-2003.

Figure 2.05 (Lapham, N., B. Zybach and A. Gruen 2003) illustrates the riparian and ridgeline trail network used for foot traffic by Alsi people and by neighboring tribes and nations visiting Alsi lands. This is the same basic Alsi foot-trail and Klickitat horse-trail network described in GLO field notes and shown on GLO maps. This location is adjacent to the Alsea River crossing on the 1856 Willamette Valley to Tidewater Indian Trail (see Appendix H.1), facing northwest, across the North Fork prairie. Location is Tsp. 13 S., Rng. 7 W., Sec. 28. Map 2.13 represents another new GIS map of Alseya Valley constructed from GLO information. However, this map is intended to be used in conjunction with historical vegetation maps, therefore the emphasis is on vegetation patterns, rather than centers and travel corridors of human use.

Map 2.13 shows a generalized reconstruction of the pre-contact "lawns, corridors, and mosaics" landscape pattern (Lewis and Ferguson 1999) that characterized Alseya Valley about 1850. It is based almost entirely on GLO notes and maps produced by the surveyors listed in Table 2.03 and incorporates the vegetation types listed in Table 3.03. For purposes of scale and display, individual glades, scatterings, meadows, patches, etc., have been collected into a single band,

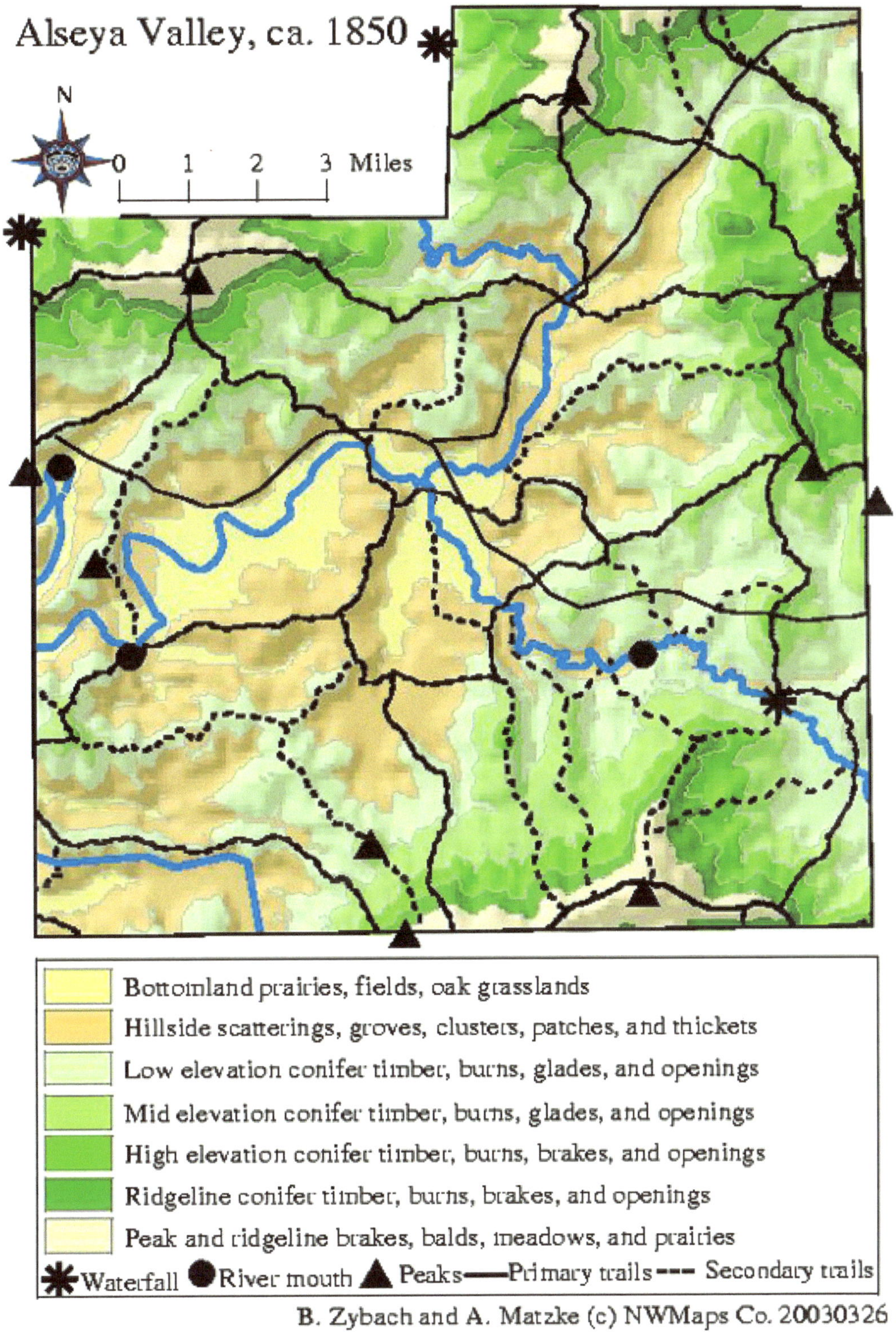


Map 2.13 "Alseya Valley" land use patterns ca. 1850 GIS map.

surrounding bottomland prairies where they were most prevalent. Similarly, ridgeline brakes and prairies are shown at the higher elevations, where they were more frequently found. This process minimizes the existence of stands of conifer forests at lower elevations and exaggerates their extent at higher elevations, eliminates the existence of openings at middle elevations by exaggerating their occurrence at lower and higher elevations, and adds a more northerly extent to the southerly prairies and balds found on peaks (Aldrich 1973: 86-88). Trail networks are likely under-represented, particularly in the lower elevations. Despite these changes, the final pattern is recognizable and representative of conditions that existed at a time preceding white settlement and survey.

## 2.3 Landscape Scales

By studying an area of tens of thousands of acres in size, it is possible to consider vegetation patterns at a landscape level. At this scale GLO data are particularly useful (Schulte and Mladenoff 2001: 7-8). The "Alseya Valley" study area (Zybach 2002b; Table 2.03; Map 2.01; Map 2.13) is about 90,000 acres in size, consisting of three full townships (Tsp. 13 S., Rng. 7 W.; Tsp. 14 S., Rng. 7 W.; and Tsp. 14 S., Rng. 8 W.) and three partial townships (Tsp. 13 S., Rng. 8 W.; Tsp. 15 S., Rng. 7 W.; and Tsp. 15 S., Rng. 8 W.) It is completely contained in Benton County, and has good GLO survey data readily available via the Benton County Surveyor's Office (Mardis 2003: personal communication).

Alsea Valley (the name "Alseya Valley" is an earlier name and refers to the same area in the 1850s) is a relatively small portion of the entire Coast Range (see Map 2.01), so more detail can be displayed graphically than a regional scale map (such as Map 2.01), but with much less detail than a one-section timber cruise map (e.g., Map 2.04). This intermediate landscape scale, between a local scale of dozens or hundreds of acres, and a regional scale of hundreds of thousands or millions of acres, can be termed a basin scale of thousands or tens of thousands of acres.

### 2.3.1 Regional and subregional

GIS was used to help assess the quality of the historical regional maps being used to produce subregional and basin scale maps for this dissertation. The regional test maps are the 1900 USDI Map (Map 2.08), the 1914 ODF Map (Map 2.09) and the 1936 PNW Map (Map 2.10). They are compared to basin scale and a local scale Alsea Valley maps for consistency and detail. This information is also useful for determining scales and vegetation types for new GIS-generated maps.

### 2.3.2 Basin and subbasin

The basin scale map used for comparison with the three regional scale maps will be the 90,000 acre "Alseya Valley" study area, constructed new from 19th century GLO survey data (see Map 2.13; Zybach 2002b). This pattern has a slightly different coloring scheme than the historical maps, but is comparable with number of vegetation types, detail of polygons, and the other type patterns.

Map 2.14 shows "Alseya Valley" basin scale GIS crops of the 1900 USDI, 1914 ODF, and 1936 PNW vegetation patterns. It also shows the 1850 GLO (Map 2.13) pattern for the area. Five Bagley (1915: map; Map 2.04) square mile (640 A.) timber cruise maps are placed on each of the four areas, in their correct geographic positions, and to scale. Using these methods to assess the quality of the regional scale historical maps at basin and subbasin scales, the following conclusions can be made:

1) Although the 1900 USDI area has representative areas of forest, burned timber, grassland, and brush, their combination is almost abstract, with Lobster Valley appearing to be three or four times as large as Alsea Valley. Also, when considered even at the scale of a single 23,000-acre township, the proportions of 1900 USDI vegetation types, when compared to the other maps, appears unrealistic.

2) The 1914 ODF area appears to be more accurate for this area than the 1900 USDI Map, but less detailed than the 1850 GLO Map and the 1936 PNW Map. Representative burns and forest seem to exist even to the local scale level for some of the section overlays. Still, the 1914 ODF polygons are far less detailed than their 1915 Bagley local scale counterparts.

3) A polygon labeled "A," drawn around the bottomland Indian prairies of the mainstem Alsea Valley on the 1850 GLO Map, seems to correlate very well with the polygon labeled "A" in the 1936 PNW Map. However, the two were drawn in entirely different fashions: the 1936 "A" was

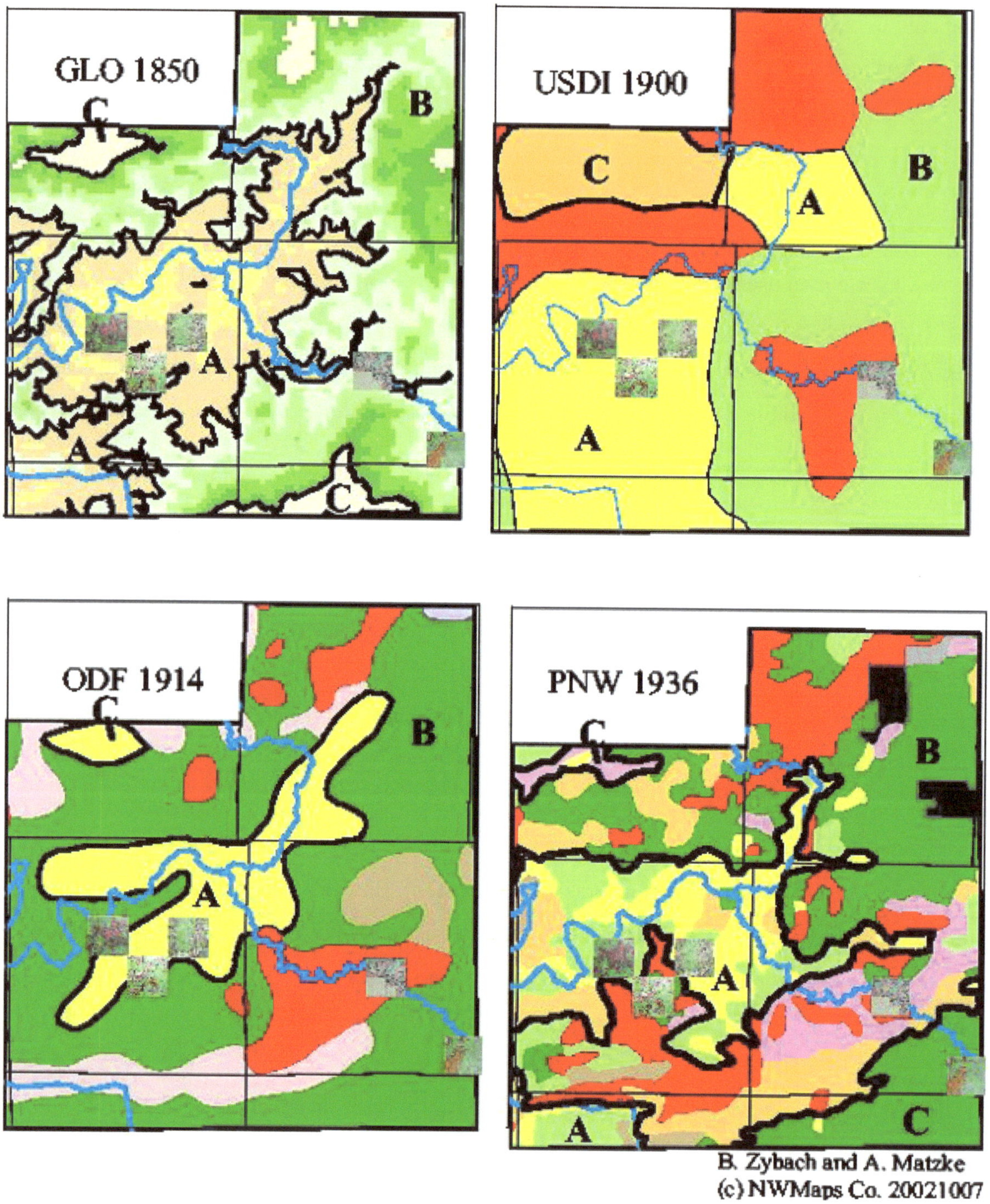


Map 2.14 Alsea Valley forest fire patterns, ca. 1850-1936.

drawn by following the innermost forest and wildfire boundaries, while the other three maps, including the 1850 "A," were drawn by following the outermost grassland boundaries. The area (the most heavily populated portion of the Alsea Valley) apparently has maintained a "cultural artifact" of precontact time that, coupled with subsequent burning, plowing, logging, grazing, and mowing activities, effectively "fire-proofed" the land for local people and most of their structures, and have helped protect it from the Great Fires and local wildfires of the past 150 years.

Alsea Valley, by happenstance, appears to be one of the few large areas in the western Coast Range that hasn't experienced a catastrophic forest fire during the past 250 years. It may not be possible to determine why--or even if--this is true. Yet it may be that it is protected from such events by the presence of unusually high Coast Range peaks in the area, including Marys Peak (see Map 2.01; Figure 2.01), Grass Mountain, Prairie Peak, and Buck Peak (see Map 2.12); or, it may also be that its buffer of historical prairies, brakes and balds form an effective barrier from fire, as apparently shown on Map 2.14. Logging practices since the 1850s may have created similar openings (Fagan 1885: 500, 525-526; Bagley 1915). It is also possible that such a fire did occur, but that rapid growth of local forests and brushlands concealed the event from early white settlers and surveyors. What is known, however, is that large scale (but non-catastrophic) forest fires had taken place in and about "Alseyah Valley" by 1853 (Webster 1853: 98), and continued at regular intervals from then until 1910 (Corvallis Gazette 1902: 2; Bagley 1915: map; Rust 1984: 25), or even more recent times (Kirkpatrick ca. 1940: 30-31; Longwood 1940: 70, 136).

### 2.3.3 Local: patch, stand, and trail

In order to test the three regional maps at a finer, local scale, a series of five Bagley (1915: map) timber cruise maps (including Map 2.05) were somewhat randomly selected for purposes of comparison. The maps represent square mile areas within "Alseya Valley," and were chosen principally on the boldness of the fire pattern or the presence of meadows, trails, and other possible artifacts of precontact time that aren't shown on the regional maps. Map 2.15 shows how the cruise maps were cropped from the larger page, and also gives an idea as to how much detail is available from these sources. Map 2.14 shows Map 2.15 and the other four selections placed on the 1900 USDI, 1914 ODF, and 1936 PNW patterns, as well as the 1850 "Alseya Valley" pattern.

Map 2.16 shows another method of using GIS to compare scale and quality of data. Each of the five 1915 Bagley sections is lined up in a vertical column, arranged in chronological order with similar columns representing the four Alsea Valley vegetation patterns in Map 2.15. These

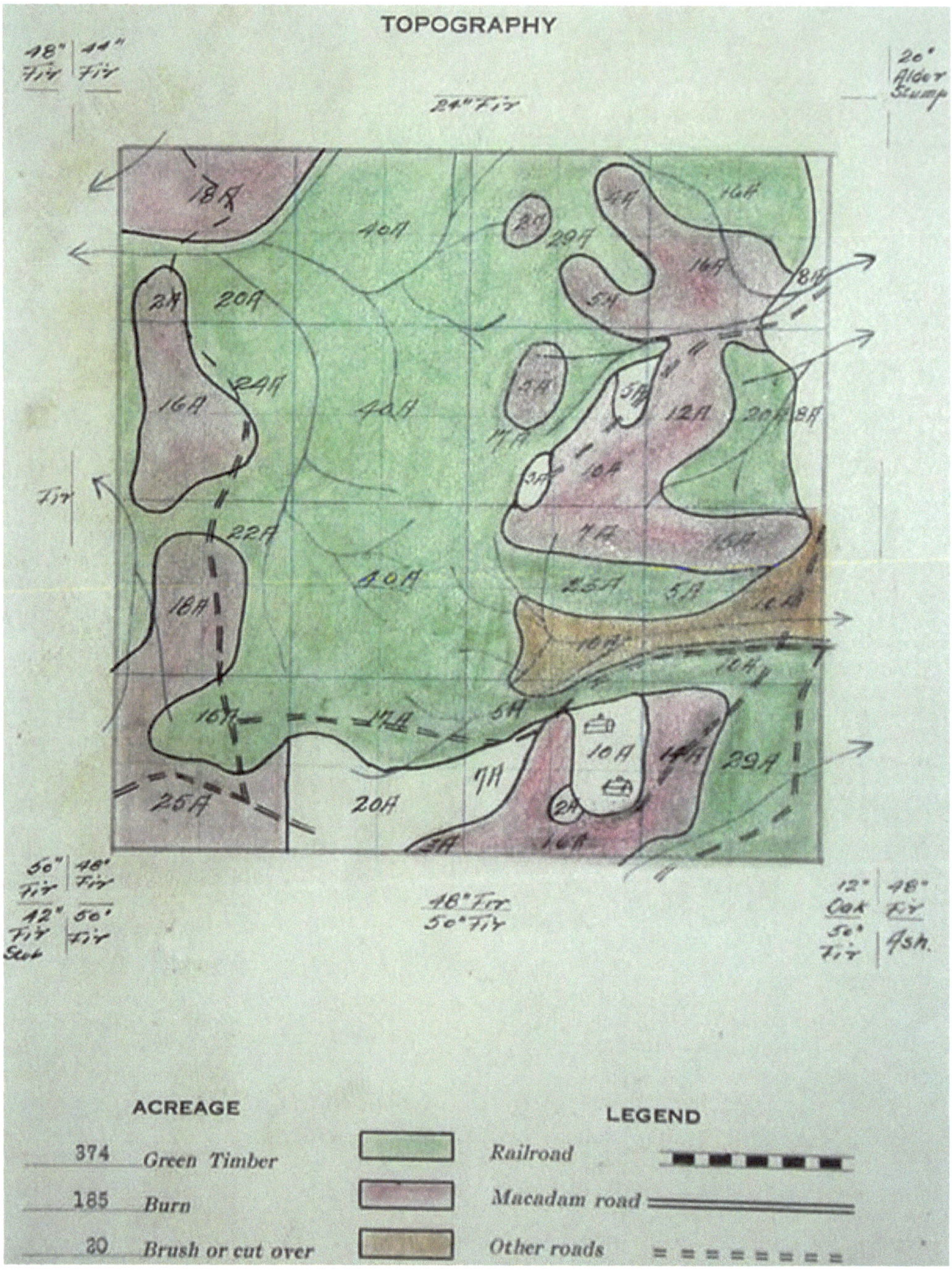


Map 2.15 T. 14 S., R. 8 W., Sec. 14 timber cruise map, 1915.

are the same five sections covered by the overlays on Map 2.14. Viewed in this manner, at a local scale rather than a subbasin scale, different nuances of the patterns emerge:

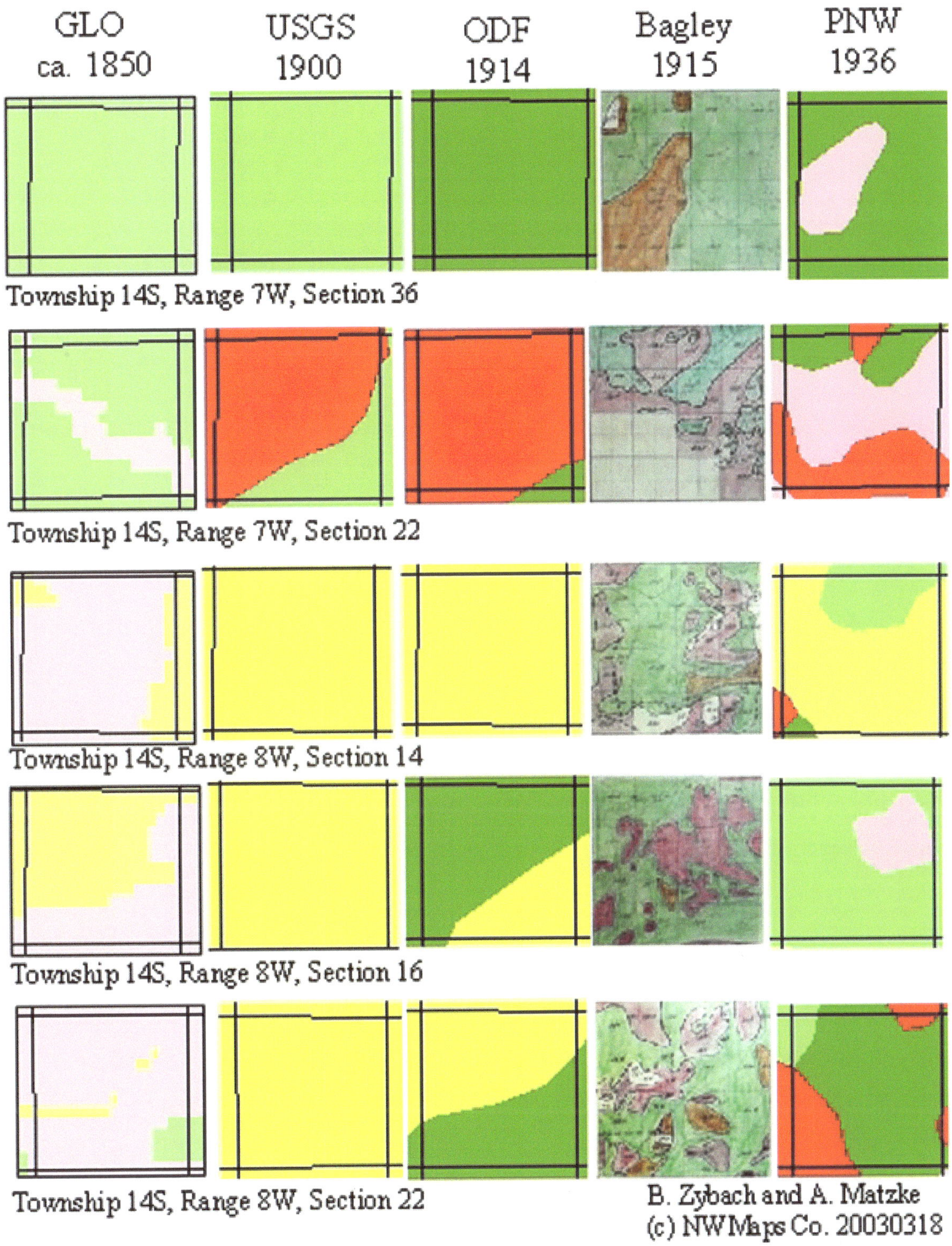


Map 2.16 Alsea Valley section-scale comparisons, ca. 1850-1936.

1) For Tsp. 14 S., Rng. 7 W., Sec. 36 (the top row), both 1915 Bagley and 1936 PNW have similar patterns of logging and brush. The other three maps are uniformly green. The 1850 GLO map, however, made no consideration of logging, so is consistent with 1915 Bagley and 1936 PNW. The 1914 ODF map is probably in error for this section, and the 1936 USDI map is possibly in error.

2) For Tsp. 14 S., Rng. 7 W., Sec. 22 (second from top row), all five maps are in fair agreement, although 1914 ODF seems to over-state the amount of burned ground by quite a bit. Again, the 1915 Bagley and 1936 PNW are in reasonable conformance.

3) For Tsp. 14 S., Rng. 8 W., Sec. 14 (the middle row), none of the four other maps comes close to matching 1915 Bagley for detail or accuracy. The 1850 GLO is the closest, given the strong elevation-based pattern that runs through both maps, as well as the description for the polygon colors that vary between the two maps.

4) For Tsp. 14 S., Rng. 8 W., Sec. 16 (next to the bottom row), only the 1936 PNW is similar to the 1915 Bagley, although the 1850 GLO again shares a more general elevation-based pattern.

5) For Tsp. 14 S., Rng. 8 W., Sec. 22 (bottom row), the result is the same as the previous section: only the 1936 PNW is similar to the 1915 Bagley, although 1850 GLO again shares a more general elevation-based pattern.

In sum, all of the regional maps use representative vegetation types that are comparable over time and at differing scales. The local scale 1915 Bagley maps are, as might be expected, the most detailed, but only represent a small portion of the landscape. The subbasin scale "Alseya Valley" pattern is generalized, but consistent with the local scale maps. The 1900 USDI map appears to be the least accurate of the three regional maps at a subbasin or local scale, but has representative portions of key vegetation types at the township (36 sections) scale. The 1936 PNW map is more detailed than the 1914 ODF map at a local scale, and is likely more accurate at a subbasin scale as well. These considerations were critical to construction of the subregional vegetation and fire pattern maps found in chapters 4 and 5.

## 2.4 Discussion and Summary

This study was compared with other similar studies throughout North America for focus and methods. It demonstrates a generally unique approach, due primarily to a lack of GIS methodology in most of the other studies. The research focus is apparently new: none of the other studies considered or discussed a comparison of long-term Indian burning patterns with subsequent catastrophic forest fire patterns. For those reasons, it is difficult to compare the principal findings of this research with the findings of others, except on the most general basis. A summary of key points in this chapter follows:

### 2.4.1 Research design

General research theory and methods. Research was conducted with the "method of multiple working hypotheses" (Chamberlin 1965): "the effort is to bring up into view every rational explanation of new phenomena, and to develop every tangible hypothesis respecting their cause and history." Research questions are answered based on the "weight of evidence," and research findings include the identification of additional "tangible questions" as they present themselves.

Research boundaries (comparisons). Chapter 1 identified subregional differences in drainage patterns, topography, weather, and vegetation types between the four regional study areas. Chapters 3 and 4 will show, as might be expected, that subregional differences also exist for precontact cultural landscape patterns, early historical land use patterns, and for catastrophic fire history patterns.

### 2.4.2 Research methods

GLO survey. This study demonstrates the potential of GLO maps and survey notes to make good estimates of precontact land use patterns and resource management practices. The detailed mapping of Indian canoe routes, foot and horse trail networks, homesites, and campsites in relation to managed fruit and nut orchards, berry patches, and extensive fields of root and grain crops cannot be duplicated by any other source of information. The location and timing of daily, seasonal, and annual practices of burning, tilling, and harvesting that took place in late precontact times can be easily and reasonably inferred from these records. The great diversity and ready access of managed environments, including grassy prairies, oak groves, meadows, and berry patches, provide insight into the lives of people and cultures that no longer exist. This combination of detailed records and reasoned insights are just as significant for considering precontact wildlife populations, habitat patterns, and food sources.

GIS methodology. New GIS maps and map layers were produced for this research in one of two ways: a method of inventory and common generalization of existing layers available via the internet or public agency, as described and illustrated with a 1900 to 1936 historical map sequence; and the production of new maps and GIS layers directly from GLO notes, as described and illustrated by the "Alseya Valley" example.

### 2.4.3 Mapping scale

Mapping for this dissertation was completed at regional, subregional, basin, and subbasin scales. Local scale data was used to corroborate details and test for accuracy, but new maps were not made at this scale.

1) The three regional historical maps (1900 USDI; 1914 ODF; 1936 PNW) are relatively accurate and complementary at a regional scale, with newer maps being more detailed and accurate at a basin scale than older maps. They are the best available source of information for reconstructing patterns of catastrophic fire and reforestation, from 1849 to 1936.

2) The GLO map and field note data was not as detailed as the timber cruise data for local scale landscape patterns, but was far better suited for interpretations of precontact environments than any of the other sources. In particular, the GLO information is best suited of all the sources for constructing precontact and early historical Indian burning patterns at all scales, especially for the 1750 to 1848 time period.

3) The local scale timber cruise maps are far more detailed than the regional scale historical maps, and slightly more detailed than GLO maps and notes. The local scale cruise maps are limited by scarcity and uneven quality. They were not made very often, and they rarely included the entire landscape. For comparative use purposes (Map 2.16) and for local scale resource management projects, early historical timber cruise maps can have great value.

INDIAN BURNING PATTERNS, 1491-1848

3. *The traveler can but imagine the numbers of these dead tribes by the mounds of clam and oyster shells, many feet in thickness and many yards in extent, which mark the site of their former camping places. The gatherers of these sea-dainties have long since passed away, and even our first records tell of a time when wars, pestilence, and the gradual pressure on these sea coast dwellers by other tribes displaced from their hunting grounds in the east and south, had already done their work.*

--David D. Fagan (1885: 320)

*All the oak timber was owned by well-to-do families and was divided off by lines and boundaries as carefully as the whites have got it surveyed today. It can be easily seen by this that the Indians have carefully preserved the oak timber and have never at any time destroyed it.*

*The Douglas fir timber they say has always encroached on the open prairies and crowded out the other timber; therefore they have continuously burned it and have done all they could to keep it from covering the open lands. Our legends tell when they arrived in the Klamath River country that there were thousands of acres of prairie lands, and with all the burning that they could do the country has been growing up to timber more and more.*

--Che-na-wah Weitch-ah-wah (Thompson 1991: 33)

This chapter provides background on the primary Indian tribes and nations that lived on the Oregon Coast Range during the late 1700s and early 1800s. It then describes the principal burning practices these people used to create landscape scale patterns of trails, forests, prairies, and openings and other habitats across the Range, and lists the plants and animals that populated these habitats. Maps, figures, and tables are used to illustrate and list local scale patterns and wildlife species. Subregional scale patterns of ca. 1800 vegetation and trail are presented as GIS maps.

## 3.1 Historical Coast Range Indian Nations: Background

Indians living in the Oregon Coast Range prior to European American contact essentially viewed the land and sea as their supermarket, hardware store, and pharmacy (Lake 2002). The area naturally offers many biologically diverse and productive habitats which people exploited to provide personal and social necessities. Every ecosystem and vegetative assemblage was likely used and managed in variable intensities over time (Pullen 1996; Boyd 1999b).

In early historical time there were at least eight Indian nations in the Oregon Coast Range and at least 26 distinct tribes. Map 3.01 shows the location of Coast Range Indian tribes during that period of time, from the late 1700s to the 1840s (Berreman 1937; Zybach et al 1995). National boundaries were determined by considering the observations of early journalists, riverine and ridgeline travel corridors, and current understanding of precontact language and cultural affiliations.

Ruby and Brown (1986) and Volume 7 of the Smithsonian's Handbook of North American Indians (Krauss 1990; Miller and Seaburg 1990; Seaburg and Miller 1990; Silverstein 1990; Zenk 1990a; 1990b; 1990c) were primary sources used to determine language names and groups. The Smithsonian publications are probably somewhat more authoritative, but the Ruby and Brown book often provides additional insights and both use a common set of references. Both authorities are in general agreement as to the languages used by Oregon Coast Range Indians, with one notable exception: Ruby and Brown consider all coastal languages south of the Siletz to be of a single stock, Yakonan (1986: 4, 79, 97, 130, 206, 275); whereas Zenk (1990b: 568, 1990c: 572) lists Alsean, Siuslawan, and Coosan. In this instance, a middle route was chosen--two languages: Yakonan (for the Yakona, Alsi, Siuslaw and Kelawatset tribes) and Kusan (for the Hanis and Miluk tribes). This intermediate position is consistent with current perceptions by modern tribal leaders (Phillips 2003: personal communication; Kentta 2003: personal communication). Hall (1992; 1995; 2001) provides details regarding the early historical mix of Athapaskan and Kusan languages in the Coquille River basin.

Whenever possible, the earliest commonly used spellings for individual tribes have been used. Newer spellings and designations are generally less accurate phonetically (e.g., Atfalati vs. Tualatin; Yamel vs. Yamhill; Killamox vs. Tillamook) and potentially confusing when using modern spellings of rivers (e.g., Marys River Indians vs. Chepenafa; Salmon River Indians vs. Nechesne; Upper Coquille Indians vs. Mishikwutmetunne). Earlier spellings also help keep references clear as to time and possible pronunciation.

Table 3.01 lists the tribal groups shown on Map 3.01 and puts them into context with current river names, counties, and cities. Note the close correlation between modern river names and the names of precontact tribes (also see Appendix B). Figure 3.01 shows a selection of tribal members as depicted by a variety of photographers and artists in early historical time. The upper

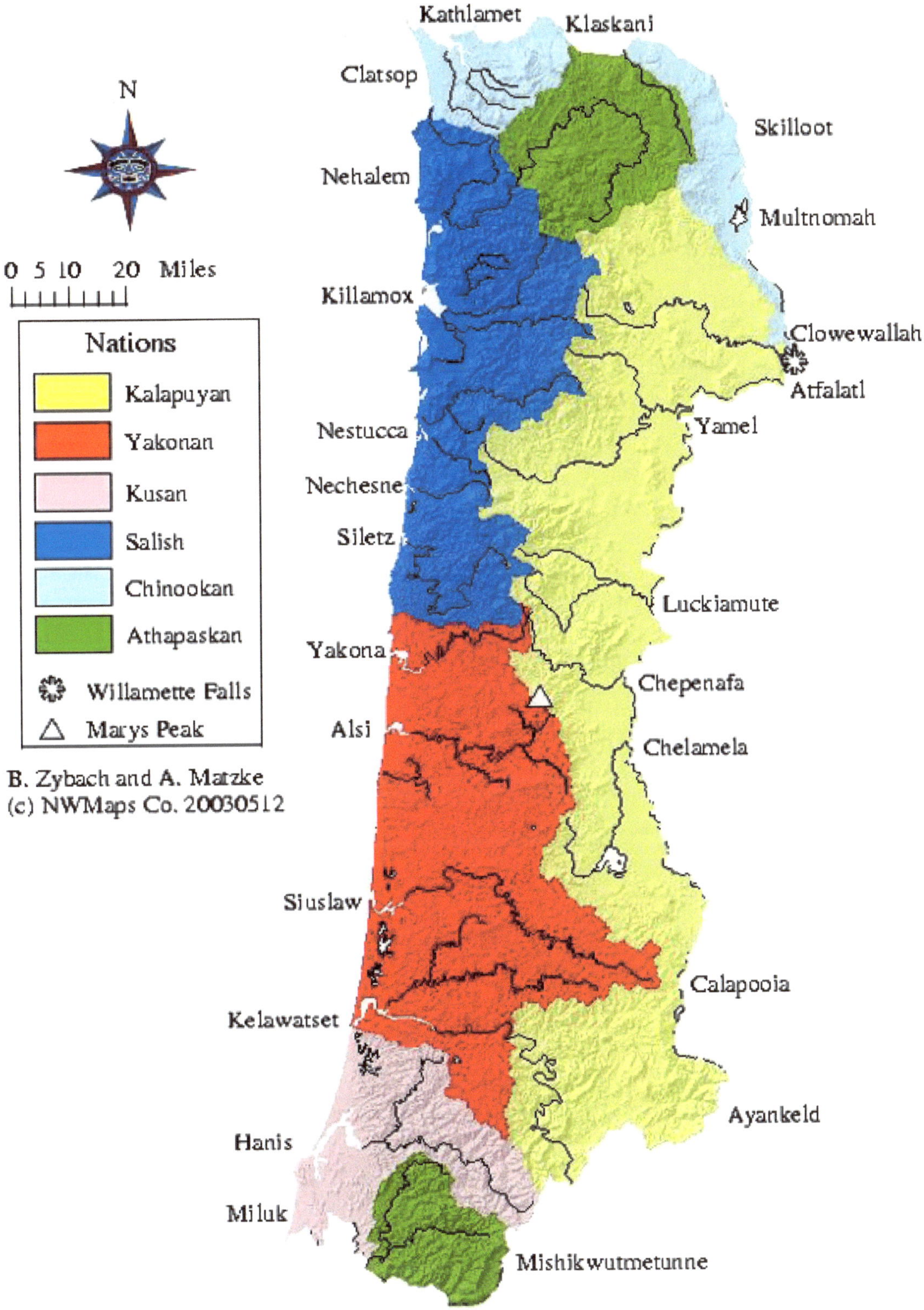


Map 3.01 Oregon Coast Range tribes and nations, ca. 1770.

Table 3.01 Oregon Coast Range tribes, rivers, and counties, 1770-1893.

| **Tribe** | **Language** | **River** | **City** | **County** |
|---|---|---|---|---|
| **North** | | | | |
| Clowwewalla | Chinookan | Willamette | Oregon City | Clackamas |
| Multnomah | Chinookan | Willamette | Portland | Multnomah |
| Skilloot | Chinookan | Columbia | Ranier | Columbia |
| Kathlamet | Chinookan | Columbia | Knappa | Clatsop |
| Clatsop | Chinookan | Youngs | Astoria | Clatsop |
| Klaskani | Athapaskan | Clatskanie | Clatskanie | Columbia |
| Nehalem | Salish | Nehalem | Nehalem | Tillamook |
| **East** | | | | |
| Atfalati | Kalapuyan | Tualatin | Tualatin | Washington |
| Yamel | Kalapuyan | Yamhill | Yamhill | Yamhill |
| Luckiamute | Kalapuyan | Luckiamute | Dallas | Polk |
| Chepenafa | Kalapuyan | Marys | Corvallis | Benton |
| Chelamela | Kalapuyan | Long Tom | Monroe | Benton |
| Calapooia | Kalapuyan | Willamette | Eugene | Lane |
| **West** | | | | |
| Killamox | Salish | Tillamook | Tillamook | Tillamook |
| Nestucca | Salish | Nestucca | Pacific City | Tillamook |
| Nechesne | Salish | Salmon | Rose Lodge | Lincoln |
| Siletz | Salish | Siletz | Siletz | Lincoln |
| Yakona | Yakonan | Yaquina | Newport | Lincoln |
| Alsi | Yakonan | Alsea | Waldport | Lincoln |
| Siuslaw | Yakonan | Siuslaw | Florence | Lane |
| **South** | | | | |
| Ayankeld | Kalapuyan | Umpqua | Yoncalla | Douglas |
| Kelawatset | Yakonan | Umpqua | Reedsport | Douglas |
| Hanis | Kusan | Coos | Coos Bay | Coos |
| Miluk | Kusan | Coquille | Bandon | Coos |
| Mishikwutmetunne | Athapaskan | Coquille | Coquille | Coos |

left photograph is of two Salish women, possibly Killamox, on a "trading trip" (Sauter and Johnson 1974: 29). The upper right drawing is of a Chinookan woman and her child. The drawing is thought to be made from a sketch by George Catlin and to depict a likely "superfluity of tattoos" (Ruby and Brown 1988: 80). Note the flattened head of the woman, denoting

Figure 3.01 Native people of the Oregon Coast Range.

"royalty" or upper caste, and the device used to create a similar effect on the child. This was a common practice among Chinookan people and certain adjacent tribes and gave rise to the general name of "flatheads" applied to several regional tribes and nations by early trappers and explorers (Carey 1971: 12). The lower left hand corner shows a family of Yakona Indians, near present-day Newport (Nash 1976: 150). This picture was drawn shortly after the Yaquina River basin had been withdrawn from the Coast Range Reservation and opened to white settlement. Note the post contact dresses, pants, and shirt combined with traditional feather headdresses, tattoos and necklaces. Kentta (2003: personal communication) speculates that modern clothing styles were inspired by early missionaries, who wanted local Indians to dress in a "more modest" fashion than provided by traditional clothing. The lower right hand drawing was made of a

Kalapuyan man, possibly of the Chelamela tribe, in 1841 by Alfred Agate, a member of the Wilkes Expedition, near present-day Monroe, in Benton County (Wilkes 1845b: 223). Note the bare hills and isolated Douglas-fir trees in the background, the forbs at his feet, and the sealskin quiver--a sign of trade or other contact with adjacent tribes. Wilkes described the occasion of the latter drawing with this account:

> Some wandering Callapuyas came to the camp, who proved to be acquaintances of Warfield's wife: they were very poorly provided with necessaries. Mr. Agate took a characteristic drawing of one of the old men.
>
> These Indians were known to many of the hunters, who manifested much pleasure at meeting with their old acquaintances, each vying with the other in affording them and their wives entertainment by sharing part of their provisions with them. This hospitality showed them in a pleasing light, and proved that both parties felt the utmost good-will towards each other. The Indians were for the most part clothed in deer-skins, with fox-skin caps, or cast-off clothing of the whites; their arms, except in the case of three or four, who had rifles, were bows and arrows, similar to those I have described as used at the north; their arrows were carried in a quiver made of seal-skin, which was suspended over the shoulders (Wilkes 1845b: 224).

Large wood products. Precontact Indian people used large wood products throughout the Coast Range over long periods of time. Figure 3.02 shows two principal uses of logs and planks by a variety of tribes. The two upper pictures show a photograph of a plank house near the mouth of the Umpqua River, taken in 1858 by an army officer stationed at Fort Umpqua (Buan and Lewis 1991: 93) and a drawing is of a similarly styled home, probably Kusan, made near the same location during the same year (Douthit 1986: 113). The lower right picture shows an 1841 drawing by Agate (Wilkes 1845a: 341) of the interior of a Chinookan lodge. Consider the amount of firewood needed to heat structures of this size--also, the amount of lumber needed for construction and maintenance:

> A single large house may have required as much as 70,000 board feet of lumber: One such structure near Portland, Oregon, was used continuously for 400 years and would have required between 500,000 and 1 million board feet of lumber during that period for maintenance and repair. And that is just one house, 55 feet wide and 120 feet long, home to forty-five to sixty people (Suttles and Ames 1997: 273).

Figure 3.02 Large wood products, Oregon Coast Range, 1788-1860.

The ocean going canoe in the lower left hand corner of Figure 3.02 (Sauter and Johnson 1974: 109) is typical of a type used by Salish people to travel and trade along the coastline. Such a canoe could easily hold more than twenty people, hundreds of pounds of seal, fish, or kelp, and facilitate the widespread trade of foods, baskets, slaves, or other items of common value. One of the earliest accounts of these canoes is by Haswell, off the mouth Tillamook Bay, in 1788 (Elliott 1928: 171): "at this time we discovered a canoe with ten natives of the country paddling towards us on there nigh approach they made very expressive seigns of friendship."

Writing from his winter encampment at Fort Clatsop on February 1, 1806, Lewis made the following observations:

> The Canoes of the natives inhabiting the lower part of the Columbia River make their canoes remarkably neat light and well addapted for riding high waves. I have seen the natives near the coast riding waves in these canoes with safety and apparently without concern where I should have thought it impossible for any vessel of the same size to [have] lived a minute. they are built of whitecedar or Arborvita generally, but sometimes of firr. they are cut out of a solid stick of timber . . . they are all furnished with more or less crossbars in proportion to the size of the canoe . . . these crossbars serve to lift and manage the canoe on land . . . some of the large canoes are upwards of 50 feet long and will carry from 8 to 10 thousand lbs. Or from 20 to thirty persons . . . 4 men are competent to carry them a considerable distance say a mile without resting . . . This form of canoe we did not meet with until we reached tidewater or below the grand rapids. From thence down it is common to all the nations but more particularly the Killamucks and others of the coast . . . In the management of these canoes the women are equally expert with the men (Thwaites 1959b: 30-32).

Disease. A common tragedy among all of the Coast Range nations was the decimation of nearly all communities and families via diseases introduced by European, African, and American explorers and fur traders, beginning in the 1770s, and perhaps dating to much earlier times. Haswell, for example, made the following observations of people near the mouth of the Siletz and Salmon rivers in 1788:

> They were armed with bows and arrows they had allso spears but would part with none of them they had both Iron and stone knives which they allways kept in there hands uplifted in readiness to strike we admitted one of them onboard but he would not come without this weepen two or three of our visitors were much pitted with the small pox (Elliott 1928a: 171).

Clark confirmed Haswell's observations--and also extended the geography of the wide-ranging effects of the ca. 1770s smallpox epidemic (Boyd 1999a) -- when he made the following observations near the mouth of the Willamette River on April 3, 1806:

> at 3 P. M. we arrived at the residence of our Pilot which consists of one long house with seven appartments or rooms in square form about 30 feet each room opening into a passage which is quit[e] through the house . . . back of this house I observe the wreck of 5 house remaining of a very large village, the houses of which had been built in the form of those we first saw at the long narrows of the *E-lute* Nation with whome those people are connected. I endeavored to obtain from those people of the situation of their nation, if scattered or what had become of the nativs who must have peopled this great town. an old man who appeared of some note among them and father to my guide brought forward a woman who was badly marked with the Small Pox and made signs that they all died with the disorder which marked her face, and which she was verry near dieing with when

> a girl. from the age of this woman this Disruptive disorder I judge must have been about 28 or 30 years past [1776-1778], and about the time the Clatsops inform us that this disorder raged in their towns and distroyed their nation (Thwaites 1959b: 240-241).

Introduced diseases inflicted a heavy toll on the social organization and infrastructure of Indian land management practices, particularly burning and trading networks. Catastrophic diseases between 1770 and 1850 removed at least 80% of the Indian population of western Oregon, including the Coast Range (Boyd 1999a). Settlement of whites followed closely behind Indian depopulation (Scott 1928). Many areas of the landscape, once commonly under Indian management and burning regimes, transitioned into "wilderness" (Anderson 1996). As a result, specific vegetation assemblages--orchards of oaks and filberts, berry patches, basketry and root gardens, and grasslands managed for hunting--missed Indian burning return intervals and forestation of prairies, brakes, balds, and meadows went unchecked (see Appendix F). What many white settlers came to witness was a transitional landscape moving from intentional management to one more influenced by natural processes (Zybach and Lake: In Review). Remnant populations of Indians focused what limited management practices they could in face of disease, genocide, and forced removal from their homelands and on to reservations. For example, writing in the 1840s, Lee and Frost (1968: 99) noted of the Klaskanis: "A clan called the Claskanios lived upon the streams which empty into the head of Young's Bay, which clan is very nearly extinct." Thirty years later, the remaining few members of the tribe were forcefully evicted from their ancestral homes and moved to a new location:

> Soldiers forced the Clatskanie Indians to leave their homes in the Upper Nehalem River Valley in the early 1870s. They had to walk over the old Salem-Astoria Military Road to their new homes in the Yamhill Valley on the Grande Ronde Indian Reservation. (Martin and Fick 2002: 95).

A chilling eyewitness account of the subsequent impact of diseases on the populations of Nechesne (see Zobel 2002) and Siletz peoples first described by Haswell (Elliott 1926: 171) in 1788 is given by Talbot as he traveled along Siletz Bay toward Salmon River (see Map 1.03) in early September 1849:

> Recrossing the horses, we extricated ourselves from this marsh and traveled down the shore of the [Siletz] bay. It was about three and a half miles long - greatest width one mile. The opposite shore was almost concealed from view by the fog, but it seemed to be heavily timbered. . . It is the custom of the Indians in

> this country to deposit their dead in canoes, and there are a great number of them along the borders of the bay.
>
> Early this morning an old Indian entered our camp. He had come in a canoe from some distance up the bay, his attention having been attracted by a large fire which we had built last evening on the southern point of the inlet. He said that himself and another man, with their families, were the only residents on this bay - the last lingering remnants of a large population which once dwelt upon these waters . . .
>
> . . . bidding our final adieau to the ocean, we struck northeast, following a small trail [present-day Highway 18] which led us over rolling hills covered with grass and a high growth of fern. About a mile to our right lay a handsome little fresh-water lake [Devils Lake], and beyond rose a succession of ridges and tall forests. Having come three miles through the hills we descended into a fine bottom lying along the banks of a stream about fifty feet wide [Salmon River] . . . There are no Indians living here (Haskin 1948: 12).

Between the evidence of small pox noted by Haswell in the late 1780s, the dramatic depopulation of northern Coast Range towns and tribes documented by Lewis and Clark in the early 1800s, and the coastal landscape nearly devoid of Indians described by Talbot in 1848, was a period of only 60 years; about three generations of people. It is reasonable (and sad) to assume that the amount of burning, tilling, trail maintenance, and other actions related to resources management declined accordingly during the same time.

### 3.1.1 North: Chinookan, Athapaskan, and Salish

The northern Coast Range was inhabited by Chinookan people (Ruby and Brown 1988) from Willamette Falls to the mouth of the Columbia River, represented by the Clatsop (Silverstein 1990: 533-546; Ruby and Brown 1986: 30-31), Kathlamet (Silverstein 1990: 533-546; Ruby and Brown 1986: 11-12), Skilloot (Silverstein 1990: 533-546; Ruby and Brown 1986: 208), Multnomah (Silverstein 1990: 533-546; Ruby and Brown 1986: 142), and Clowwewalla (Silverstein 1990: 533-546; Ruby and Brown 1986: 31-32) tribes. The Salish Nehalem tribe (Seaburg and Miller 1990: 560-567; Ruby and Brown 1986: 240-243) lived along the seacoast, to the south of Tillamook Head, an historical boundary between them and the Clatsop. Klaskani (Krauss 1990: 530-532; Ruby and Brown 1986: 29) people largely occupied the forested Tualatin Hills that bordered the Columbia River and the headwaters of the Nehalem River, to the north and west of Willamette Valley Kalapuyans (see map 3.01).

Some of the earliest and most detailed accounts of the Chinookans and their local landscape were by Lewis and Clark, during their travels of late 1805 and early 1806. Clark, for example, described a Skilloot town and nearby prairies and woodlands on November 4, 1805:

> on the Main Lard. [larboard, or left] Shore [Oregon side of the Columbia] a Short distance below the last Island we landed at a village of 25 *houses*: 24 of those houses we[re] thached with Straw, and covered with bark, the other House is built of boards in the form of those above, except that it is above ground and about 50 feet in length (and covered with broad split boards) This village contains about 200 Men of the *Skilloot* nation I counted 52 canoes on the bank in front of this village maney of them verry large and raised in bow. we recognized the man who took us over last night . . . he invited us to a lodge in which he had Some part and gave us roundish roots about the Size of a Small Irish potato which they roasted in the embers until they became Soft, This root they call *Wap-pa-to,* the *Bulb* of which the Chinese cultivate in great quantities . . . we purchased about 4 bushels of this root and divided it to our party,
>
> at 7 miles below this village passed the upper point of a large Island nearest the Lard. Side, a Small Prarie in which there is a pond opposit on the Stard. [starboard, or right] here I landed and walked on Shore, about 3 miles a fine open Prarie for about 1 mile, back of which countrey rises gradually and wood land commences Such as white oake, pine of different kinds, wild crabs [crabapples] (Thwaites 1959a: 196-197).

Note the references to, and availability of local bulbs (wapato), nuts (oak), and fruits (crabapples). Several early residents of Astoria, including Gabriel Franchere (Franchere 1967), Ross Cox (Cox 1957), and Alexander Ross (Ross 1923), also provided detailed accounts of Chinookan tribes during the 1810-1814 time period.

#### 3.1.2 East: Kalapuyan

The eastern slopes of the Coast Range were inhabited by Kalapuyan people, who maintained tens of thousands of contiguous oak savannah acres through the practice of annual broadcast burns (Collins 1951; Boyd 1986; Gilsen 1989). As with other tribes of the Coast Range, most Kalapuyans lived in tribes closely associated with a particular river (see Table 3.01; Map 1.03). The Atfalati (Zenk 1990a: 547-553; Ruby and Brown 1986: 5-6) occupied the mouth and headwaters of the Tualatin River; the Yamel (Zenk 1990a: 547-553; Ruby and Brown 1986: 274-275) lived to their south; along the Yamhill River; the Luckiamute (Zenk 1990a: 547-553; Ruby and Brown 1986: 109-110) lived to the south of the Yamel, along the Luckiamute River, the

Chepenafa (Zenk 1990a: 547-553; Ruby and Brown 1986: 18-19) lived along the Marys River, to the south of the Luckiamute, although they likely shared common grounds, such as Soap Creek Valley, a tributary to the Luckiamute. The Chelamela (Zenk 1990a: 547-553; Ruby and Brown 1986: 17) resided along the Long Tom River to the south of the Chepenafa. The Calapooia (Zenk 1990a: 547-553; Ruby and Brown 1986: 10-11) were more wide-ranging than other Kalapuyans during early historical time and apparently occupied much of the woodlands along the headwaters of the Willamette, Long Tom, and Siuslaw rivers, and traveled southward into the Umpqua basin.

The Kalapuyans of the Willamette Valley are one of the best documented nations of the Coast Range (Williams 2003). Early observers include Alexander Roderick McLeod (Davies 1961) and David Douglas (Douglas 1906) in the 1820s, Methodist missionaries (Lee and Frost 1968) and beaver hunters (Pipes 1934) in the 1830s, and numerous early settlers (e.g., Neall 1977; Phinney 2000) and others (e.g. Wilkes 1845) during the 1840s.

### 3.1.3 West: Salish and Yakonan

Salish and Yakonan speaking tribes dominated the western Coast Range (see Map 3.01 and Table 3.01). The Killamox (Seaburg and Miller 1990: 560-567; Ruby and Brown 1986: 240-243) lived in the vicinity of Tillamook Bay, to the south of the Nehalem and to the north of the Nestucca (Seaburg and Miller 1990: 560-567; Ruby and Brown 1986: 240-243), who lived between the Killamox and Cascade Head, in north Lincoln County, which separated them from the Nechesne (Seaburg and Miller 1990: 560-567; Ruby and Brown 1986: 240-243; Zobel 2002), also known as the Salmon River Indians. To the south of the Nechesne were the Siletz (Kent 1977; Seaburg and Miller 1990: 560-567; Ruby and Brown 1986: 202), who lived along Siletz Bay and extended southward as far as Yaquina Head, to the north of Yaquina Bay. All four tribes spoke Salish, the southernmost nation of people to use this language in historical time.

To the south of Yaquina Head, and extending southward to Tenmile Lake between the Umpqua and Coos rivers, were the Yakonan speakers. The Yakona (Zenk 1990b: 568-571; Ruby and Brown 1986: 275-276) lived along the Yaquina River, the Alsi (Zenk 1990b: 568-571; Ruby and Brown 1986: 4-5) along the Alsea River, and the Siuslaw (Zenk 1990c: 572-579; Ruby and Brown 1986: 206-207) along the Siuslaw River. Haswell was the first to note differences in the Yakona and Salish people, when he observed in 1788:

> The long boat in the evening returned alongside they had seen nothing remarkable except vast numbers of the [probably Alsi] natives they appeared to be a very hostile and warlike people they ran along shore waving white skins these are the skins of moose deer three or four thicknesses compleatly tanned and not penetrable by arrows these are there war armour they would some times make fast there bows and quivers of arrows to there spears of considerable length and shake them at us with an air of defyence every jesture they accompaneyed with hideous shouting (Elliott 1928: 169-170)

The following day Haswell sailed further north with his ship, where they encountered an entirely different response:

> Made sail along shore at 11 A M there came alongside two Indians [likely Nechesne or Siletz] in a small canoe very differently formed from those we had seen to the southward it was very sharp at the head and stern and Extremely well built to paddle fast they came very cautiously toward us nor would they come within pistol shot untill one of them a very fine look'g fellow had delivered a long oration accompaneying it with actions and jestures that would have graced a European oritor the subject of his discorse was designed to inform us they had plenty of Fish & fresh water on shore at there habitations which they seemed to wish us to go and partake of (Elliott 1928: 170)

One line of speculation is that the Alsi people had already been subjected to European diseases and had determined the source of their problem as having arrived by sea, in the same manner and from the same direction as Haswell and his shipmates.

#### 3.1.4 South: Kusan, Athapaskan, and Kalapuyan

The southern Coast Range contained a mix of cultures, weather, and landscape patterns that reflected similarities to each of the other three subregions. The Kelawatset (Zenk 1990c: 572-579; Ruby and Brown 1986: 97-99) lived along the lower Umpqua River mainstem, and were the southernmost speakers of the Yakonan language. Their territory extended south to Tenmile Lake, which the Hanis (Zenk 1990c: 572-579; Ruby and Brown 1986: 79-81) are said to have claimed for the wapato that grew there. They also lived along the Millicoma River, although it was named for the Miluk tribe (Zenk 1990c: 572-579; Ruby and Brown 1986: 130-133), whose territory included land south of Coos Bay to the mouth of the Coquille River, and then inland, to the present-day site of Myrtle Point. Both Hanis and Miluk spoke Kusan, although the Miluk are thought to have been largely bilingual (Hall 1992; Hall 1995: 25-38); if so, a likely result of being bordered on the north by the Hanis, and to the east and south by the Mishikwutmetunne (Miller and Seaburg 1990: 580-588; Ruby and Brown 1986: 64-66) and other Athapaskan speaking tribes. To the east of the Yakonan and Kusan people was the southernmost tribe of Kalapuyans, the Ayankeld (Zenk 1990a: 547-553; Ruby and Brown 1986: 276-277), who lived near present-day Yoncalla, south to present-day Roseburg or Winston. These people maintained the central Umpqua Valley in nearly the same manner that other Kalapuyan tribes maintained the Willamette Valley: that is, they used fire annually to broadcast burn great expanses of oak savannah and grasslands. A key difference with the Willamette Valley people is that the climate was drier, so grass and ferns grew less profusely and acorns were harvested from mostly black oak and tanoak, rather than white oak.

The Indian communities of the southern Coast Range (in common with the other nations of western Oregon) were decimated by diseases a generation or more before their lands were claimed and occupied by white immigrants (Boyd 1999a). Writing in 1856, Dr. John Milhau noted:

> The Kalawatsets and Coos indians . . . subsist chiefly on fish, berries, roots, and seeds, but are fond of whales and seals . . . the Indians of these Tribes have at one time been very numerous, the number of [people] and varieties of habitations showing that every stream and nook was once populated (Younker et al. 2000: 2).

## 3.2 Types of Burning Practices

Fire was one of the most energetically efficient tools available to manipulate local ecosystems to produce or induce ecological qualities and derive socially desired products from the land (Kimmerer and Lake 2002). Through time, observations of natural processes, and experience, people learned to use fire to maintain areas of biological diversity and to enhance the productive capabilities of the land (Anderson 1997; Turner 1991). As a result, they were able to consistently obtain a wide variety of foods, construction materials, medicines, and other products from known locations during certain seasons throughout the year. Such practices also reduced the likelihood of wildfire, and ensured personal and community safety when wildfires did occur (Williams 2000).

Indian burning patterns, by definition, are caused by people, and are the result of purposeful actions. Occasional fire escapements were probably a significant part of the landscape pattern created by daily firewood storage and use, situational patch burning, and seasonal broadcast burning. Trails would have been regularly cleared by fire and routinely harvested for firewood along their routes (Norton et al 1999). The same would likely have been true of canoe routes, at least seasonally, along stretches of most low gradient streams and rivers throughout the Range.

The use of fire in the landscape varied from culture to culture over time, and according to circumstance. Differing climates, topographies, and plant assemblages led to--and resulted from--such differences throughout the region (Boyd 1999b). All tribal groups used firewood, wove baskets, or manipulated vegetation that affected fuel structure and composition. Indian people managed oak savannas for acorns, and prairies primarily for seed, camas or other bulbs, or root crops (Norton 1979b). Indians on the coast used fire along coastal headlands, affecting the production of berries, ferns, and facilitating more open wildlife habitat (Pullen 1996). Local knowledge of fire and fire's effects on different ecosystems increased predictability and certainty of such seasonally available resources (Turner 1991). Increased diversity, predictability, and certainty resulted in improved individual quality of life and social security for local communities (Anderson 1996).

Table 3.02 (Lake and Zybach: In Review) describes three major types of Indian burning practices that affected landscape patterns of vegetation and provided definition to local wildlife habitat conditions: firewood gathering and burning, patch burning, and broadcast burning. Firewood gathering and burning involves the movement of fuels to specific locations, resulting in areas containing relatively little (or stockpiled) large, woody debris and spots of repeated, intense, and prolonged heat. Patch burning is defined as having a specific purpose and involves fuels

Table 3.02 Oregon Coast Range Indian burning practices, pre-1849.

| Type of burning | Products and purposes | Timing |
|---|---|---|
| Firewood gathering and burning | 1-2 purposes: heat, light, cooking, boiling, cleaning, fuel stores, celebration, ceremony, security. | Daily: concentrated near homes, trails, settlements and campgrounds. |
| Patch burning | 1-2 purposes: hunting, berry patch maintenance, root fields harvesting, pest control, weaving materials, trail maintenance. | Seasonal and situational. |
| Broadcast burning | Multiple purposes: stable wildlife habitat; curing seeds; hunting; viewing; transportation; weaving materials; acorn harvest. | Seasonal: late summer, early fall for grasslands; late winter, early spring for bracken fern. |

within a bounded area, such as burning an older huckleberry patch, maintaining a trail, or clearing a field of weeds. Broadcast burning is the practice of setting fire to the landscape for multiple purposes and with general boundaries, such as burning a prairie to cure tarweed seeds, eliminate Douglas-fir seedlings, expose reptiles and burrowing mammals, and harvest insects (Lake and Zybach: In Review).

### 3.2.1 Firewood

Firewood gathering and use was probably a daily process for most families, hunters, gatherers, and travelers for hundreds and thousands of years throughout the Coast Range. Principal locations were probably located along the shores of estuaries and at the mouths of major tributaries. Low gradient riverbank floodplains were also likely locations of homesites and campgrounds. Springs, peaks, waterfalls, meadows, berry patches, root fields, filbert orchards, oat fields, camas patches, pea fields, and other favored locations were also the likely sites of seasonal camping and food processing activities that required intensive, localized firewood gathering activities. The value of firewood to families and communities--at least in areas of scarcity--was illustrated by Clark's April 17, 1806 assessment of "the 2nd Chief" at the Skillute trading village near Celilo Falls:

> I was envited into the house of the 2nd Chief where concluded to sleep. This man was pore nothing to eat but dried fish, and no wood to burn. Altho' the night was cold they could not rase as much wood as would make a fire (DeVoto 1953: 355).

Clark was gauging the man's wealth by the amount of fuel he had available, rather than food or housing: despite the apparent poverty of the 2nd Chief, Clark also "observed maney stacks of fish remaining untouched on either side of the river" (DeVoto 1953: 353).

The likelihood of most bonfires, campfires, oven fires, and sweathouse fires resulting in wildfire events was probably very low. Fires left unattended for the purpose or desire of being spread were probably fairly common (Minore 1972; French 1999). Such fires were intended to spread when possible and cannot be considered escapements. The cumulative results of widespread and systematic firewood gathering over time undoubtedly had a major impact on the location, distribution, and quantity of fuels consumed during wildfire, field clearing, or crop management processes.

### 3.2.2 Patches

Daily and seasonal trail clearing activities, combined with occasional and seasonal brush clearing, hunting, seed curing, and sprout-inducing (for food and weaving materials) burns, made year-around open field burning a likelihood. Areas most likely to be burned in this manner included ridgeline trail segments, hilltop balds, bracken fern prairies, berry patches, filbert orchards, and other travel corridor segments or croplands (Zybach 2002b). The escapement potential of such fires was probably moderate, depending on weather, the fuels being burned, and the condition of burn boundaries.

Many areas (specific habitats or patches) across the landscape within different ecosystems were nationally, family or individually owned (Thompson 1991). Ownership of productive areas across the landscape was viewed as a care-taking socio-ecological responsibility. Indians managed many of the most productive hunting and gathering areas with fire. Parcels of land that could provide productive, abundant, and predictable natural resources provided foods, medicines and material goods for Indian people. A productive and diverse landscape reflected a wealthy and healthy social community. In sum, fire was an ubiquitous tool used by Indian people to perpetuate ecological goods and services necessary for survival and trade (Kimmerer and Lake 2001).

### 3.2.3 Broadcast

Seasonal broadcast burning activities varied from firewood and patch burning actions in two important ways: fire boundaries were not so clearly defined, and there were multiple objectives for burning. Large grass or fern prairies and extensive oak savannahs were maintained by seasonal broadcast burns for a wide variety of purposes, including land clearing, hunting, seed processing, weeding, insect harvesting, and enjoyment. Escapement likelihood from these actions was, as with patch burning, probably moderate. Indians viewed the application of broadcast burning as essential to maintain diversity and productivity of the landscape. The scale of such broadcast burning varied but could result in much larger expanses of land base if climate or weather intensified fire behavior (Lake and Zybach: In Review).

## 3.3 Native Foods and Fire

The development and maintenance of transportation corridors, extensive oak savannahs, prairies, berry patches, filbert groves, camas fields, lawns, and balds by Indian burning practices also resulted in beneficial habitat to a number of plant and animal species, providing sunlight, abundant food, ready transportation corridors, and certain types of cover. During wildfire events, these areas were not prone to being burned, or burned at relatively low temperatures, and could also function as "refuges" for threatened wildlife species (Robbins 1988: personal communication; Krech 1999: 112). Areas that were regularly burned produced a number of food plant species (see Appendix A), as well as plants that could be used for other purposes, such as weaving materials, medicines, dyes, and construction materials (Zobel 2002: 307-308). By providing stable breaks in the landscape with little long-term fuel build-ups, these areas also protected adjacent land areas that produced firewood, large wood products, and provided long-term refuges for big game animals, forest-dwelling species, or other useful plants, such as mosses (Kimmerer 2003: 108-110) and mushrooms (Lake 2003: personal communication). The use of fire across the landscape, therefore, provided benefits for people that were shared by a wide range of plant and animal species, many of whom were apparently dependent on the practice to maintain habitat or food.

### 3.3.1 Plants

Coast Range Indians used fires on the landscape principally for the purpose of obtaining food. Oak, filberts, crabapples, and chokecherry would become dominated and replaced by conifers if not for regular disturbances provided by fire (Thilenius 1964; Stewart 2002). Shrubs, flowers, and grasslands are quickly invaded and replaced by trees when prescribed fire (or other regular disturbance, such as grazing or mowing) is removed from the environment (e.g., Moravets 1932). These plants provided much of the basic food stores for precontact people in the forms of fruits, nuts, seeds, roots, stems, and bulbs (Todt and Hannon 1998; Boyd 1999b). Many of the same foods were important to the survival of such animals as bear (Wilkes 1845b: 184), deer (Whitney 2001), and butterflies (Schultz and Crone 1998). Coblentz (1980: 348), for example, notes, "acorns are among the most important fall and early winter foods for numerous wildlife species."

Table 3.03 (Zybach 2002: 177) lists a range of native plant environments encountered by GLO surveyors in Alsea Valley (see Appendix D) between 1853 and 1897 (see Table 2.03). These designations were common with the remainder of the Coast Range and were required to be used as stipulated by federal regulations until the 1900s (Moore 1851: 32-34; Moore 1855: 17-18). Note that some descriptions (belt, cluster, forest, glade, grove, meadow, patches, and scattering) were only used in later surveys. As a result, it is reasonable to assume that the terms opening and thicket of the earlier surveys were simply supplanted by the more descriptive glades, groves, meadows and patches of the 1890s. Most of the surveyors consistently noted other vegetation types and environments throughout the 45-year survey period, no matter when the survey was conducted: brakes (usually called fern prairies, fern patches, and fern openings), burns (called "deadenings" by some surveyors), prairies, and trails.

Table 3.03 Native plant environments of "Alseya Valley," 1853-1897.

| Name | Years | Townships | Burning |
|---|---|---|---|
| Belt | 1893 | 14-7 | Rare |
| Bottom | 1853-1893 | 13-7; 14-7; 14-8 | Situational |
| Brake | 1856-1897 | 13-7; 13-8; 14-7; 15-7 | Spring |
| Burn | 1853-1897 | 13-7; 13-8; 14-7; 14-8; 15-8 | Fall |
| Cluster | 1891 | 15-8 | Situational |
| Forest | 1893 | 14-7 | Rare |
| Glade | 1893-1897 | 13-8; 14-7; 15-7 | Situational |
| Grove | 1893 | 14-7; | Situational |
| Meadow | 1891-1893 | 14-7; 14-8; 15-8 | Situational |
| Opening | 1856-1893 | 13-8; 14-7; 14-8 | Situational |
| Patches | 1893 | 14-7 | Fall |
| Prairie | 1856-1897 | 13-7; 13-8; 14-7; 14-8 | Fall |
| Scattering | 1878-1891 | 14-7; 14-8 | Fall |
| Swamp | 1856-1878 | 13-7; 14-7; 14-8 | Rare |
| Thicket | 1856-1893 | 14-7; 14-8 | Situational |
| Timber | 1856-1897 | 13-7; 14-7; 14-8 | Rare |
| Trail | 1853-1897 | 13-7; 14-7; 14-8 | Situational |

A column titled "Burning" has been added, to provide an approximate idea as to how often--and what time of year--the landscape needed to be burned in order to remain free of tree growth. Typically, brakes were burned in late winter and grasslands burned in late summer. There is a biological reason for this timing. As Ross notes:

> The name "brake" or bracken, which traces back into a number of European languages, may have come from the broken appearance of the fern cover after the first heavy frost. The leaves lie collapsed like a miniature forest hit by a tornado (Ross 1971: 2).

Prior to being desiccated by frosts in the fall, bracken fern plants simply held too much moisture to burn effectively. Following a burn, roots (and animals) were left exposed, making harvest an easier task. New sprouts in the spring ("fiddleheads") could also be easily located and harvested once the heavy plant cover had been removed. Conversely, grasslands became desiccated through late summer heat and drought and were ready to be burned at that time. And, also unlike bracken fern, which often existed in relatively pure stands dominated by this single species, the components of grassy prairies, meadows, glades, and balds usually offered a wide variety of food resources. For example, Aldrich's 1972 study of the grass balds on Prairie Peak (Aldrich 1972: 110-115) and Grass Mountain (ibid: 105-110) provides a table of associated plants found on the balds (ibid: 148-153) that included at least 16 species considered to be major food plants for Indian communities in southwest Washington (Leopold and Boyd 1999: 159-162). The Alsi and their neighbors used many of these plants for food as well, including strawberries, blackberries, raspberries, chocolate lilies, tiger lilies, tarweed, fireweed, thistle, wild onions, and yampah (see Appendix A).

These same records can be used to look at tree species diversity as well as food plants. Bottomlands, for example, were noted as containing alder, ash, balm (black cottonwood), cedar (redcedar), cherry (probably both wild cherry and chokecherry), chittam, fir (Douglas-fir), (wild) crabapple, (bigleaf) maple, and willow tree species. Two swamps in Alsea Valley were described as a "cedar swamp" and a "willow swamp". The lowest elevation and highest elevation grassy prairies along the Alsea River and on Prairie Peak contained (white) oak bearing trees, although oak was not mentioned in any other locations. In addition to Douglas-fir, redcedar, alder, maple, cherry, and chittam, higher elevations also contained chinquapin, dogwood, hemlock, "mountain balm" (probably madrone), myrtle (possibly a mistake, see Gesner 1891a: 252), spruce (possibly Sitka spruce, but maybe a "true" fir), white fir (possibly grand or noble fir), and yew.

Table 3.04 (Zybach 2002b: 175) is a list of food plants identified by the same surveyors, encountered in the various environments listed in Table 3.03. The combination of environments other than forests, prairies, and savannahs can generally be considered "patches." The timing of patch burns can be reasonably inferred by the length of time it took for fruits and berries to set, ripen, and be harvested, or the appropriate time to clear land for root digging or fiddlehead picking. Frachtenberg (1920: 204), for example, gives the Alsi word for May as "the month for picking salmonberries" and the word for July as "the month for picking salal-berries." These factors also influenced the time and volume of trail use and the structure of adjacent forested areas. Dozens or hundreds of people moving into bracken fern prairies, filbert groves, or salmonberry fields to camp, burn, or pick crops, whether daily or on a seasonal basis, must have contributed to the lasting definition of local trails. The daily use of firewood in these locations would have resulted in annual clearings of forest debris resulting from wind, ice, fire, or other forms of tree and limb mortality. Lower limbs would likely have been removed from many trees adjacent to trails and campsites, and certain trees, such as willow, chittam, yew and redcedar, would have borne the marks of peeling, carving, and the removal of bow staves. Local weather conditions (see Table 1.01) would have further dictated burning times.

Table 3.04 Seasonal locations of Alsea Valley Indian fires.

| Food | Species | Years | Townships |
|---|---|---|---|
| Berries | Blackberry | 1891-1897 | 13-7; 13-8; 14-7 |
| Berries | Gooseberry | 1891 | 15-8 |
| Berries | Huckleberry | 1891-1897 | 13-7; 14-7; 14-8; 15-7; 15-8 |
| Berries | Oregon Grape | 1856-1891 | 13-7; 14-7; 14-8; 15-7; 15-8 |
| Berries | Salal | 1856-1897 | 13-7; 13-8; 14-7; 14-8; 15-7; 15-8 |
| Berries | Salmonberry | 1865-1891 | 13-7; 13-8; 14-8; 15-8 |
| Berries | Thimbleberry | 1891-1893 | 13-7; 13-8; 14-7; 14-8; 15-8 |
| Fruits | Choke Cherry | 1856-1897 | 13-8; 14-7; 14-8; 15-8 |
| Fruits | Crab Apple | 1856 | 14-8 |
| Grains | Grasses | 1856-1893 | 13-8; 14-7; 14-8; 15-8 |
| Nuts | Filbert | 1853-1897 | 13-7; 14-7; 14-8; 15-7; 15-8 |
| Nuts | White Oak | 1856-1891 | 13-7; 13-8; 14-7; 14-8; 15-8 |
| Peas | Legumes | 1856-1893 | 13-7; 14-7; 14-8 |
| Roots | Bracken fern | 1853-1897 | 13-7; 13-8; 14-7; 14-8; 15-7 |

Some of the entries in this table require further explanation. White settlers often referred to wild legumes as "Indian peas", and "fern" was, without doubt, mostly bracken fern. Both plants were maintained, harvested, and processed as important foods, providing protein and starch to many--if not all--western Washington and western Oregon Indian communities in precontact and early historical time (Frachtenberg 1920: 129-133; Norton 1979). The US government surveyed these lands for the purpose of being sold to private landowners--mostly farmers and ranchers--and livestock subsistence was critical to purchasers in the pre-automobile era. Grass and legumes were important feed crops for horses and cattle used for transportation, work, beef, and milk, and camas, acorns and fern roots were used to fatten hogs for market (Krewson 1955: 86, 95). Fagan (1885: 498), for example, noted that the Alsea basin had become an important area of pork and butter production by the 1880s, a fact supported by Kirkpatrick more than 50 years later (Kirkpatrick 1939: 14).

Tables 3.03 and 3.04 demonstrates that GLO survey records can be used to determine the location of precontact food plants, the environments in which they existed, and the seasons in which they were harvested (picking, cutting, tilling) and burned. From this information we can infer much about the locations and occupations of local people in late precontact time. Because desirable food animals, such as bear, deer, and rabbits, used many of the same foods at the same times of year and because people used firewood for a variety of purposes wherever they went, we can begin to draw a fairly accurate picture of the daily lives and seasonal use of the landscape by people who have left few additional records.

Table 3.05 (Lake and Zybach: In Review) is derived from tables 1.03, 3.03, and 3.04, combining Coast Range seasonal weather patterns with fuel environments and types (species) of fuel. Because many of the native plants managed by precontact people persist in the environment to this time, Table 3.05 contains information of value to modern resource managers that work with fire in the environment.

Table 3.05 Oregon Coast Range weather and burning patterns.

| Mo. | Season | Weather | Temperature | Plant Fuels | Burning |
|---|---|---|---|---|---|
| Jan. | Winter | Wet | Freezing | Dormant | Firewood |
| Feb. | Winter | Wet | Freezing | Dormant | Patches |
| Mar. | Spring | Wet | Freezing | Budburst | Patches |
| Apr. | Spring | Mixed | Cool | New Growth | Patches |
| *May* | *Transition* | *Mixed* | *Warming* | *Growing* | *Projects* |
| Jun. | Summer | Dry | Warm | Growing | Firewood |
| Jul. | Summer | Dry | Warmest | Growing | Firewood |
| Aug. | Late Summer | Dry | Warmest | Dormant | Broadcast |
| Sep. | Late Summer | Dry | Warm | Dormant | Broadcast |
| *Oct.* | *Transition* | *Mixed* | *Cooling* | *Fall Growth* | *Patches* |
| Nov. | Fall | Wet | Freezing | Dormant | Firewood |
| Dec. | Fall | Wet | Freezing | Dormant | Firewood |

Burning seasons can be further described as follows:

Winter patch burning (February to April). Patches of old berries, unburned grasses, and bracken fern could be readily burned after a few killing frosts had created enough fuel. The principal limitation was moisture: regular winter storms from the west keep most fuels sopping wet during these months. The exception was continental east winds, which could dry fuels within a few hours or days and tended to drive fires (including escapements) westward.

Spring project burning (May). Spring weather is unpredictable, and it can be difficult to plan for outdoor burning practices as a result. Abandoned plank houses could be burned in the rain and old fields could be burned on east winds, for example, depending on conditions and desired results. Areas burned in the winter would be showing results in the forms of edible sprouts, weaving materials, and firewood gathering strategies. Spring burns of filbert orchards, for example, will tend to produce sprouts useful for a number of manufactured products, while fall burns of the same species tend not to create sprouts, and can be used to weed, hunt, or more easily gather nuts instead (Buckman 1964).

Late summer broadcast burning (August to September). Most Indian burning on record (e.g., Boyd 1999b) regards the burning of grassy prairies and oak savannah from July until the beginning of fall rains.

Fall patch burning (October). Fall weather, much like the spring, is unpredictable and landscape burning was opportunistic as a result. Patches and special projects could be burned as weather allowed.

The principal result of maintaining a diverse environment through regular burning practices was, as has been stated, a predictable and abundant supply of a wide diversity of food plants. Table 3.06 is a listing of "signature" foods found throughout the Coast Range, often identified with specific tribes and nations (e.g., wapato with Chinookans and Atfalati Kalapuyans; camas with Kalapuyans; cranberries and myrtle nuts with Kusans). Many of these foods were critical to survival at different times of the year. Louisa Smith, a Siuslawan, for example, stated in 1911:

> They had dried salmon, and likewise (dried) fern-roots, which they ate during the winter. They ate fern-roots (mostly). Thus the people did during the winter . . . Such was the food of the people belonging to the past (Frachtenberg 1914: 81-83).

The column marked "Fire" denotes whether plants were dependent on regular disturbance for their survival--such as provided by fire--or whether they were merely tolerant of such actions. "XX" denotes plants largely dependent on fire for their existence and "X" denotes plants tolerant of fire; blank spots denote plants in which the relationship to fire is unclear. Note that no plants are intolerant of fire: all have either depended on its use, or become tolerant of its existence.

Table 3.06 Principal native food plants of the Oregon Coast Range.

| Food Type | Food Name | Fire |
|---|---|---|
| **Berries** | Blackberry | XX |
| | Huckleberry | XX |
| | Salmonberry | XX |
| | Strawberry | XX |
| | Thimbleberry | XX |
| **Bulbs** | Camas | XX |
| | Lily, Chocolate | XX |
| | Lily, Tiger | XX |
| | Onion | XX |
| | Wapato | X |
| **Fruits** | Crabapple | X |
| | Chokecherry | XX |
| | Manzanita | XX |
| | Rosehips | XX |
| **Grains** | Grass seed | XX |
| | Indian peas | XX |
| | Sunflower | XX |
| | Tarweed | XX |
| **Greens** | Dock | XX |
| | Miner's Lettuce | XX |
| | Nettles | XX |
| | Seaweed | X |
| **Mushrooms** | Chicken-in-the-woods | |
| | Morels | XX |
| | Puffballs | XX |
| | Shaggy Manes | |
| **Nuts** | Acorns | XX |
| | Filberts | XX |
| | Myrtle nuts | XX |
| | Pine nuts | X |
| **Roots** | Bracken fern | XX |
| | Mountain carrot | XX |
| | Yampah | XX |
| **Stalks** | Cat-tail | X |
| | Fiddleheads | XX |
| | Skunk cabbage | X |
| | Thistle (Edible) | XX |

Other forms of disturbance related to plant management, such as tillage, picking, and pruning, were also present in all environments but seem far more likely to have occurred in regularly burned areas.

Acorns, filberts, camas, wapato, tarweed, huckleberries, blackberries, bracken fern, nettles, tobacco and other signature food crops (see Table 3.06) were often managed in select areas over long periods of time. Crops were maintained and harvested in discrete locations in which the dominant species—usually the crop species itself—had been established or rejuvenated within a few weeks or months' time. This approach creates a condition that is called "even-aged" management. Foot trails or canoe traffic, depending on location, provided access to croplands.

A few native Coast Range food plants, such as seaweed or wapato, grew independently of most burning strategies. Tidal actions and seasonal floods provided the regenerative disturbances needed for these plants. The harvest of these foods was not insignificant, and may have pointed to a much larger human population prior to the advent of disease. Clark, for example, noted in 1806, near the mouth of the Willamette:

> the man who took us over last night . . . gave us a roundish roots about the Size of a Small Irish potato which we roasted in the embers until they became Soft, This root they call *Wap-pa-to* the *Bulb* of which the Chinese cultivate in great quantities called *Sa-git ti folia* or common arrow head, . . . it has an agreeable taste and answers verry well in place of bread (Thwaites 1959a: 196-197).

Lewis and Clark were so impressed with the great quantities of wapato grown, harvested, and traded along the Columbia River between the Cascades and Oak Point that they named it "Wap-pa-too Valley" Thwaites 1959a: 202). Darby (1996: 94) has estimated that enough wapato grew on Sauvies Island alone to feed between 18,270 and 36,777 people year around. Add those numbers to other wapato growing areas of the region--such as Columbia Slough and Wapato Lake in the Tualatin drainage--and the potential population of precontact Coast Range people that could survive on wapato, camas, acorns, bracken fern, seals, salmon, clams, venison, etc., would seemingly be well over 100,000. This is far more people than can be accounted for in historical time (Boyd 1999a), and might even exceed many of the more liberal and controversial estimates of Denevan (1992), Mann (2002), and others. Certainly, the human "carrying capacity" of the Oregon Coast Range must have been at least tens of thousands of people during precontact time.

### 3.3.2 Animals

Energetically, many of the plants and wildlife species used and managed by people were also important to other plants and animals for habitat, cover, or forage (Norton et al 1984; Todt and Hannon 1988), as has been stated. Table 3.07 lists important animal food groups for Coast Range peoples native to the area in precontact time. This table indicates the wide variety and abundance

Table 3.07 Principal native food animals of the Oregon Coast Range.

| Food Type | Food Name | Fire |
|---|---|---|
| **Crustaceans** | Crabs, Dungeness | O |
| | Crawdads | X |
| | Shrimp | O |
| **Fish** | Eels, Lamprey | X |
| | Eulachon | X |
| | Halibut | X |
| | Salmon, Chinook | X |
| | Salmon, Coho | X |
| | Sturgeon | X |
| | Trout, Cutthroat | X |
| **Fowl** | Ducks | XX |
| | Grouse, Ruffed | XX |
| | Geese | XX |
| | Pigeons, Band-tailed | XX |
| **Insects** | Grasshoppers | XX |
| | Stoneflies | X |
| | Yellow jackets (larvae) | XX |
| **Red Meat** | Bear, Black | XX |
| | Deer, Blacktail | XX |
| | Elk | XX |
| | Gray Diggers | XX |
| | Seals | O |
| | Sea Lions | O |
| | Squirrels, Gray | XX |
| | Whale, Grey (occasional) | O |
| **Shellfish** | Clams, Butter | X |
| | Clams, Geoduck | X |
| | Clams, Razor | O |
| | Mussels, saltwater | O |
| | Mussels, freshwater | X |
| | Oysters | X |

of important animal foods and trade items available to local people, as well as the response of favored species to regular fire management practices. As with Table 3.06, the column marked "Fire" denotes if animals benefited by regular burning practices (XX), had adapted to such practices (X), or were independent of such practices (O). Only animals that lived in the ocean, or on sandy beaches, rocks, and in tide pools adjacent to the ocean were not directly affected by Indian burning practices. Even anadromous fishes and freshwater animals had to adapt to influxes of carbon and changes in solar energy caused by fire.

This relationship of Coast Range Indian burning practices to wildlife habitat--especially habitat for such food animals as birds, ungulates, rabbits, and squirrels--was first noted by Haswell as he sailed along the southern Oregon Coast near Coos Bay in August, 1788:

> . . . this Countrey must be thickly inhabited by the many fiers we saw in the night and culloms of smoak we would see in the day time but I think they can derive but little of there subsistance from the sea but to compenciate for this the land was beautyfully diversified with forists and green veredent launs which must give shelter and forage to vast numbers of wild beasts most probable most of the natives on this part of the Coast live on hunting for they most of them live in land this is not the case to the Northward for the face of the Countrey is widly different (Elliott 1928: 167-168).

## 3.4 Cultural Landscape Patterns

The combination of widely diverse landscape conditions and differing Indian cultures throughout the Coast Range led to significant differences in local and subregional landscape patterns. The following maps and eyewitness accounts describe and locate some of the principal differences that existed throughout each of the Coast Range subregions.

### 3.4.1 North: firewood and flood

The northern Coast Range seems to have been heavily populated by people who trafficked almost exclusively along the Columbia River, with small prairies and floodplains serving to provide most needs for a land base. Trade with coastal seagoing peoples and upland Kalapuyans who managed plant foods for subsistence probably provided most of the necessary means needed to acquire most regional products. Figure 3.03 (Spencer 1950: 62) shows pre-1941 seasonal flooding of Sauvies Island, at the mouth of the Willamette River. Between 1938 and 1941, 12,000 acres [of 24,064 total] of Sauvies Island were diked. Willow and cottonwood groves were cleared, and lakes drained. According to Spencer:

> Until recent years annual freshets from the Columbia and Willamette have covered most of its area for parts of each year. Its soil is deep and very fertile . . . Most of the land within the dike had always been used in a state of nature with wild meadows producing pasture and hay. This had to give way because wild meadows depended on annual freshets for their luxurious growth (Spencer 1950: 3, 83).

A map of the northern Coast Range, prepared from GLO survey notes (see Appendix D) and early vegetation maps (see Chapter 2.2.6) was constructed to display local landscape patterns as they likely appeared in early historical time (ca. 1800). Map 3.02 depicts the wetland prairies of

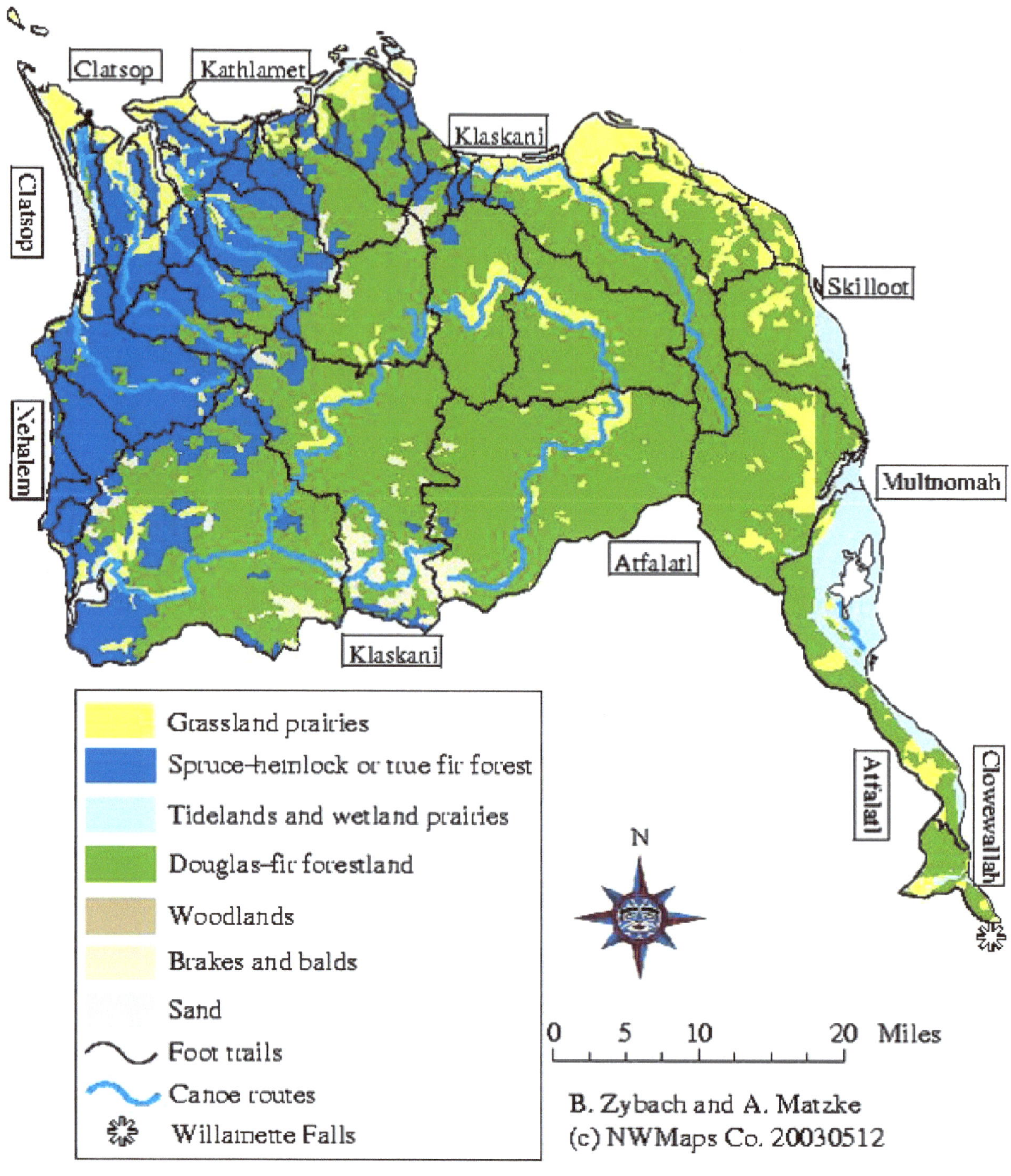


Map 3.02 Northern Coast Range landscape patterns, ca. 1800.

Figure 3.03 Sauvies Island seasonal flood, pre-1941 dike completion.

the Multnomah, the tidal wetlands and adjacent prairies inhabited by the Clatsop and Nehalem, and a well-developed foot trail network between the Chinookans and the Salish that was used and maintained by inland Klaskani. A significant string of prairies line the banks of the Nehalem River that was probably shared by both Nehalem and Klaskani, with Nehalem tending more toward the ocean and the Klaskani nearer the Columbia and Willamette rivers. The area is mostly dominated by an extensive stand of conifer forestlands, with hemlock, cedar, and spruce typifying the lands of the Salish people, and Douglas-fir populating most of the remainder.

### 3.4.2 East: oak savannah

Precontact Kalapuyan tribes maintained the Willamette Valley as an oak savannah, through the annual use of extensive broadcast burning projects that likely took place every fall (Boyd 1986; Gilsen 1989). The result was a white oak savannah covering tens of thousands of contiguous acres, with hundreds of camas prairies, berry patches, root fields, and tarweed fields interspersed with "islands" of conifer trees and "gallery forest" riverine floodplains dominated by cottonwoods, bigleaf maple, ash, pine, true fire, and Douglas-fir.

Although Charles Wilkes did not travel south through the Willamette Valley with other members of his expedition, he had access to their daily journals, from which he assembled his final report in 1845. He did travel extensively along the Columbia River, however, and made a brief trip into the northern part of the Willamette River:

> . . . on the 14th [of September] we took leave of Vancouver. After proceeding down to the mouth of the Willamette . . .
>
> . . . we were a good deal annoyed from the burning of the prairies by the Indians, which filled the atmosphere with a dense smoke, and gave the sun the appearance of being viewed through a smoked glass (Wilkes 1845b: 141-142).

Earlier that month a party of his expedition had started south from Fort Vancouver, heading overland toward San Francisco. On September 7th or 8th, the following view was reported from somewhere around the Salem area:

> The country in the southern part of the Willamette Valley, stretches out into wild prairie-ground, gradually rising in the distance into low undulating hills, which are destitute of trees, except scattered oaks; these look more like orchards of fruit trees, planted by the hand of man, than groves of natural growth, and serve to relieve the eye from the yellow and scorched hue of the plain. The meanderings of the streams may be readily followed by the growth of trees on their banks as far as the eye can see (Wilkes 1845b: 221-222).

On September 9 the party experienced an unseasonal "severe frost." The following day, near present-day Corvallis, Wilkes reported:

> On the 10th, the country was somewhat more hilly than the day previous, but still fine grazing land. During the day they crossed many small creeks. The rocks had now changed from a basalt to a whitish clayey sandstone. The soil also varied with it to a grayish-brown, instead of the former chocolate-brown colour, which was thought to be an indication of inferior quality. The country had an uninviting look, from the fact that it had lately been overrun by fire, which had destroyed all the vegetation except the oak trees, which appeared not to be injured (Wilkes 1845b: 222).

In contrast to the northern subregion, the eastern Coast Range contained very little conifer forestland. This was in part due to the annual burning practices of local Kalapuyan families, but also due to the relative fall drought experienced by the eastern part of the Range compared to the northern and western subregions (see Table 1.01 and Map 1.04). Talbot noted the difference in

September 1849 as he traveled along the current route of Highway 18, passing the watershed boundary from the westward-flowing Salmon River, to the eastward-flowing Yamhill:

> Near the Couteau, or summit line of the range there are many open spots, all covered with luxuriant crops of fern. Descending into the valley of the Willamette, we camped on a fork of the Jam Hill [Yamhill] river . . . We were much struck by the contrast in the appearance of the vegetation on this side of the mountain, parched and withered by the long droughth, while on the west slope we had left it fresh and green as in the early spring (Haskin 1948: 14).

Despite this difference, once Indian burning was stopped in the late 1840s, much of the Willamette Valley brakes and grasslands began to develop Douglas-fir forests (see Appendix F). Figure 3.04 shows the first steps in this transitional process. In 1845, Neall made the observation:

> The leading features of the Willamette Valley and Tualatin plains were peculiar and strange to me as compared with any other country I had seen. Among the striking peculiarities was the entire absence of anything like brush or undergrowth in the forests of fir timber that had sprung up in the midst of the large plains, looking at a distance like green islands here and there dotting the vast expanse of vision. The plains covered with rich grasses & wild flowers looking like our vast cultivated fields, and where the rolling foothills approached the level valley these spurs would be sprinkled with low spreading oak trees, frequently with a seeming regularity that would seem unlike nature's doing, and at a distance like orchards of old apple trees (Neall 1977: 44).

In 1885, forty years after the Willamette Valley had passed from Kalapuyan hands to American agriculturists and ranchers, the effect was strikingly similar. A typically florid writer of that time, writing of Benton County history, wrote:

> . . . a feast for the eye presented itself as the fertile prairie of the Willamette valley was espied from the far-off height of a crag or a mountain pass. And what was it like? For mile upon mile and acre after acre, tall wild grasses grew in wonderful profusion--one great, glorious green of wild waving verdure--high over the backs of horse and ox and shoulder high with the brawny immigrant. Wild flowers of every prismatic shade charmed the eye, while they vied with each other in the gorgeousness of their colors and blended into dazzling splendor.
>
> One breath of wind and the wide emerald expanse rippled itself into space, while with a heavier breeze came a swell whose rolling waves surged over the foot-hills, beat against the mountain sides, and, being hurled back, were lost in the far away horizon. Shadow pursued shadow in one long merry chase; the air was filled with the hum of insects, the chirrup of birds and an overpowering fragrance weighted the air. The river's bank was clothed in its garment of green foliage,

Figure 3.04 Willamette Valley, 1845-1888.

while, the dark green forest trees lent relief to the eye. The impenetrable jungle of to-day, at this time was not, the smaller growth being kept low by Indian fires, while the timber land presented a succession of tempting glades open to movements on foot or on horseback (Fagan 1885: 328).

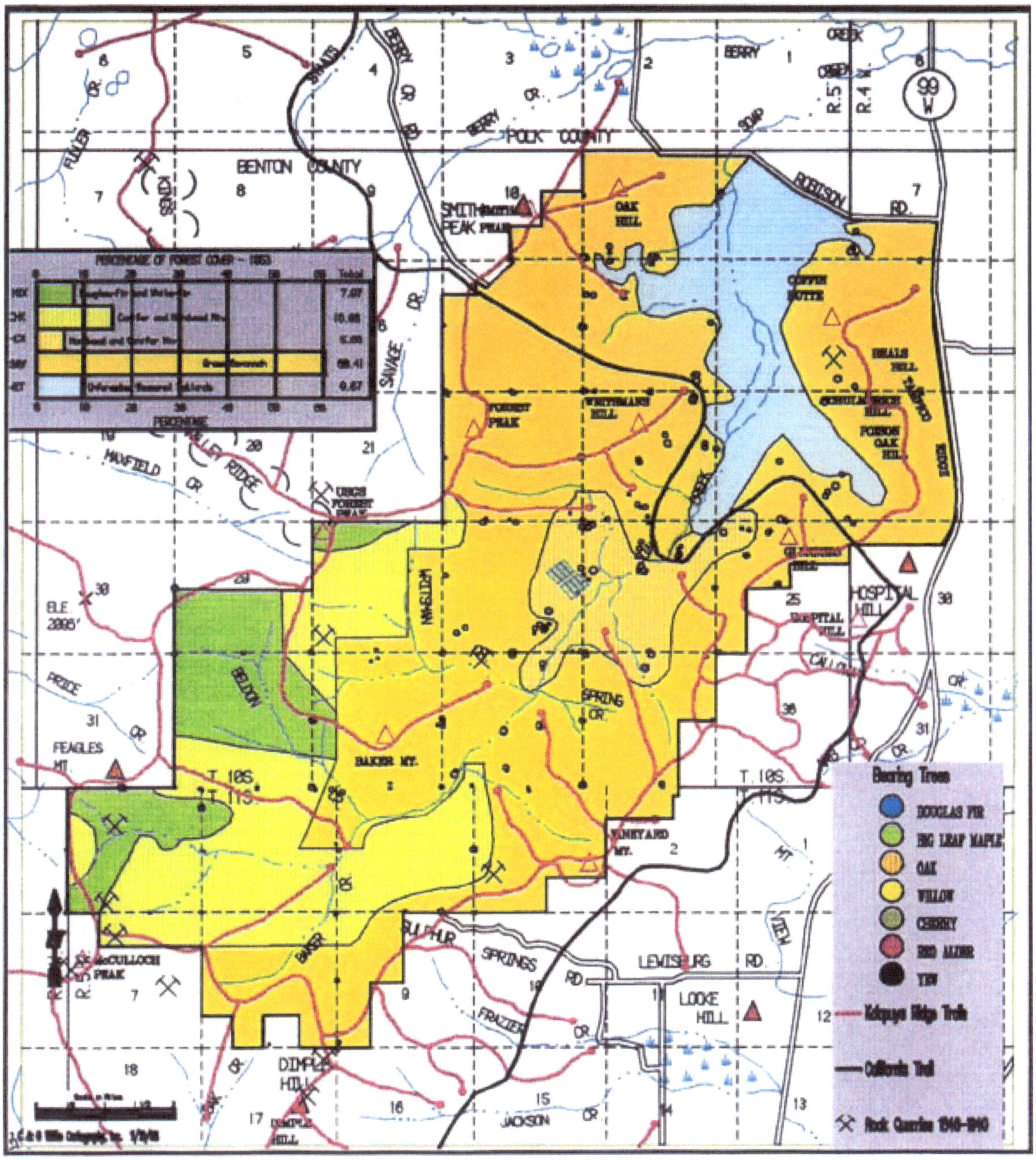


Map 3.03 Soap Creek Valley, ca. 1826 GIS map.

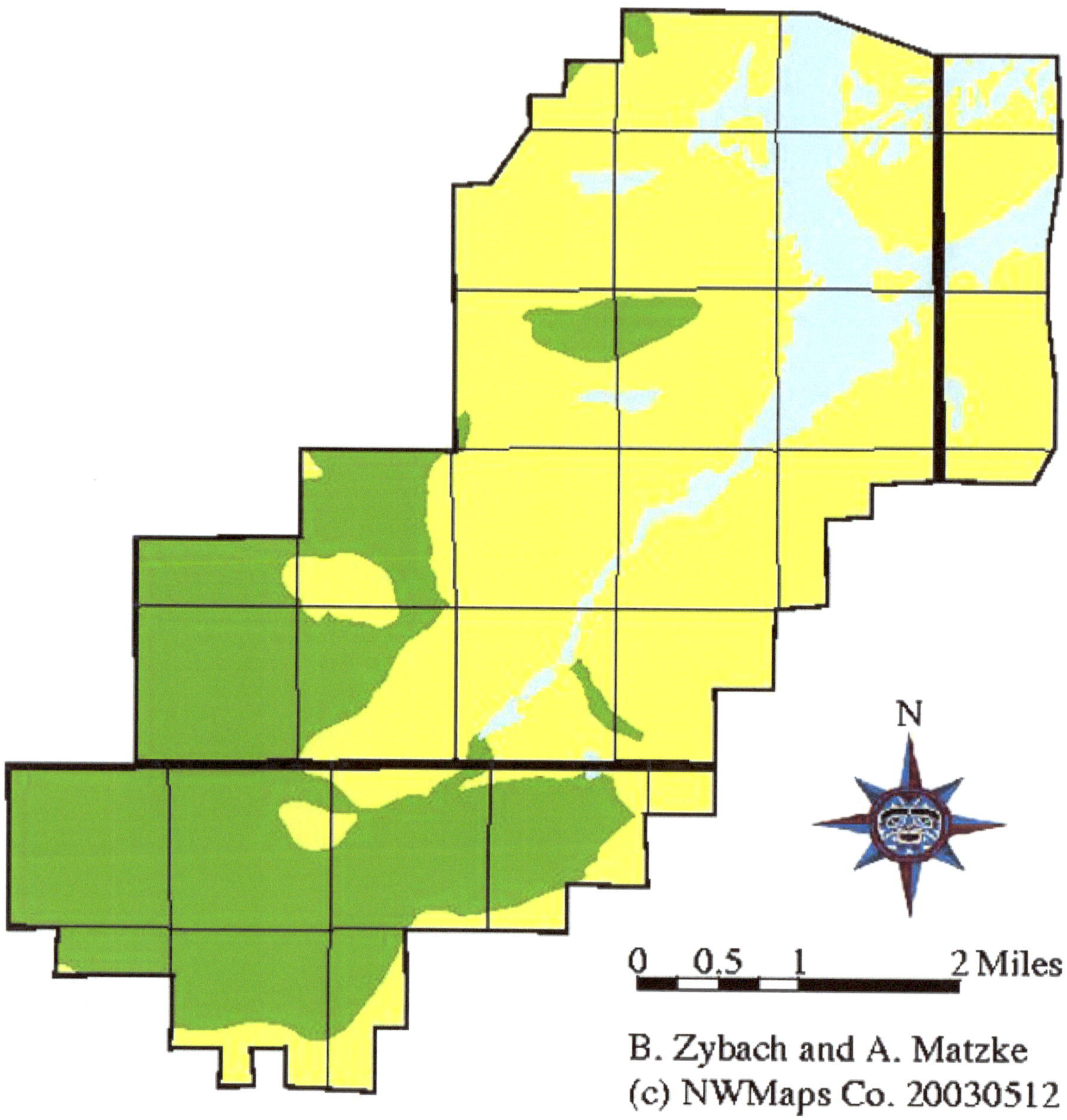


Map 3.04 Soap Creek Valley, ca. 1850 GIS map.

By 1885, then, significant brush had grown up and into the "peculiar islands" noted by Neall. Further, the survey lines of immigrant landowners turned into property lines and then roads and fence lines, imposing straight lines into the landscape (see Figure 2.03). To further document this process, the 15,000-acre Soap Creek Valley area can be used as an example. Map 3.03 is a map of the Valley constructed by plotting every GLO subdivision and DLC bearing tree noted by the original land surveyors (Zybach 1999: 275-292). Inferences drawn from tree spacing, species,

Figure 3.05 Soap Creek Valley, 1914-1989.

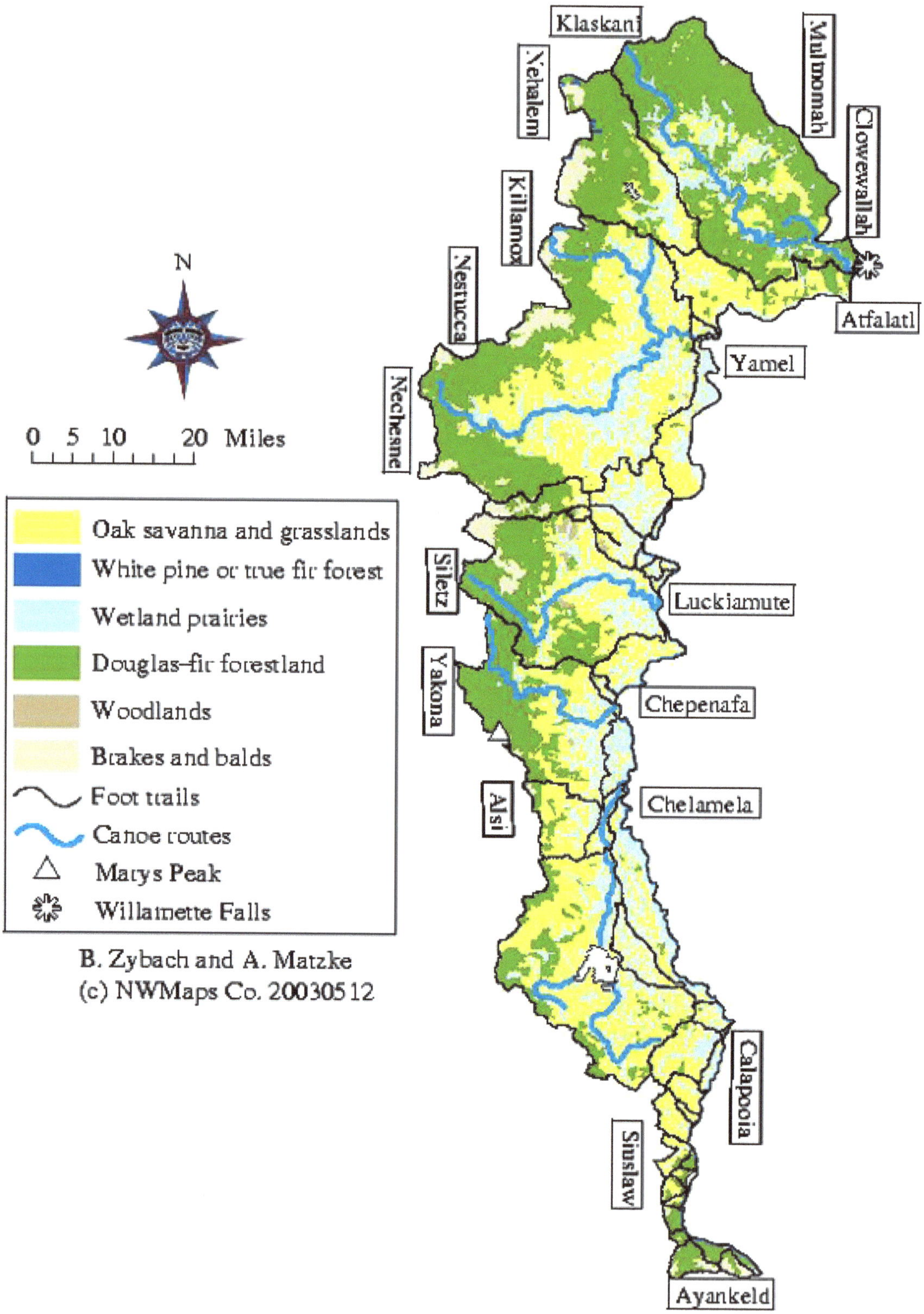


Map 3.05 Eastern Coast Range landscape patterns, ca. 1800.

diameters and growth rates were used with a map of 1939 aerial photographs to derive a fairly accurate picture of local vegetation patterns in 1826. A more general use of the same notes was used to construct an 1850 pattern of the same area (Christy and Alverson 2003). Map 3.04 was constructed without plotting individual trees and without using DLC survey data. USGS 7.5 minute quadrangle maps were used for vegetation patterns instead of aerial photos, and a much wider range of vegetation patterns was developed. Despite these differences, the two patterns are very similar and further support the use of GLO data for interpreting past environments. The two maps also show little change and a fairly stable vegetation pattern that can be mostly attributed to regular burning practices that took place between 1826 and 1850. Figure 3.05 (Zybach 1999: 112-113) includes a photograph taken of Soap Creek Valley in 1914, showing that a combination of post-contact farming, grazing, and slashing has somewhat retained the "cultural legacy" of earlier generations of Kalapuyans. This is similar to the differences noted in Figure 3.04. However, a 1989 photograph taken from the same perspective paints an entirely different picture. Farming practices that replaced burning practices have themselves been replaced with forestry practices. The oak savannah of the Kalapuyans and the open grazing lands and fenced pastures of early white settlers has been replaced by Oregon State University Research Forests' timber: the McDonald and Dunn forests (see Appendix F).

Does the Soap Creek pattern hold true for the entire Willamette Valley? Map 3.05 was constructed in the same manner as Map 3.02, but the landscape is entirely different. Wetland prairies dominate the lowlands, and oak savannah dominates the uplands, with only ridgelines, occasional riparian areas, and steep valleys showing any significant amounts of conifer trees.

### 3.4.3 West: "lawns, corridors, and mosaics"

In precontact time, the western Coast Range landscape was dominated by the ocean, tides, and bays, rather than a major river and seasonal flooding, as in the north, or by regular burning, as in the east. Saltwater, rather than freshwater, plants and animals formed an important part of daily and seasonal diets and recreation. Large canoes capable of sophisticated international trade excursions traveled up and down the coast for the entire length of the Coast Range, and along the Columbia River from the coast eastward, to the Cascades.

The northern part of the western Coast Range was opened to white settlement in the early 1850s, where white immigrants coexisted with small bands of Killamox that had survived the small pox and malaria epidemics that had devastated their own families and had caused the extinction or near-extinction of many neighboring tribes. One such immigrant was Warren Vaughn, who apparently kept a journal and used it as the basis for his memoirs, written in the 1880s. Speaking of the early 1850s, when he first arrived at Tillamook Bay, he observed:

> At that time, there was not a bush or tree to be seen on all those hills, for the Indians kept it burned over every spring, but when the whites came, they stopped the fires for it destroyed the grass, and then the young spruces sprang up and grew as we now see them (Vaughn 1923: 40).

The first person to accurately describe the mix of trails, brakes, balds, and bottomland prairies that characterizes much of the western Coast Range is Talbot (Haskin 1948), who traveled along the Siletz, Yaquina, Alsea, and Salmon rivers in late August and early September 1849. Talbot was sent on a military expedition from Fort Vancouver, to report on the largely uninhabited area that would become part of the Coast Range Indian Reservation in 1856 (Kent 1977). As Talbot followed a Klickitat horse trail westward from Kings Valley, near the headwaters of the Luckiamute River in the Willamette Valley, he made the following notes:

> 2 miles below our camp of last night we struck the main fork of the Celeetz [Siletz; Talbot was actually following present-day Rock Creek] river flowing from the N.E. . . . Crossing it we ascended the bank into a handsome prairie, extending several miles along the north side of the river, which from the junction of its forks takes a nearly west course [near present-day Logsden, at the fork of Rock Creek and the mainstem Siletz River]. The soil of the river bottom is very rich; grass growing most luxuriantly where not completely choked up by the fern - this plant usurping possession of nearly every open spot of ground. It grows here from eight to ten feet in height, and is quite serious empediment to travel. We encamped in an open prairie bottom about a mile long and a half mile in width, just where the river, changing its course, makes an abrupt bend to the north. We are surrounded on all sides by tall forests of pine, fir, spruce, hemlock, etc., which gave quite a sombre [sic] appearance to this sequestered valley . . . There are no Indians residing permanently on this river, and no trails going further down; the one which we have followed thus far crossing the river here and striking south [toward the Yaquina River, near present-day Toledo] (Haskin 1948: 4-5).

Talbot subsequently went as far south as Alsea Bay, before returning north along the coast. He reached Yaquina Bay on September 3:

> Moving camp we came two miles along the shore of the bay; thence striking north, traveling three miles through an open rolling country covered with fine grass and some smal[l] patches of fern and thistles. The soil here appeared to be very rich, and was well-watered by numerous little springs (Haskin 1948: 10).

From Yaquina Bay he continued northward, toward Siletz Bay:

> Our road gradually improved as the mountains, receding, left a beach of open land extending from the top of the precipices bordering the ocean to the foot of the steep timbered acclivities, a space varying from one-fourth to half a mile in width, well watered, with rich soil, bearing a luxuriant crop of clover, grass, and their usual concomitant of fern. . . we came to the upper part of Celeetz [Siletz] bay, where we encamped on a small prairie covered with fine bunch-grass and clover.
>
> . . . We soon constructed a small raft for ourselves and baggage, the shore being strewn with thousands of drift-logs. . . we were glad to substitute in its stead a fine large canoe which we found concealed among the bushes on the opposite bank. It was after night before all had crossed, and we camped a hundred yards from the shore, at the edge of a pretty grassy prairie which borders the bay (Haskin 1948: 11-12).

In the early 1850s, at the time of initial white settlement in the area, "Alseya Valley" existed as a series of prairies, brakes, balds, openings, patches and meadows connected by a network of foot trails, horse trails, and canoe routes, and bounded by stands of even-aged forest trees, burns, seedlings and saplings (Zybach 2002b). This condition has been described as "yards, corridors, and mosaics" (Lewis and Ferguson 1999). Lewis and Ferguson initially used the phrase to describe a cultural landscape pattern maintained by Native people who lived in the boreal forests of Canada and Alaska (ibid: 164, 172-178), but determined that similar management patterns were also used by people in the conifer forests of the Rockies and Sierra Nevada (ibid: 164), northwest California (ibid: 167-168), western Washington (ibid: 168-169), Australia (ibid: 169-170), and Tasmania (ibid: 170-171). They found that in each instance, fire was the tool most commonly used to establish and maintain grasslands and other openings ("fire yards"), bounded by stands of trees and open transportation routes ("fire corridors"). Fire was also the agent that entered unmanaged forested areas, whether by human cause or lightning, and caused burns that regenerated to a shifting mosaic of even-aged stands of seedlings, saplings, and trees (ibid 164-165; Stewart 2002: 250-255).

The existence of significant western Coast Range meadows and prairies in early historical time is shown in Chapter 2 and in Appendix D with the "Alseya Valley" example. The transformation of grasslands to forestlands during historical time is shown with the Soap Creek Valley example in the previous section of this chapter. Figure 3.06 (photos by B. Zybach and N. Lapham 2003) shows persistent vegetation patterns that still exist in Alsea Valley to this time: the persistent cultural legacy of past Alsi Indian land management practices. These pictures were taken in spring, 2003 (see Appendix H). The upper left picture is near the single "Indian Trail" crossing of the Alsea River that connected the Willamette Valley to Alsea Bay in 1856 (see Figure 2.05), and shows an early oak prairie that was first surveyed in 1856 (Table 2.03). The lower left picture is a long riparian meadow (or glade) adjacent to Alsea River, directly behind an isolated Douglas-fir old-growth and just south of a popular campground and boat launch. There is no evidence that trees ever became established here (except along the riverbank to the right) at any time since the 1850s. The upper right photo and the lower right photo were taken about five miles' distance from one another, in two locations believed to have been Alsi (or their predecessors') townsites in precontact time. In all four instances, the fire-dependent landscape patterns of the 1850s have been largely maintained by subsequent management actions of farming, grazing, and logging.

Figure 3.06 Alseya Valley prairie relicts, April 14, 2003.

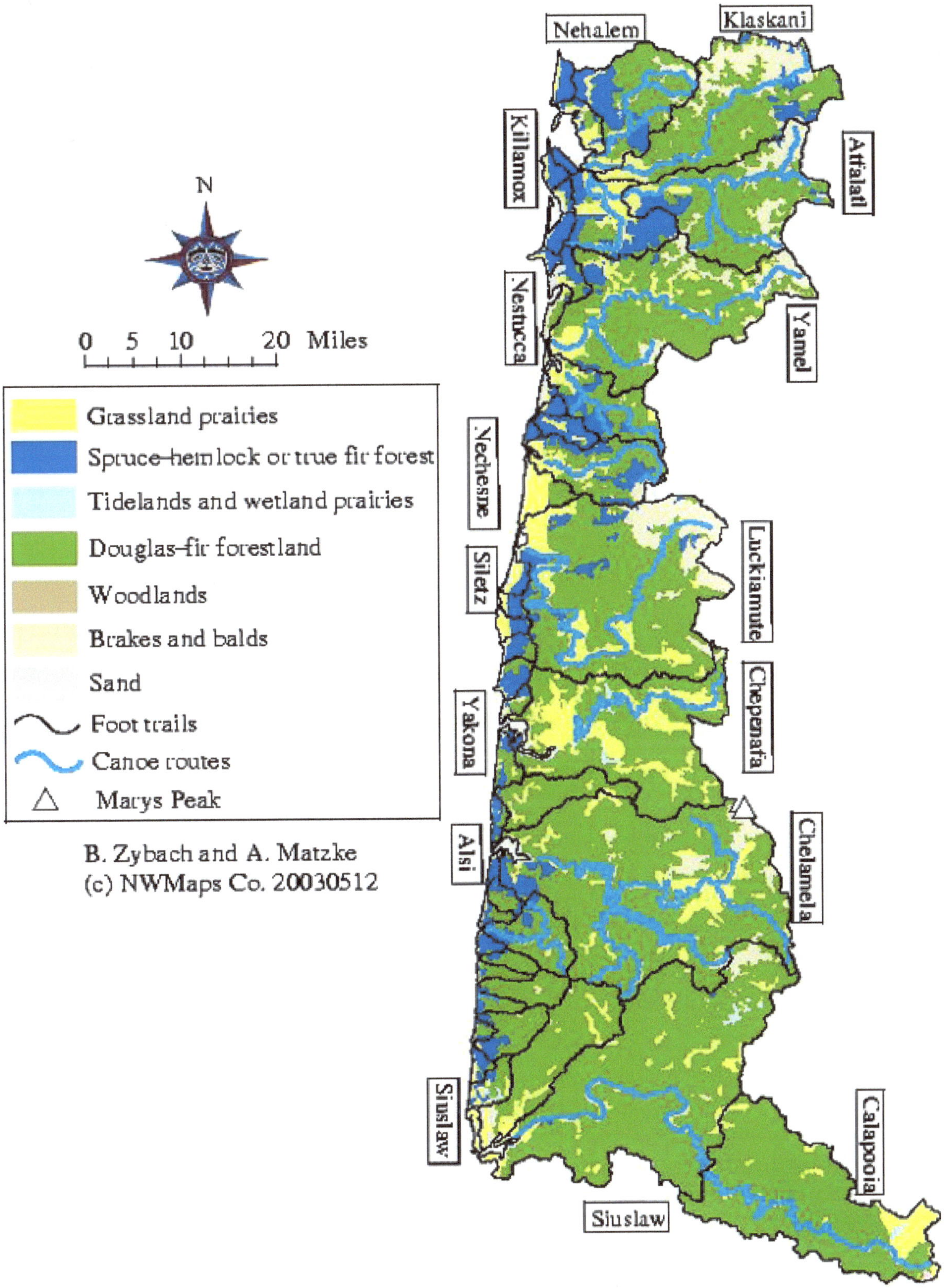


Map 3.06 Western Coast Range landscape patterns, ca. 1800.

Map 3.06 was constructed in the same manner as Maps 3.02 and 3.05. The principal differences are: 1) a better developed ridgeline trail network than North (more foot traffic; better access to inland resources; better ridge alignments) and East (flat topography; seasonal wetlands; few impediments to foot travel in any direction); 2) more brakes, balds, and bottomland prairies than the North, far less savannah and oak woodlands than the East; 3) far less canoe usage than the North, but far more than the East; 4) a transportation and trade focus on coastal estuaries that establishes a radiant trail network from those locations, which hardly exists in the North, and doesn't exist at all in the East (because of no estuaries or coastal shoreline); and 5) there is a much more extensive and pure Douglas-fir forest component than in the North (which is smaller--see Table 1.03--and contains a relatively large amount of fog belt spruce and hemlock) and the East (very little conifer forest of any kind).

### 3.4.4 South: mixed

Precontact vegetation of the southern Coast Range reflected the cultural and topographical features of the landscape--that is, it was a distinct mix: combining the wapato and flood lands of the north, with the oak savanna and camas of the east, and the even-aged conifer forests, inland prairies, and coastal grasslands of the west. Landscape management actions were also mixed, when compared to the other three subregions: the use of wetland prairies of the Coos and Coquille estuaries were reminiscent of the uses of Columbia River tidelands; the central Douglas-fir forests and inland prairies connected by foot trails were similar to the western Coast Range subregion; and the black oak and tanoak savannah of the Umpqua were managed with annual fall broadcast burns, similar to the way Kalapuyans managed the white oak savannah lands of the Willamette. Wilkes describes the process of periodic, low-intensity fires that were used to maintain such savannah conditions on the Umpqua in 1841, although he seems to have misinterpreted the basic purpose of these fires:

> During the day they passed over some basaltic hills, and then descended to another plain, where the soil was a fine loam. The prairies were on fire across their path, and had without doubt been lighted by the Indians to distress our party. The fires were by no means violent, the flames passing but slowly over the ground, and being only a few inches high (Wilkes 1845a: 228-229).

Figure 3.07 is a picture of the town of Coos Bay in 1888, when it still went by the name of Marshfield. The cultural legacy of low-lying prairie lands and sparsely timbered hillsides show a

Figure 3.07 Coos Bay (Marshfield), 1885.

disturbance history based on floods, similar to the northern Coast Range, and fire, similar to the eastern Coast Range or the interior valleys of the western Coast Range. The taking of prime Indian lands by subsequent white settlers is also similar to the history of the remainder of the Coast Range.

Map 3.07 of ca. 1800 southern Coast Range vegetation patterns also presents the "mixed" picture of landscape patterns that characterizes the southern Coast Range today. The eastern oak savannah/woodlands pattern is similar to that of the eastern Coast Range, but likely had better developed canoe routes and foot trails for reasons related to topography. The central portion of the area is largely an even-aged stand of Douglas- fir, with limited trail access; similar to Athapaskan lands to the north. And, as with the north, the Athapaskan speaking people maintained a direct trade connection to the coast via riverine townsites. One major difference

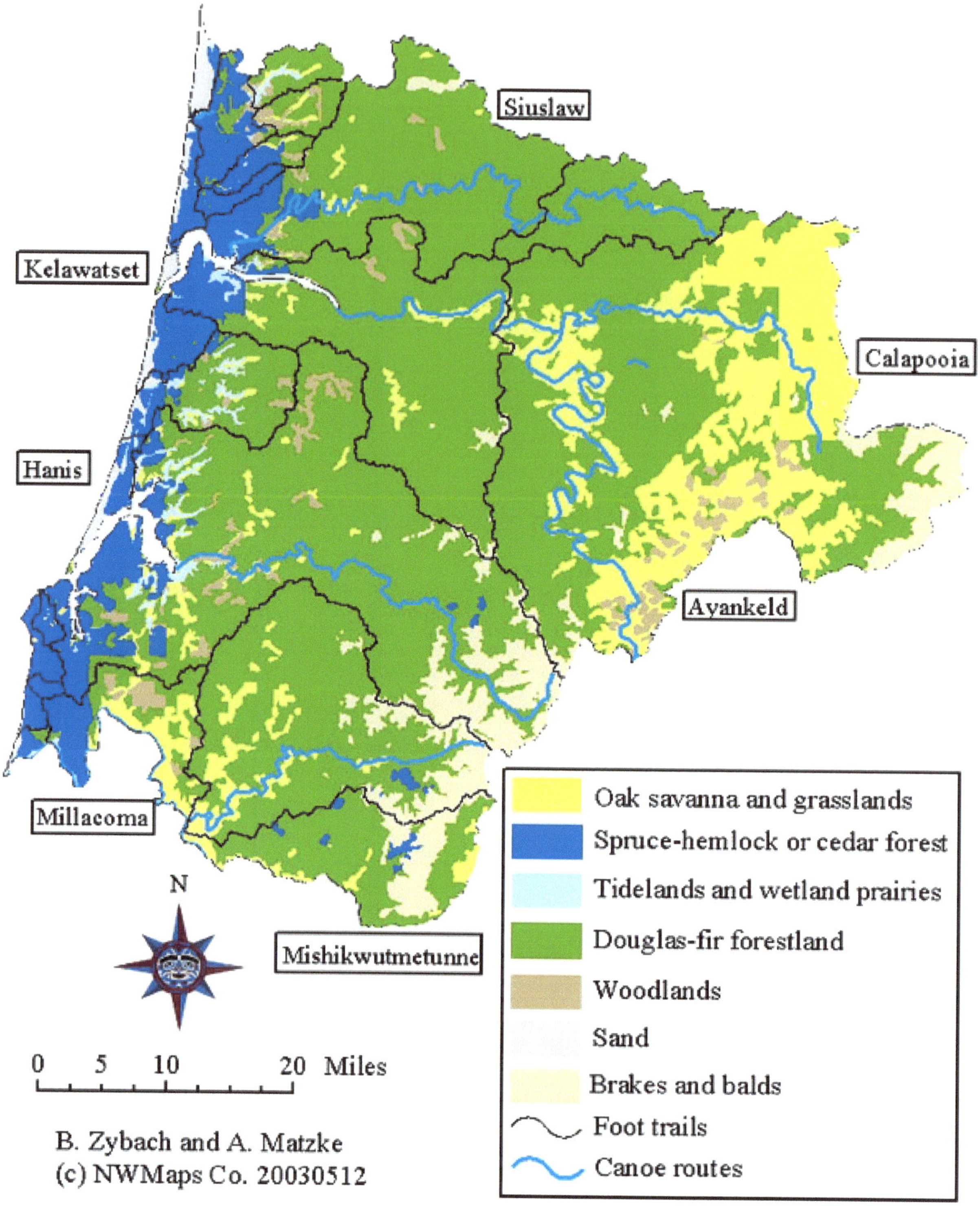


Map 3.07 Southern Coast Range landscape patterns, ca. 1800.

from the other subregions, though, is the extensive series of coastal sand dunes that exists between Coos Bay and the mouth of the Siuslaw; however, the total amount of land taken by these dunes is relatively small when compared to the entire area. The southern Coast Range also contains large cranberry bogs and stands of myrtle and Port Orford whitecedar that are not found in other areas of the Coast Range.

## 3.5 Discussion and Summary: Cultural Legacy

In precontact and early historical time, local Indian communities systematically managed native plants in all river drainages of the Oregon Coast Range. Firewood burning was a daily occupation by many precontact Indians. Patch burning practices were more likely to take place during seasonal periods, but also could be performed in almost any weather if fuel conditions permitted.

All historical accounts of broadcast burning activities in the Coast Range occurred during two fire seasons: late winter/early spring "fern burning" and late summer/early fall "field burning." In this manner, seasonally desiccated ridgeline brakes and bald peaks could be burned whenever a drying east wind came up for a few days; anytime from late February to early May. Valley grasslands, coastal headlands, oak woodlands, and tarweed fields were more likely to be burned in August or September, after vegetation had been dried by summer drought. East winds (see Map 1.08) were a factor that increased the possibility of late summer/early fall grassland burns in the eastern Coast Range entering Douglas-fir forests and developing into wildfires. White (1999: 47) makes a point that has bearing to this research when he states: "Indeed, tribal divisions did not differentiate the villages of the Puget Sound region as well as the cultural divisions of inland, river, and saltwater—divisions first mentioned by American settlers and later adopted by anthropologists." These divisions also hold true for the Oregon Coast Range. Both eyewitness accounts and historical maps show significant differences in the use of fire by river (Chinookan), saltwater (Salish, Yakonan, and Kusan), and inland (Kalapuyan and Athapaskan) peoples (see Map 3.08). Further, there is a marked difference between the "inland" burning practices and resulting landscapes of Athapaskans, who tended to live in dense Douglas-fir forestlands, and of Kalapuyans, who maintained open oak savannas.

Native food plants typically existed in relatively pure, even-aged patches or fields, and often bordered fallow mosaics (or "stands") of conifer trees; which can also be generally characterized

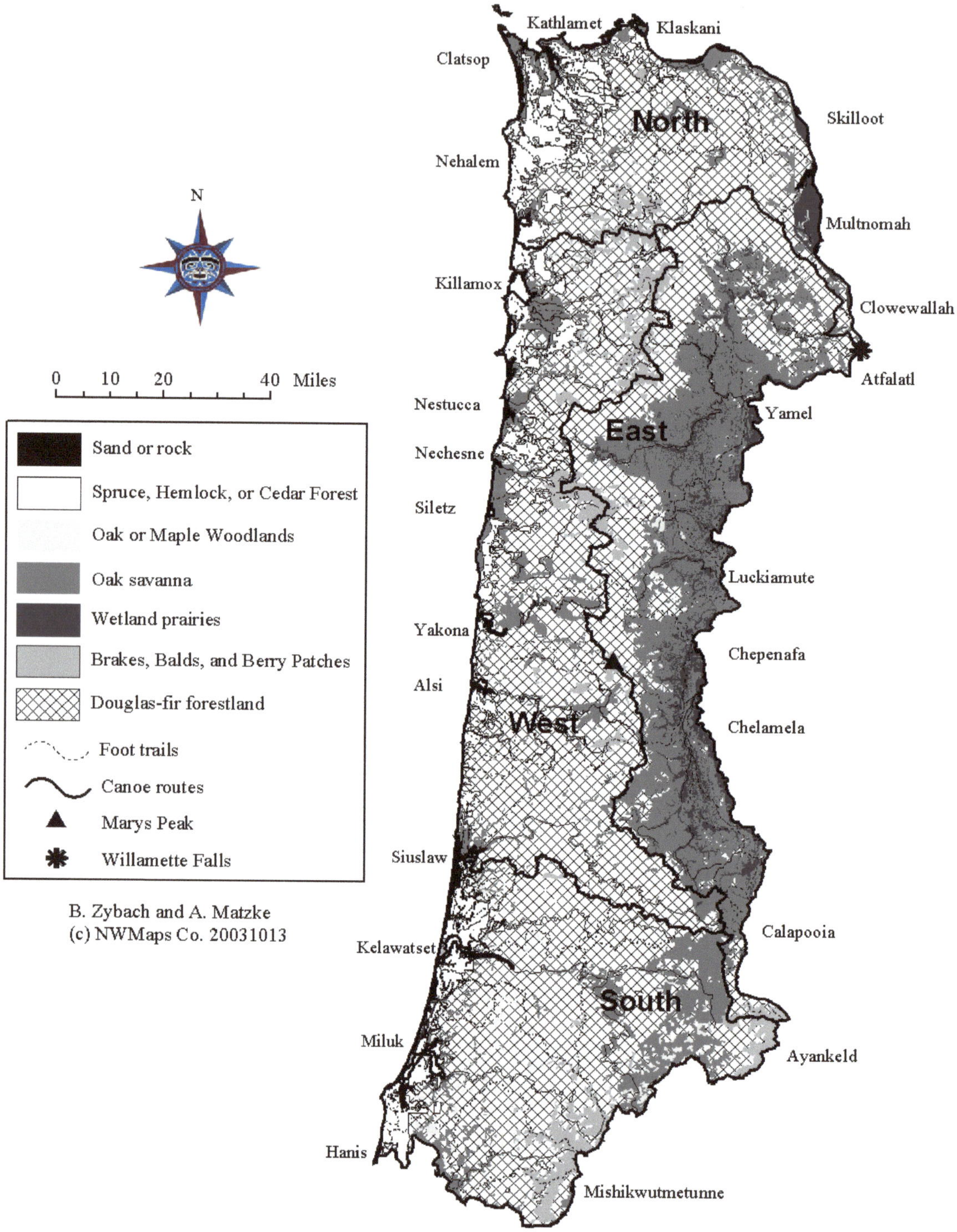


Map 3.08 Oregon Coast Range cultural landscape patterns, ca. 1800.

as existing in even-aged groupings of a single, dominant species. Stands of shorepine, Sitka spruce, Douglas-fir, noble fir, and western hemlock differed from managed areas in that they were probably not established or maintained for a distinct purpose, such as food, fiber, or tobacco production. Because of the even-aged nature of these stands, it is likely they were established by seeding, following either a forest fire or the abandonment of a fire-dependent crop (see Appendix F). Animals that benefited by the maintenance of large tracts of food plants, such as deer, bear, birds, and rabbits, were, in turn, often used as food by people. The coincidence of usage by both people and animals that occurred when sprouts appeared or fruits ripened likely made hunting activities more efficient.

The role of people in Coast Range forest and fire histories should not be underestimated. Lightning has been commonly assumed as the principal source of historical fire on the Oregon Coast Range landscape (Agee 1993: 54-55). The role of lightning fire vs. Indian burning in shaping fire regimes that produced varying vegetation patterns on the Coast Range has only been generally described (Boyd 1986; Boyd 1999c). The role of Indian burning is most commonly attributed to valleys and lowland areas, and little evidence has been put forward until now as to when, why, or how Indians used fire in mountainous areas (Agee 1993: 56; LaLande and Pullen 1999: 266). The lack of understanding by many scholars of the specific uses and application of fire by Indians in particular vegetation types has contributed to the dismissal of Indian burning as an important factor in shaping the composition, structure, diversity, and productivity of coastal forests and prairies (e.g., Tappenier et al 1997: 638; Franklin et al 2002; Whitlock and Knox 2002). Climate is a significant driver in potential vegetation assemblages at long time scales (Hansen 1947), but human actions often add detail and stability to those patterns (Pyne 1982). Such is the case with the Oregon Coast Range.

This research has found that lightning has not been a significant factor in recent Coast Range fire history, and that dominant patterns of precontact vegetation--including vast tracts of even-aged Douglas-fir forests--are probably a result of Indian burning practices (Stewart 2002: 250-255; Weisberg and Swanson 2003). The cultural legacy of those practices remains evident today in the general locations of forest plant species assemblages, and the persistent existence of relict balds, meadows, brakes, prairie fragments, and berry patches.

# 4. CATASTROPHIC FOREST FIRE PATTERNS, 1491-1951

> *The voyage from San Francisco is almost all the way in sight of land; and as you skirt the mountainous coast of Oregon you see long stretches of forest, miles of tall firs killed by forest fires, and rearing their bare heads toward the sky in a vast assemblage of bean poles - a barren view, which you owe to the noble red man, who, it is said, sets fire to these great woods in order to produce for himself a good crop of blueberries. When, some years ago, Walk-in-the-Water, or Red Cloud, or some other Colorado chief, asserted in Washington the right of the Indian to hunt buffalo, on the familiar ground that he must live, a journalist given to figures demolished the Indian position by demonstrating that a race which insisted on living on buffalo meat required about 16,000 acres of land per head for its subsistence, which is more than even we can spare. One wonders, remembering these figures, how many millions of feet of first-class lumber are sacrificed to provide an Indian rancheria with huckleberries.*
>
> --Charles Nordhoff (1874: 338)

> *The paths of the great forest fires of the last century or two are plainly marked by even-aged stands, consisting to the extent of at least 90 per cent of Douglas fir (if within the preferred habitat of this tree), regardless of the proportion of Douglas fir in the original fire killed stand.*
>
> --Thornton Munger (1940: 1)

This chapter provides background on the "Great Fires" and catastrophic "fire years" of the Oregon Coast Range, and compares the cause, timing, size, and boundaries of these events to the precontact Indian burning practices and resulting landscape patterns presented in Chapter 3. Maps, figures, and tables are used to depict spatial and temporal fire patterns, and to illustrate the severity and other dramatic environmental changes represented by these patterns of "stand replacement" events (Winter et al 2002: 1039).

## 4.1 Catastrophic Fires: Background

Catastrophic forest fires, by definition, are wildfires that cover more than 100,000 acres of contiguous forestland during the course of a single event. Catastrophic fire years are calendar years in which more than 100,000 acres of forestland within a given area are burned by wildfire, whether during the course of a single event, or as a result of numerous events.

Map 4.01 shows the locations and boundaries of the Great Fires of the Oregon Coast Range. Indians likely caused the ca. 1765 Millicoma Fire; whites probably caused all, or most all, of the remainder (including the Nestucca Fire, generally blamed on a black settler clearing land near

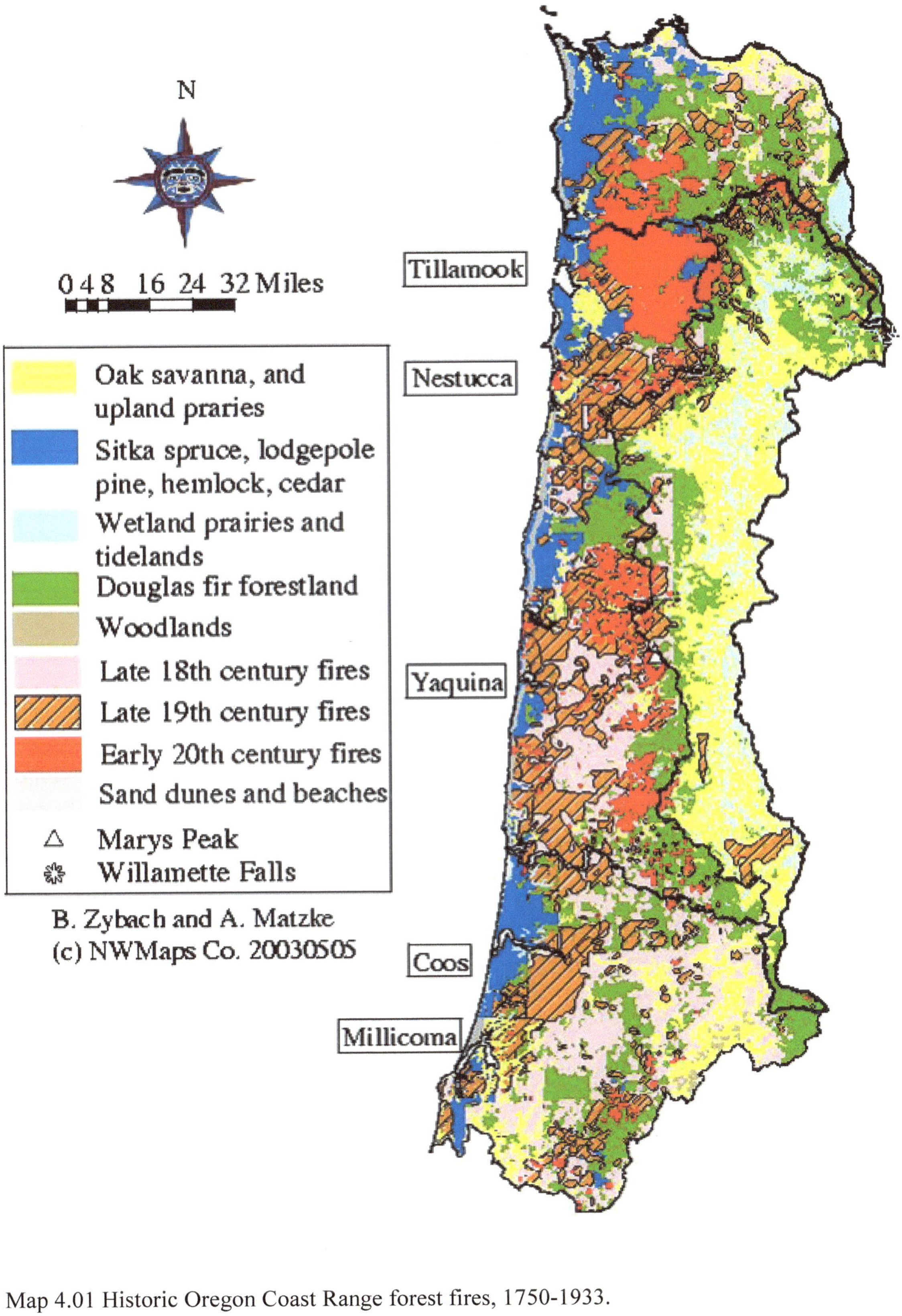


Map 4.01 Historic Oregon Coast Range forest fires, 1750-1933.

Table 4.01 Historic Oregon Coast Range forest fires, 1750-1951.

| Name | Year | Season | Acres | Authority |
|---|---|---|---|---|
| Millicoma | ca. 1765 | Unknown | 200,000 | Phillips 1988: map; Smyth 2002 |
| Yaquina (1) | 1849 | Aug./Sep. | 450,000 | Gannett 1902; Morris 1934b |
| Nestucca | ca. 1853 | Aug./Sep. | 350,000 | Gannett 1902; Munger 1944 |
| Yaquina (2) | 1868 | Aug./Sep. | 300,000 | Fagan 1885; Kirkpatrick 1940 |
| Coos | 1868 | Sep/Oct | 300,000 | Kirkpatrick 1940; Phillips 2003 |
| Tillamook (1) | 1933 | Aug. | 311,000 | Dague 1934; Fick & Martin 1992 |
| Tillamook (2) | 1939 | Aug./Sep. | 225,000 | ODF 1951a; Fick & Martin 1992 |
| Tillamook (3) | 1945 | Jul./Aug. | 110,000 | ODF 1951a; Fick & Martin 1992 |
| Tillamook (4) | 1951 | Apr.-Sep. | 49,500 | ODF 1951b; Chen 1997: 4 |

Champoeg -- see Morris 1934b: 318). Table 4.01 lists the Great Fires, their likely or known year of occurrence, and their approximate size. Drawings and photographs are used in the following pages of this chapter to show how the burns and reforestation resulting from these events must have appeared to Nordhoff (1874: 338), Nash (1976), Munger (1916), and others in the late 1800s and early 1900s. These illustrations also indicate the dramatic change in local wildlife habitat conditions and populations that occurs when tens of thousands of acres of forestland are burned within a few hours or days' time. Brief individual profiles of each of the Great Fire events, as well as a general background on the fires' fuel histories, are also provided.

### 4.1.1 Mystery of the 16th Century

Something happened in the Coast Range during the 1500s that resulted in widespread forestation into the 17th century, which in turn likely formed a majority of fuels consumed in subsequent wildfire events throughout the time of this study. Common speculation is that catastrophic wildfires deforested much of western Oregon (Andrews and Cowlin 1940; Berkley 2000; Weisberg and Swanson 2003) and the Pacific Northwest (Morris 1934b; Schmidt 1970; Agee 1993) during that time, including Coast Range forests (Juday 1976; Tappeiner et al 1997; Ripple et al 2000). This speculation is partly based on the assumption that 15th and 16th century fires burned vast tracts of old-growth forests (200 years or greater age), but little evidence exists to support that belief. If such fires did occur, it seems as likely--or possibly even more likely--they burned saplings or young second-growth as older trees. Other possibilities include a cessation of widespread Indian burning, with subsequent afforestation of abandoned prairies, meadows, and fields (see Appendix F). Lightning fires are rare in the region (Towle 1974: 37; Chapter 1.4.3), and people caused all historical Coast Range fires of a similar magnitude (see Appendix G). Were human populations greater in the 1500s than the 1600s, leading to more forest fires or more widespread land management practices? A subsequent decline in human populations might reasonably lead to an expansion of existing forestlands, as occurred in the early 19th century. Or could other factors, such as catastrophic winds, diseases, or insect infestations, have been a cause? These hypothetical questions are examined in more detail in Chapter 5.

It is known that widespread forestation took place in the Coast Range during the 17th century because, over the course of the last century, hundreds of people systematically counted and recorded the rings of thousands of trees throughout the entire region (e.g., Kimmey and Furniss 1943; Weyerhaeuser Company 1947; Impara 1997). Tax revenue agents, timber cruisers, forest scientists, resource managers, and graduate students, among others, have done this work. Reasons for counting rings have varied: to measure tree growth; tax timber harvest, record fire events; track seasonal climate changes; or document disease. A common result of these efforts is tree ages have become known and recorded for a significant number of locations throughout the region (see Appendix G). Very few of these trees have exceeded 400 years age, and none are known to have achieved 600 years; that is, almost all Coast Range trees described, measured, burned or logged in the last 200 years began to grow sometime after 1600--and the majority of those trees likely began to grow sometime after 1800. In other words, most of the trees known to have lived in the Coast Range during the past 400 years have been second-growth (less than 200 years of age) rather than old-growth, and the majority of trees that have achieved old-growth status spent the majority of their lives as second-growth. For example, over 90% of the Teensma (1987) and Connelly and Kertis (1992) study areas in the western Cascades became forest after 1535, Impara (1997) has only one Coast Range date before 1530, and Weyerhaeuser Company (1947) has no dates in Coos County before 1565 (see Appendix G). Similar findings are documented for the Siuslaw River (Poage 2001), Siletz River (Grantham 1953; Boyce and Wagg 1953), Luckiamute River (Zybach 1999), and other Coast Range basins. This fact is of secondary interest to this dissertation, but should have important policy (e.g., FEMAT 1993), scientific (e.g., Agee 1993), and forest and wildlife management (e.g., Shea 1963) implications.

That second-growth trees have probably dominated Oregon Coast Range forests for most of the last 500 years is an interesting finding for several reasons. The Coast Range contains many of the largest and fastest growing species and specimens of forest trees in the world (Franklin and Dyrness ca. 1988). However, Douglas-fir, Sitka spruce, western hemlock, redcedar, black cottonwood, red alder, and perhaps even white oak all tend to live much longer in other regions of the Pacific Northwest, often taking many additional decades or centuries to achieve similar heights, diameters, or volumes in size as those in the Coast Range. Douglas-fir, for example, is known to reach ages of 700, 1000, or even more years in areas of the western Cascades (e.g., Franklin 1981: 94), Olympic Peninsula (Henderson et al 1989), or Vancouver Island (Schmidt 1970: 108), yet the height, diameter, and/or volume of those trees is not significantly different than the largest, though much younger, Coast Range trees (Franklin and Dyrness ca. 1988).

Figure 4.01 shows examples of old-growth Coast Range trees established in precontact time. The upper left hand drawing of sailors measuring a giant Douglas-fir near Astoria was made in 1841 by "Mr. Drayton," who "took a camera lucida drawing" of the event (Wilkes 1845b: 116). This method uses a lens to reflect an image onto paper, which is then traced, producing a drawing that can be nearly as accurate as a photograph, particularly in regard to scale. The tree was "one of the largest" in the "primeval forest of pines" and measured 39 feet, six inches in diameter, with bark 11 inches thick at eight feet above the ground. The tree was "perfectly straight" and estimated to be "upwards of two hundred and fifty feet" tall (ibid: 116). In 1825, Douglas had measured a bark-less, three-foot high stump behind Fort Astoria that was 48 feet in circumference that, he noted, had been logged "to give place to a more useful vegetable, namely potatoes" (Meany 1935: 53). Douglas' stump may have belonged to the tree described by Cox in 1814:

> The largest species [of "fir-trees"] grow to an immense size, and one immediately behind the fort at the height of 10 feet from the surface of the earth measured 46 feet in circumference! The trunk of this tree had about 150 feet free from branches. Its top had been some time before blasted by lightning; and to judge by comparison, its height when perfect must have exceeded 300 feet! This was however an extraordinary tree in that country, and was denominated by the Canadians *Le Roi de Pins* (Cox 1957: 71).

The upper right photograph (B. Zybach 2003) shows an old-growth Douglas-fir located along the 1856 ridgeline "Indian Trail to Tidewater" (see Map 2.11). The heavy lower limb structure indicates that the tree grew in the open, thereby reducing risk of crown fire. The leeward side of such trees provides good protection from wind and rain, and likely served such a purpose from time to time for Alsi Indians in the 1700s and early 1800s, local travelers and residents in the

Figure 4.01 Precontact Douglas-fir and white oak old-growth, 1841-2003.

1800s, and loggers or treeplanters in the 1900s. In precontact time this tree would have been clearly visible from much of Alsea Valley and the surrounding landscape, adding potential value as a landmark or meeting site. The lower left picture is of old-growth oak (photograph by B. Zybach 2003), located near the town of Alsea (see Figure 2.03) in an area containing numerous precontact artifacts (personal observation and informal interviews with owners and adjacent residents). Oaks were prized by precontact Indians for many reasons, including their ability to produce acorns. In Coast Range areas where oak are left undisturbed by fire, grazing, mowing, plowing, logging, or other management processes, they are soon replaced by competing conifers (Sprague and Hansen 1946; Zybach 1999: 102-117). The lower right hand corner (photograph by N. Lapham 2003) shows another, nearby old-growth oak being threatened by second-growth

Douglas-fir, several centuries younger in age. The outstretched arms of the man at its base signify a diameter of nearly six feet. Many Indians prized oak trees and groves highly, and viewed most conifers as invasive weeds (Thompson 1991: 33). These trees are typical of certain precontact old-growth forms and species and demonstrate potential "functions" in earlier landscape patterns, dating to the beginnings of this study period.

Figure 4.02 (Williams 1912: 137) shows the change of values toward old-growth trees that accompanied the replacement of Indians with whites across the Coast Range landscape. Large, sound conifer trees with a minimum of lower limbs were highly prized by early loggers and sawmill owners, particularly if they were located near a major waterway or could be readily accessed via railroad construction. These trees, as with those pictured in Figure 4.01, likely date to the 16th or 17th century; a key difference is they were allowed to grow in an absence of regular human disturbance, rather than be dependent on it, as evidenced by stems per acre (average distance between trees) and lack (or presence) of lower limbs.

Figure 4.02 Logging of Clatskanie River basin, ca. 1900.

### 4.1.2 Millicoma Fire, ca. 1765

Shortly after World War II, in 1945 and 1946, Weyerhaeuser Company did an inventory of its land and timber holdings in Coos County, principally within the Millicoma River basin (see Map 1.03). Foresters traversed grids across the landscape, bored about 1500 trees, mostly Douglas-fir, and counted the rings, over more than 100,000 acres of forestland (Weyerhaeuser Company 1947). They also noted species, diameter, the length of clear boles, and other measures of interest to timberland managers. Most of the trees, covering a swath more than 90,000 acres in size, were large second-growth of the same age--about 180 years old (see Figure 4.03; Smyth 2001: 53; used by permission of A. Smyth; Weyerhaeuser Company Archives). One of the foresters who worked on the project, in apparent concert with his peers, concluded the stand dated to a single catastrophic fire that occurred about 1765, and that may have been 200,000 acres in size (Smyth 2000: 2-4; Smyth 2003: personal communications). A few older clumps of trees, dating to the 1720s, were also included in the inventory, and assumed to be possible seed sources for the younger stand. Adjacent stands of timber, notably the so-called 28,000 acre Pillsbury Tract (Shea 1963), were also possible seed sources, and dated to the 1600s (see Appendix G).

The Weyerhaeuser timber cruise provides the first compelling evidence of a Coast Range catastrophic fire that dates to precontact times, although it remains possible the trees became established following some other form of disturbance. Figure 4.03 shows a portion of the Millicoma stand as it appeared to foresters in the 1940s. Note that snags and large woody debris in this area appear to be from younger trees in the stand, rather than provide evidence of an earlier population. It is also possible that the ca. 1720 trees represent an initial fire in the stand, and that the 1765 component represents a reburn over much of the original fire's extent. Either--or both--date(s) would satisfy Morris' observation:

> When land in western Oregon and western Washington is deforested by fire or cutting and then left unmolested, within a few years it is clothed with a new stand of trees almost uniform in age (Morris 1934b: 314).

The Weyerhaeuser cruise covered most of the southern part of the Millicoma stand; the northern part extended into land that became Elliott State Forest, most of which forest became established following the 1868 Coos Fires. Jerry Phillips, who worked on the Elliott for 33 years, mapped the northern part of the ca. 1765 stand to an approximate boundary between the two Great Fire events

(Phillips 1988: map). The combined Weyerhaeuser cruise and Phillips map conform fairly well to the 1936 PNW Forest Type map (see Map 2.10; Andrews and Cowlin 1936: map) as an extensive stand of large second-growth Douglas-fir, apparently 68 to 170 years of age (see figures 1.05 and 4.01) [in 1936].

Figure 4.03 Millicoma Fire second-growth reforestation, ca. 1946.

Smyth (2000: 2) gives a figure of about 200,000 acres for the fire, but doesn't acknowledge 28,000 acres of older trees (Shea 1963: 4), or directly account for another 80,000 acres that wasn't included in the 1945-1946 Weyerhaeuser cruise. Phillips shows the ca. 1765 fire as bordering directly on the 1868 fire, so there is no telling how much of the original boundary was burned again at that time. Phillips thinks the line is fairly distinct, in that the 100-year old Douglas-fir forest that had grown in the wake of the Millicoma Fire, was essentially "fire-proof" of ladder fuels, hanging mosses, and dead, pitchy snags that characterize older Douglas-fir forests (Phillips 2003: personal communications). A reasonable size estimate for the Millicoma Fire, whether occurring in ca. 1720, ca. 1765, or both years, is about the same size range as currently given for the 1868 Coos Fire: 125,000 to 300,000 acres (Pyne 1982: 338).

Phillips (2003: personal communications) hypothesizes that the bulk of the Coos Fire was 1500s-era Douglas-fir, with a component of hemlock and redcedar, similar to relict stands that survived the 1868 event, and that mosses, dead snags, and other ladder fuels were principal causes of spread wherever they occurred within the fire boundary. If the same pattern holds true for the boundaries of the Millicoma Fire, the fuels would have been principally older second-growth and younger old-growth Douglas-fir trees between 150 to 250 years old, with a cedar and hemlock component, and ladder fuels that included hanging mosses and pitchy snags.

It is reasonable to speculate that the year 1765 is an average age for the cruised stand, and that the year of the principal fire is actually sometime between 1710 and 1760. One condition that would help resolve the issue of dating is whether the 1720 component existed in clusters, or was somewhat evenly distributed throughout the entire 1760 stand--one condition would indicate a possible seed source for the younger trees, and the other would indicate possible relicts of a reburn. Smyth (2000: 3-4) further speculates the fire was caused by Indians and traveled from the northeast to southwest. Phillips' (1988: map) and Weyerhaeuser's (1947) maps seem to agree with the burn pattern and with detailed 1936 vegetation patterns (Andrews and Cowlin 1936: map), and general Coast Range fire history supports the likely cause (people) and direction (east wind).

#### 4.1.3 Yaquina Fire, ca. 1849

The Yaquina Fire may be the largest of the Great Fires. Morris (1934b: 322) says it extended southwesterly from the South Fork (Rock Creek) of the Siletz River to the mouth of the Siuslaw

River (see Map 1.03). William Smith (see Appendix C), an Alsi Indian, seems to be the only reliable eyewitness of record to this event (Frachtenberg 1920: 213-219), and he agrees with Morris on the time (about 1849) and southern extent (Siuslaw River). Phillips (2003: personal communication) speculates the southern extent may be the divide between the Smith and Umpqua rivers, which he believes to be the northern-most extent of the 1868 Coos Fire. Talbot (Haskin 1948) was in the Yaquina Fire area in late August, 1849--when he noted heavy smoke moving west from the area of the Marys and Yaquina river headwaters--through early September, when he left Alsea Bay, crossed the Yaquina and Siletz estuaries along the coast, and returned to the Willamette Valley via the Salmon River. If the fire did occur in 1849, which seems likely, it must have covered most of its extent in September, after Talbot had traveled through its interior--which also seems likely. It may have also crossed the Siuslaw River and ended near the Umpqua River; if not in 1849, then perhaps in 1857 or 1868 (Morris 1934b: 322).

Both Talbot and Albert, who preceded Talbot along the same general route in 1848--but in reverse direction--noted fairly large burns that had already occurred near the northeast extent of the Yaquina Fire:

> [Albert] The Celeetse [Siletz] prairie is level, rich, and the fern is eight or ten feet high. The prairie is surrounded by timber . . . Coming up the Celeetse river, we passed prairie bottoms or plains . . . The Celeetse river has many forks, but only two principal ones. The main river is larger than Pudding river . . . the timber, however, is not very good, having been badly burnt for some miles around the N.E. portion of it. The path from the Celeetse prairie is rugged, full of logs, &c. (Albert 1848: 1).
>
> [Talbot] Our road today [August 26, 1849] . . . [was] much obstructed by fallen trees and thick brush. We passed through one tract of burnt forest several miles in extent where the little trail which we followed, indifferent at the best, was often completely broken up, and we were compelled to have recourse to our axes to make a way through the heaps of charred logs. We descended, after a toilsome day's journey, into a grassy valley, about half a mile in length, watered by a fork [Rock Creek] of the Celeetz [Siletz] river, in which we encamped, having made nine miles (Haskin 1948: 5).

The burned logs may have resulted from clearing fires made by Klickitat Indians as they constructed a horse trail from Kings Valley (Luckiamute River) to the Siletz and Yaquina rivers in the late 1840s (Haskin 1848: 4). This route follows the northern boundary of the Yaquina Fire (see Map 4.01; Braman 1987). Both journalists noted extensive burns, brakes, prairies, downed logs, and big trees, but no evidence of the subsequent Great Fire. Munford quotes James Blodgett:

> As a boy, I knew this country well. A great fire in 1850, while my great-grandfather was in California, destroyed about 150 square miles [100,000 acres] of virgin forest. The fire stopped along a line approximately from Hoskins to Blodgett to Marys Peak. What is again dense forest was, in 1877, open fern-covered ground interspersed with the skeletons of burned trees. Burnt Woods [see Appendix B] takes its name from this feature (Nash 1976: 281).

This story was confirmed with Blodgett on August 23, 1984. He conceded the fire may well have occurred in the fall of 1849. His ancestors had been among the first to travel to the gold fields in California, having left in July 1848. There, his grandfather (also named James) was born on August 12, 1850. The family returned to Oregon in time to be counted in the October 7, 1850 Oregon Territory census ["Dwelling #148"], when they had been "amazed" to see that "the entire area from Marys Peak to the ocean had been burned clean by fire" and had become a "sea of snags." They were also gratified to note their homestead, "Blodgett's Valley," to the west of Philomath, had been spared. Blodgett claimed his father had told this story to him in the 1920s, and whose parents, in turn, had told him the same story in the 1890s. In sum, the Blodgetts had returned to their homestead on the eastern boundary of the Yaquina Fire sometime between late August and early October, 1850, when they found a fire had destroyed almost all of the timber between their home and the ocean. Talbot had noted large fires in the area in the fall of 1849, when the Blodgetts were still in California, and the Blodgett family history made no mention of smoke or fire in 1850--only burned land and snags. Therefore, an August to September 1849 date and an east wind burning pattern seem to be the most likely attributes of the fire.

Figure 4.04 (Fagan 1885: 104) shows results of the Yaquina Fire in 1885, near Toledo, along the Yaquina River, a few miles east of Yaquina Bay. Note the location of Marys Peak in the background. This area likely burned twice; first in September 1849, and the second time in September 1868. On September 2, 1849, Talbot made the following notes:

> Should it be satisfactorily ascertained that ships may come in [to Yaquina Bay] with safety, this harbor will become exceedingly valuable, as it is surrounded by a country covered with forests of the finest kind of timber, has good mill-seats, and roads could be constructed which would afford a near market for the produce of the upper Willamette (Haskin 1948: 9).

Figure 4.04 Yaquina Burn, Toledo, Lincoln Co., Oregon, 1885.

While traveling through the same area as Talbot, in 1848, Albert had noted:

> The Yac-quin-na harbor, they [some local Indians] said, was long and deep, and also deep at the mouth . . . The land from the Yac-quin-na is rolling--in fact, mountainous, with heavy timber, and is really the prettiest building timber we have ever seen--containing many springs, and small prairies covered with fern (Albert 1848: 1).

Because a second (or possibly third or fourth) major fire occurred throughout the Yaquina Burn in 1868 (Morris 1934b: 326-332; Kirkpatrick ca. 1940: 33), the "rolling" land behind Toledo in Figure 4.04 remained largely unforested into the 1880s, as shown in the drawing. Nash described the Yaquina Burn in 1877:

> The next day we camped near the new farmhouse of Mr. Meade, a settler of twelve years history. His house stood in the heart of a large tract [his house was in Nortons, upriver from Toledo, along the Yaquina River, between Eddyville and Nashville] which had been swept by a forest fire forty years ago. For fifteen or twenty miles we passed among the huge, black standing trunks which had

> survived the fire, and resisted the slow decay which had brought down many of their neighbors, now rotting into red and yellow soil among the thick fern and wild pea-vines. From a hill-top, looking on a wide prospect of this fire-ravaged land, the distant stems reminded us of a forest of masts standing in crowded docks. It was only by drawing close to one of these scarred trunks that one could realize its size--five or six feet thick, and ninety or a hundred feet high (Nash 1976: 143-144).

A related factor to the lack of rapid reforestation of the Yaquina Burn is the "Yaquina Tract," a 20-mile wide tract of land from Corvallis to Yaquina Bay, which was removed from the Coast Indian Reservation and opened to white settlement in 1865 to provide land for Civil War veterans, among others (Kent 1977: 18-20). A major portion of the Yaquina Tract had been burned during the 1849 fire, and most of that land reburned in 1868. The burned land was subsequently used for grazing cattle, sheep, and mohair goats, a general use that extended well into the 1930s (Wakefield 1989: personal communication). Local grazing practices included seasonal "fern burning" to increase pasturage, as well as seasonal browsing and trampling to further eliminate or delay the process of reforestation. Gannett (1902) observed:

> Since this region, especially that portion of it west of the Cascades, where the great burns have occurred, is abundantly watered, and in all other respects extremely favorable for tree growth, all of the burns are in some stage of reforesting, and in most of them reforesting has gone forward rapidly and very favorably. Much of the burned country is now covered with a dense stand of young trees. The species following the fires are, in nearly all cases, similar to those destroyed. Fir follows fir, yellow pine follows yellow pine . . .
>
> In some cases, however, and especially in the largest burns, the work of reforestation has made little progress, owing to the difficulty in reseeding large burned areas. Since over many square miles all the trees were killed, the seeds of a new crop have had to come from outside the region, and hence the seeding process has been slow. Areas are reported which were burned twenty-five to fifty years ago on which there is no vegetation larger than brush and ferns, trees of any species not yet having obtained a foothold (Gannett 1902: 12)

The southernmost extent of the fire, as noted above, was possibly its western-most extent as well, near the mouth of the Siuslaw River. Smith (see Appendix C) repeats the story of his father and uncle, who had traveled south from Heceta Head to the mouth of the Siuslaw during the aftermath of the fire:

> Then they started. Where there was a mountain [possibly Cape Mountain], that place there did not burn. So they two kept on going on that trail, and they arrived at where there was a place (covered) with grass. And only there did the fire reach. Then they two ascended. Now it was gradually getting light all over [the smoke was beginning to clear, after darkening the skies for about 10 days]; just a little (light) showed far away. At last they came below, whereupon they two started to walk on that beach. Then they two kept on going along the beach. Everywhere even the blossoms of the highest trees had burned down, (as could be seen) after the water came with them to the beach. Now not long (afterward) they two saw a bear walking along the edge of the water, just partially burned. At last they two arrived at the mouth of the Siuslaw River. All the pine trees (there) were partially burned. Only ashes (could be seen) all over, because all the pine trees had caught fire (Frachtenberg 1920: 217).

Smith's description conforms very well with subsequent maps of the area (see Map 4.01), which show the fire burning all the way to the coastline--right to the strip of shorepine noted by Smith that borders the beach--at the mouth of the Siuslaw River. Whether the fire continued southward to the Umpqua River, or whether the burned land between the Umpqua and Siuslaw dates to the 1868 Yaquina reburn or to the Coos Fire of the same year, is unknown.

#### 4.1.4 Nestucca Fire, ca. 1853

The Nestucca Fire is one of the least documented and most storied of the Great Fires. Munger (1944: 342) states there was only a single eyewitness accounting of the fire ("reported many years later to a now anonymous writer by Dick Harney, chief of the Nestucca tribe") that "certainly spread from east to west" and covered "300,000 or 375,000 acre expanse of the country's finest timber." According to Morris (1934b: 321), the fire extended south from the Nestucca-Trask river basins divide to Siletz Bay, with a marked difference between the northern and southern portions of the resulting burn (see Map 4.10). Munger (1944) attributes the differences to subsequent land management history of grazing and land clearing, but the presence of precontact Indian prairies, balds, brakes, and berry patches undoubtedly affected fire extent and subsequent afforestation and reforestation patterns. The 1858 US Coast Survey, for example, describes the southern face of Cape Lookout (to the north of Nestucca Bay) as "destitute of trees" (US Coast Survey 1859: 384) and Cape Falcon (to the north of Tillamook Bay) as: "Like some other points in this latitude, the southern face of the cape is destitute of trees, but covered with a thick growth of grass, bushes, and fern" (ibid: 385).

The oldest conifers on the 1935-1988 Cascade Head growth and yield study were established ca. 1852 (USDA Siuslaw National Forest ca. 1988). In 1940, the eleven "practically even-aged, uniform in stocking" growth study plots averaged 89-years old (Munger 1944: 344); that is, they were also established sometime around 1850. Whether the new forest plots dated to a catastrophic fire or to a cessation of Indian burning is difficult to determine. There is no way to ensure the accuracy of the ring counts without taking further samples. Several of the study plots still exist, but none of the literature mentions the existence of older trees, snags, stumps, or logs within their boundaries. Arthur (1848: 1), Talbot (Haskin 1948), and Vaughn (1923) were through the area in the late 1840s and early 1850s and made no mention of a fire. Ballou (2002) gives a date of 1853, but Vaughn probably traveled through the area a year or two after that time, without mentioning a large-scale burn. Morris (1934b) lists 1857 as a year of much smoke, and it seems likely the principal Nestucca fire may have occurred at that time, with boundaries being extended in 1868, along with the Coos and Yaquina fires of that year.

Figure 4.05 shows photographs of portions of the great Yaquina and Nestucca burns, as they appeared around 1900 (Gannett 1902: 12). Note the scattered "parent trees" that survived both fires and were likely seed sources to the reforested stands of the 20th century. Also, note the cleared area of land claimed by an early settler in the lower picture. The lack of snags in the clearing indicates a likelihood of regular precontact use by local Indians, and the presence of young saplings in the same area indicates the likely cessation of such use. These patterns of widespread snags, scattered surviving trees, tree-less prairies and openings, and the presence of young trees throughout are typical of much of the Coast Range. Also, note the similarity in size and distribution of Yaquina Burn snags in figures 4.04 and 4.05.

Remarking on the cause of the Nestucca Fire, Munger notes:

> How it started will never be known. It is unlikely that lightning was the cause for thunderstorms are rare in the Coast Range of Oregon; it my have been from an Indian campfire of from fire that Indians to the east set, as they are known to have done customarily to enlarge areas of "prairies" and open grass and browse land in the valleys. Or possibly it may have spread from the stump fires of some covered wagon pioneers or Hudson's Bay voyageurs who were camping on the eastern edge of this forest wilderness. There were no white people living in this part of Oregon (Munger 1944: 342)

U. S. GEOLOGICAL SURVEY PROFESSIONAL PAPER NO. 4 PL. III

A. THE GREAT NESTUCCA BURN.

B. THE GREAT YAQUINA BURN.

Figure 4.05 Yaquina and Nestucca burns, ca. 1902.

Munger further notes that the fire "certainly spread from east to west" from to "headwaters of the Nestucca to the Salmon River, and perhaps further . . . to tide line" (Munger 1944: 342). He attributes reforestation of the Nestucca Burn to scattered surviving trees, but:

> At any rate, a new forest of conifers sprang up on the burn, and has flourished . . . These trees soon formed such a dense canopy that brush, bracken fern and weeds were shaded out . . . So, the trees on Cascade Head that were born in the ashes nearly a hundred years ago have grown and grown until at the present time [1944] this is what we find: A forest of 200 trees an acre, mostly from twelve to forty inches in diameter and from 120 to 160 feet tall. It is practically even-aged, uniform in stocking, but varying in its proportion of spruce, hemlock, and fir (Munger 1944: 344)

Munger paints a different picture for the reforestation of the northern ("Hebo country") Nestucca Burn, however. For that area, he theorizes that the land had initially reforested with conifer seedlings in much the same manner as Cascade Head, but "fire after fire spread from the clearings along the creeks" by "settlers" had "annihilated the tender [conifer] seedlings" and:

> So, the hills from the Nestucca River to the top of Mt. Hebo were periodically baptized in fire, the down tree trunks were gradually burned up, some of the snags were burned down and the surface converted to a sea of bracken fern, salal, thimbleberry and other worthless plants. Only here and there did a Douglas fir or spruce sapling or a clump of red alders gain a foothold and survive long enough to be immune to the periodic surface fire (Munger 1944: 345).

Munger further noted that the "periodic scorching that the white man's fire gave it" was responsible for turning "this whole section of the Coast Range, which once bore such a magnificent forest" into "a waste of worthless brush and weeds" (Munger 1944: 345).

In sum, the Nestucca Fire took place sometime between 1845 and 1857, with ca. 1853 being a fairly reasonable guess. It probably took place in August or September, was started by people to the east, probably in the Willamette Valley or eastern Coast Range foothills, and traveled west until it reached "tide land." The entire fire is estimated to have been 300,000 to 375,000 acres in size.

#### 4.1.5 Coos Fire, 1868

In his 19th-century history of Coos and Curry counties, Orvil Dodge noted: "in 1868 the face of the country was changed by a great forest fire which swept over miles, reaching Port Orford, and leaving destruction and desolation in its wake" (Dodge 1898). The fires of 1868 burned large areas in Coos County. The most extensive continuous fire area--named the Coos Fire--was indicated by even-aged forest, reaching from Scottsburg in Douglas County, southeast to Coos Bay and nearly 300,000 acres in size (Phillips 1988: map; Phillips 2003: personal communications). Kirkpatrick (ca. 1940: 33) also claimed the Coos Fire was 300,000 acres in size, but he may have been including the entire southwest Oregon complex for that year: second-growth stands in 1936 near the coast from Coos Bay to Port Orford (Map 2.07; Andrews and Cowlin 1936: map) show where numerous smaller fires also burned during 1868.

In 1868 forest fires burned throughout the Pacific Northwest. In addition to the Coos and Port Orford fires, the snags from the 1849 Yaquina Fire re-burned (Kirkpatrick ca. 1940: 33), and the 1868 Yaquina Fire may have extended tens of thousands of acres into more green timber in the Yaquina, Alsea, Marys, and Siuslaw river basins. Fires were reported in other areas of western Oregon and southwest Washington as well:

> Forest fires along the Washington side of the Columbia River were raging about this time [ca. September 16] at several points below Portland . . . The woods for miles around were a mass of flames. Evidently high winds had spread the fire there, for "the wind created by the fire tore up fruit trees by the roots" and the fire traveled at least five miles per hour. Large pieces of bark were carried fully one mile (Morris 1934b: 330-331).

Morris (1934b: 332) summarizes the regional fire weather for the 1868 fires as: the precipitation at the mouth of the Columbia showed the driest June, July, August, and September in a 58-year record [as of 1934] for the station; Vancouver, Washington had the second driest measures (after 1866) for the same four months in a 72-year period; "practically no rain fell" in September; strong east winds began to be reported at the mouth of the Columbia on September 6, at the Vancouver weather station on September 15, on ships along the southern Oregon coast from September 22 to 25, and along the southern Washington coast from October 3 to 12. In sum, Oregon Coast Range fire weather from September 6 to October 12, 1868 can be described as: five weeks of dry, east winds, following a record three-month drought:

> The year 1868 is known as the year in which the whole coast was afire, the year in which the sun was darkened for weeks and people had to burn candles in the daytime. In this year occurred the great burn extending from Coos Bay southward along the coast to Rogue River Mountains. In this year also the area between the forks of Coquille River was reburned (Gannett 1902: 13).

The following account of the 1868 Coos Fire and its general reforestation history is based on notes taken during interviews with Jerry Phillips that took place between 1988 (Phillips 1988: map) and November 2003 (Phillips 2003: personal communication):

The 1868 Coos Fire is believed to have started with a land clearing fire in Green Acres, about two miles east of Scottsburg, where Highway 38 crosses Wells Creek. From there it spread in a northerly and southeasterly direction on east winds. The northern boundary was the ridge between Smith and Umpqua rivers. This divide may have developed as a result of wind and topography, or may also have been the southern boundary of the 1849-1868 Yaquina Fire events and simply ran out of fuel. The southeastern boundary was a 20-mile "fire-proof" stand of 100-year-old Douglas-fir second-growth that dated to the Millicoma Fire of ca. 1765. Phillips considers this age less likely to burn because at such a young age, the canopy maintains a dark and humid shade covering, and heavy mosses, pitchy snags, and other volatile ladder fuels have not begun to develop to a great degree. A typical stand of this nature existed along Elk Ridge. Another boundary type was a 28-year old stand of Douglas-fir and hemlock, apparently dating to an 1840 fire. A curious characteristic of the 1840 age group is that it contains hemlock, whereas, all of the 1868 fire reforestation was nearly pure Douglas-fir or alder, with no redcedar, Sitka spruce, or hemlock present. An example of the 1840 type, including hemlock component, was in Steeler Creek, Tsp. 23 S., Rng. 10 W.

The western boundary of the Coos Fire was "sand dunes and spruce," where the fire either ran out of fuel, or couldn't penetrate the moist fog belt spruce-hemlock forests along the coast. One such boundary is Wind Ridge, above Scholfield Creek, in sections 18 and 19, Tsp. 22 S., Rng. 11 W., and sections 24 and 25, Tsp. 22 S., Rng. 12 W. The southern boundary is more indeterminate, partly because of logging history (see Map 2.06), and partly because of wildfire history (see Map 2.07). Most of the 1868 burn reforested to nearly pure stands of red alder or Douglas-fir. Phillips (2003: personal communication) believes the principal conifer seed source may have been a belt of Douglas-fir along Soup Mountain, Rainy Peak and Old Blue, to the east of Loon Lake in Tsp. 23 S., Rng. 9 W. and Tsp. 23 S., Rng. 8 W. Phillips' thought is the seed was borne southwesterly from these sources on the same type of late summer east winds that originally carried the Millicoma and Coos fires. Another possibility is seed from the pockets of Douglas-fir and hemlock that survived the fire, such as those along Fish Creek, Glen Creek, or the headwaters of Schofield Creek. However, Phillips notes that many of these pockets of older conifers are surrounded--curiously enough--by nearly pure five to 10-acre red alder stands. Alder is also the dominate tree along the western boundary of the 1868 fire, covering perhaps half the landscape, where it exists as pure stands in Tsp 23 S. Rng. 11 W., Tsp 23 S., Rng. 12 W., and Tsp 24 S., Rng. 12 W. Despite the predominance of hardwood in these stands, Phillips claims they replaced a ca. 1570 stand of Douglas-fir killed in the 1868 fire. The ca. 1570 stands contained a hemlock and redcedar component missing from post-1868 stands. When the Douglas-fir and alder from the 1868 burn do mix with other species, it is usually with each other, and often in about equal volumes (Phillips 2003: personal communications). In instances where the two species mix, the Douglas-fir is typically present in "stringers" and there are no spruce, hemlock, or redcedar.

Phillips (2003: personal communication) speculates the alder stands may signal the abandonment of fields and pastures by farmers and ranchers in the late 19th and early 20th century, or that they may be a consequence of massive boomer ("mountain beaver") colonies in the area. The boomer is particularly destructive to young conifers, including Douglas-fir and Sitka spruce, and Phillips estimates ODF may have spent more than "a million dollars" trapping the animals and protecting planted seedlings against them with vexar tubing. Other vegetation types that persisted following the 1868 Coos Fire included a 70 -acre brackenfern prairie that existed along Burnt Ridge on Trail Butte in Sec. 18 of Tsp 24 S., Rng. 11 W. It was maintained into the 1950s with regular burning for the purpose of attracting deer. Smaller brakes of a fewer acres each were also noted in the Glen Creek area, in Sec. 6 of Tsp. 24 S., Rng. 10 W. ("Big Saddle"). The Glen Creek meadows were burned into the 1930s, to provide pasturage for local landowners. Only one stand of myrtle was noted, a 15 acre ca. 1890 patch on Roberts Ridge that was scheduled to be replaced with planted Douglas-fir trees.

Figure 4.06 (courtesy Jerry Phillips) shows an aftermath of the 1868 Coos Fire. The subsequent landslide, whether fire-related or simply coincidental, created a lake ("Gould Lake") that is now part of the Elliott State Forest (Phillips 2003: personal communication). Note the young Douglas-fir saplings, likely dating to the 1868 event. Eighty years after this picture was taken, the slide had stabilized and this area supported a thriving stand of second-growth Douglas-fir trees. Also note the apparent small diameter and close spacing of the snags: they would have been 100-years old if they were the same age as the Millicoma Fire regeneration; 300-years old if they were the same age as the Pillsbury Tract old-growth (Shea 1967: 4).

Figure 4.06 Coos Burn, Gould's Lake landslide, 1894.

4.1.6 Tillamook Fires, 1933-1951

The August 1933 Tillamook Fire was the best documented and most publicized of the Great Fires. Part of the reason for this result is the enormity of the event, part is because it occurred in nearly full view of a large majority of Oregon and southwest Washington residents, and part is because it was technically possible to do so. Wireless communications, moving pictures with sound, and aerial photography were all new or developing technologies at the time, and the spectacle of a catastrophic fire provided an ideal stage to use these media for news reels, newspapers, and magazines. As early as 1934 Morris was able to write:

> The history of the Tillamook fire was set down and published in great detail. Forest protective agencies recorded its daily spread and behavior; photographers recorded many of its successive visual aspects both from the ground and from airplanes; later, timber owners cruised the fire-killed timber and mapped the area

> burned; and the newspapers made the results of all these activities available to the public (Morris 1934b: 313).

The first Great Tillamook Fire started between 12:20 and 12:45 PM on August 14, 1933, in Section 17, of Tsp. 2 N., Rng. 6 W. (Morris 1935: 3). The reported cause of the fire was "friction of a three-foot Douglas fir log with bark, being dragged across a windfall having no bark, in a logging operation" (ibid: 2). The operation was Gales Creek Logging Company, a Crossett Western subsidiary, salvage logging 20 to 40 inch diameter second-growth, in a 200-acre October 1932 burn. The earlier burn took place along "a broad spur ridge" of Gales Creek Canyon; the new fire began moving southerly, toward "cut-over land logged before 1925" (ibid: 2-2a).

At 1:00 PM, the Hoffman Lookout, 10 miles from the Gales Creek logging fire, reported a "big smoke" in Section 17. By 1:30 PM--about "fifty minutes" after the fire had first started--Fire Warden Kyle arrived to find:

> . . . thirty-five men fighting the fire, and it had spread to 30 acres in a southerly direction through the fresh slashings and was spotting fast. At 2:30 P. M., it was forty to sixty acres in size and at 3:30 the head of the fire had burned and spotted south through the recently felled timber and along the spur ridge to the 300-foot Gales Creek Gulch, one-half mile distant, crossed the gulch into the old cut-over land in section 20 and up the slope to the railroad grade in the middle of section 20, nearly one mile from the starting point. (Morris 1935: 3).

Spot fires were recorded two, four, and five miles south of the main fire during that afternoon of the first day. The lookout on Saddle Mountain, six miles to the south of the fire, reported being "showered with cinders in the evening of the first day" and having "watched the fire spotting ahead" (Morris 1935: 3-4).

The fire weather on August 14 was typical for wildfire conditions. Except for sprinklings from August 3 to 5, there had been no rain since June 9--nearly five weeks. The relative humidity had fallen gradually and steadily the previous week, from August 7 to 13, until a low point of 20 to 30 percent was reached on August 14 at the five reporting stations nearest the fire (Morris 1935: 4-5). There was reportedly little wind as the fire first started: "the smoke seemed to go straight up and then flatten out to the south at a high level"; that night the Saddle Mountain lookout reported a strong wind from the north (ibid: 5).

On the second day of the fire, a strong wind began blowing from the east; Portland recorded a high of 102° and Forest Grove, nearer the fire, recorded 105° (Morris 1935: 5). At 1:00 PM, with the arrival of the east wind, the fire crowned rapidly to the west; spot fires joining into a single fire, 3 1/2 miles long (ibid: 6).

Then, from August 16 to 21, the weather changed and the fire halted. It was foggy on the 16th, winds became light and variable, and minimum relative humidity was 20 % greater than the previous day (Morris 1935: 6). Weather conditions remained similar through the 21st, and the fire remained about the same size. A moist westerly wind of about 10 miles an hour blew for nearly two days, further dampening the fire and keeping it contained. An inspection of the fire's perimeter on the 17th found:

> . . . the fire smoldering quietly with no flames visible [and] no spread of the ground fire beyond the crown fire which had been driven by wind at least 38 hours previously. The timber was mature Douglas fir with some hemlock, and bore much moss which was a very important factor causing the ground fire to flare quickly into the crowns (Morris 1935: 7).

By the 17th, 565 men had been assigned to fight the fire (Morris 1935: 8). On the 19th, the weather was reported as "windy, cloudy, foggy, and cold," and rain was recorded along the coast. On the 20th "a Stimson [Lumber Company] representative reported this part of the fire was dead at 3:30 P. M. and no smoke visible from Saddle Mountain." About 4:00 PM the Gales Creek Logging Company fire warden reported, "the wind had shifted to the east." At 7:15 PM the Mt. Hebo fire lookout reported the fire "smoking us good." That night the fire moved southward from Saddle Mountain (ibid: 8-9). The morning of the 21st saw the beginning of 40 hours of high-velocity, low-humidity winds from the east.

On August 21, northern and southern fires merged into a single fire, burning 15,000 acres (Morris 1935: 9). On the 23rd and 24th, weather became cooler and moister, and the fire stalled. At 4:00 PM on the 24th, someone started a fire of "undetermined cause in a 300-acre block of 3-year old unburned slashing, 8 miles north of the Tillamook Fire"; at 11:45 PM, a 3 to 4 mph east wind was recorded "on the ground in the open timber" (ibid: 14). The new fire, driven on the wind measuring "26 miles per hour from the southeast," swept "beyond control" at 10:00 AM on the morning of the 25th, traveling northwesterly and crossing the Nehalem River at midnight, eleven

Figure 4.07 Tillamook Fire mushroom cloud, August 24, 1933.

miles from its start. Between midnight, August 24 and 9:00 AM August 26, more than 75% of the total fire area of the 1933 Tillamook and Wolf Creek fires had burned (ibid: 17). By noon, August 26 both the Tillamook and Wolf creek fires were reduced to burning within their interiors, and no further advances in fire boundaries were made after that time. At 11:00 PM, August 28 a hard rain fell at Mt. Hebo, followed by several days of clearings and showers. Morse notes of the two fires:

> Like the Tillamook Fire to the south, the weather was primarily responsible for stopping the Wolf Creek fire, and, except in small areas, the efforts of men were inconsequential after the fire gained a running start (Morris 1935: 21).

People who were in northwest Oregon or southwest Washington on August 24 or 25, 1933 remember the Tillamook firestorm of those days in much the same manner that later generations recall the Columbus Day Storm 0f 1962 or the eruption of Mt. St. Helens in 1982. These, too, were natural, unexpected, catastrophic-scale stand replacement events in full view of the Portland and Vancouver urban populations, that captured national and international attention, and that were well documented. Figure 4.07 (Tillamook Burn Collection/ODF; Zybach 1984: 14; Fick and Martin 1992: ix; ODF 1997) shows a 40-mile wide and eight mile high pyrocumulus mushroom cloud erupting over the 1933 Tillamook Fire. The vantage is looking west from an airplane, with the Willamette Valley in the foreground. The picture was taken by Eyerly Photography on August 24, 1933, during a day in which the fire suddenly erupted and burned more than 200,000 acres of mostly roadless, old-growth Douglas-fir forest; called a "blow-up" or a firestorm (Peterson 2000: 7-8). A total of 12 billion feet of timber was killed, of which eight or nine billion feet was killed in that single 20-hour period (Cronemiller 1961: 29). This compares to the 500,000-acre Biscuit Fire of 2002, in which a total of only 3.6 billion feet of timber was killed during the entire event (Sessions et al: 2003: 53). Figure 4.08 (Tillamook Burn Collection/ODF; Zybach 1984: 14; Fick and Martin 1992: ix; ODF 1997) gives an indication of the condition of the trees following the burn. About 82 percent of the timber killed in the 1933 Fire was Douglas-fir; 15 percent hemlock; two percent cedar; and one percent other species. About 82 percent of the fire took place in Tillamook County; 16 percent in Washington County; and two percent in Yamhill County (ODF 1933: 1, 4). In his August 27, 1934 "Memorandum Regarding Condition of Tillamook Burn" (Morris 1935: 23-24), Munger noted:

> Here and there were single green trees, even in the midst of the heavy burn. We concluded that in all the country we saw there was not a square mile that did not have some living timber. Here and there along the river are groups and groves of alder and maple, which were not killed by the fire, with not infrequently single trees or groups of trees both Douglas fir and cedar which were alive (Morris 1935: 24)

The "six-year-jinx" of Tillamook Fires began to be officially noticed in the spring of 1951 (ODF 1951: 3), when an April 23-26 wildfire was extinguished after reaching 7500 acres in size. Catastrophic fires had occurred in Tillamook County in 1933, 1939, and 1945--every six years since the 1933 blow-up (Heinrichs 1983: 442-443). The second fire, in 1939, was 225,000 acres in size, including 28,000 acres of green timber. The fire spread among the dry snags of the 1933 fire and destroyed most of the natural conifer reproduction throughout the old burn (Isaac and

Figure 4.08 Tillamook Burn snags, ca. 1938.

Meagher 1936; 1938; Grefe et al 1943). On July 9, 1945, the third Great Tillamook Fire got started. This fire lasted six weeks, burned at least 110,000 acres of old snags, conifer reproduction, and timber, and caused three deaths (Fick and Martin 1992: 89). This third fire is blamed for destroying the remaining seed and seedlings from the 1933 and 1939 fires, necessitating reforestation of the Burn through state-funded hand planting and aerial seeding projects (ODF 1945; Zybach 1982; Fick and Martin 1992). Finally, in 1951 the six-year jinx came to an end. For the first time since the jinx started, less than 100,000 acres burned and none of it was in green timber. The 1951 fires were unusual, in that they began in April, flared up again in July, and went out of control with "strong east winds on the 20th and 21st of September." Tillamook wildfires had been occurring off and on for five months when, in the course of two days in late September, they burned more than 40,000 acres. As with the 1933 Fire, which was about five times larger, most of the acreage that burned during the 1951 fires did so in a matter of hours, and, "despite the efforts of the men the fire could not be brought under control until late on the evening of the 24th when the first rain dampened the forest" (ODF 1951b: 3).

## 4.2 Catastrophic Fire Years

For purposes of this dissertation, the "catastrophic fire" measure of 100,000 of burned forestland is also used as a standard for catastrophic fire years; i.e., any year in which 100,000 acres or more forest burned in western Oregon, is considered a "fire year" for the region. To be conservative, all ODF records for years in which at least 75,000 acres burned are listed (Ballou 2002). This is due to inconsistencies in acreage measures and because there is uncertainty whether the ODF records include fires on federal lands for all years and, if so, what federal lands are involved (Ballou 2003: personal communication). For precontact fire years, tree ring studies are used to identify "severe" fire years. Particular attention is given to Impara's 1997 study because: 1) it is located in the Coast Range, 2) he generalizes his findings for the entire Coast Range, and 3) his study extends well into precontact time, including the period covered by this dissertation (see Appendix G).

More than 99% of all forest fires worldwide are started by either people or lightning (Agee 1993: 25-26). In the Oregon Coast Range, there is no record of a catastrophic fire or catastrophic fire year caused by lightning (Keeley 2002). People, rather than lightning or other causes, have been responsible for the vast majority of wildfires in this region (Kirkpatrick 1940; Bonnicksen 2000: 392; Stewart 2002: 255), whether measured by acreage or number of events (Ballou 2002). Lightning has been rare in the region throughout recorded history (see Chapter 1.4.3), and fires started by lightning are even more unusual (see Map 1.07; Kirkpatrick 1940; Morris 1934a). When lightning does occur, it is often accompanied by heavy rains and tends to put fires out, rather than start them (Morris 1934b). Thus, wildfire patterns preceding white settlement in the mid-1800s can reasonably be attributed almost entirely to Indian burning practices. Subsequent wildfires--particularly those occurring after 1843 (Carey 1971: 311), when Indian populations were low and restricted to a few wandering bands and minor communities--can likewise be attributed almost entirely to fires set by whites.

### 4.2.1 Precontact evidence: 1491-1775

The principal scientific evidence of Coast Range catastrophic fire years in precontact time are Impara's (1997: 212) "high severe fire years" based on tree ring studies, Long's (1996) Little Lake "charcoal peaks," and Andrews and Cowlin's (1936: map) detailed mapping of fire boundaries, Douglas-fir forest stands, conifer seedlings, and other vegetation types. Note that Impara's two greatest fire years, 1852 and 1871 (see Appendix , are both three years after the historic fire years of 1849 and 1868. The events were either a lot more localized than thought (bringing the general value of the pre-1851 and 1880s fire dates into question), or the methodology tends to add about three years to the findings. This is an important point, because he uses the absence of personal findings in the early 1900s to conclude that federal fire suppression policies all but stopped fires (except for Tillamook, which he didn't find a record for) after 1910.

### 4.2.2 Early historical accounts: 1776-1848

Although there are no specific mentions of catastrophic forest fires in the historical literature before 1849, there are numerous accounts of forest and grassland fires throughout the literature by Douglas (1905), Hines (1973), Wilkes (1845b) and many others. Many such references are made throughout this dissertation, and are also cited by others (e.g., Morris 1935B; Boyd 1986; Whitlock and Knox 2002). A typical example is the October 8, 1826, observation by Douglas in the eastern Coast Range oak savannah:

> Thus we live literally from hand to mouth, the hunters all declaring that they never knew the animals of all kinds to be so scarce and shy, which is attributable to the great extent of country that has been burned (Douglas 1905: 80).

### 4.2.3 The Great Burns: 1849-1910

Background on the Great Fires was given in the first part of this chapter. Historical quotes and photographs were used to document specific events, and also to demonstrate the relative poor quality of information available on the topic for historical time. One consequence of poor record keeping is that it is difficult or impossible to identify specific years in which fire was widespread throughout the region, but in which no single event dominated. A combination of public awareness regarding forest fire destruction, modern technology (photographs, movies, telephones), and creation of large federal forest reserves in the 1890s, specifically intended to be protected from fire, soon changed the quality and focus of both news gathering and record keeping. In 1902, and then again in 1910, forest fires occurred throughout the entire Pacific Northwest, including the Coast Range (Morris 1932b). News media reported on these fires in detail, and state and federal organizations soon formed for the specific purpose of reducing wildfires and/or "battling" to control them (Pyne 1982: 239-321). The fires of 1910, and particularly the fires in the Rocky Mountains of Idaho and Wyoming that year in which millions of acres burned and dozens of firefighters died, focused attention on the US Forest Service and led to a national resolve to intensify wildfire control strategies (ibid; 239-241). In Oregon, county fire control associations were established, funded at first by voluntary contributions of affected landowners. The Oregon Department of Forestry (ODF) began consolidating records kept by the county crews in 1911 (Ballou 2002) and, in 1914, State Forester Elliott authorized and directed the assemblage of a statewide Fire Map (see Map 2.05).

Appendix G includes a summary of the forest fire record assembled for western Oregon by the Oregon Department of Forestry from 1911 to 2002, derived from state and county records. Table 4.02 summarizes Appendix G. Note that approximately seven times as many fires for this time period were started by people, rather than lightning. Because the large majority of western Oregon lightning strikes, lightning storms, and lightning-caused fires in western Oregon from 1925-1996 are recorded in the Klamath-Siskiyou mountains of southwest Oregon and in the western Cascades, the ratio of human-caused to lightning-caused forest fires in the Coast Range is likely much larger than the 7:1 ratio shown for the entire area. Accounts of the 1902 and 1910 fires were routinely carried in local newspapers and document the widespread occurrences of wildfire during those years. Typical is the account from the Triangle Lake/Little Lake subbasin:

> In 1910, fire raged over many miles of the [Lake Creek] valley and mountains [in the Siuslaw River headwaters of Lane County], destroying thousands of acres of timber. Settlers saved themselves by lying in the creek and throwing water on each other. Some who tried to escape had their wagons and belongings burned in the road after the horses had been cut loose (Rust 1984: 25)

Recent occurrences of catastrophic fires in southwest Oregon have been blamed, in part, on dangerous fuel build-ups resulting from federal and state fire control policies adopted following the 1910 fire year (Sessions et al 2003: . Although fire suppression policies may be a factor in such fires, there is little evidence to support this claim for Coast Range forestlands. Table 4.02 shows 16 catastrophic fire years in the three decades immediately following 1910 (1911-1940), or more than 50% of the years during that time period. From 1912 to 1951 the total is 18 fire years, or 45% of the 40 years following adoption of these polices. Then, from 1952-1986, there is a string of 35 years without a single catastrophic fire or catastrophic fire year for all of western Oregon, at which time (1987) the Silver Complex fire took place in southwest Oregon. For the Coast Range, the sudden reduction in catastrophic fire seasons and events is even more striking--

Table 4.02 Historic western Oregon forest fire years, 1849-2003.

| Year | Acres | People | Lightning | Authority |
|---|---|---|---|---|
| 1849 | 500,000+ | Yes | | Morris 1934b: 322 |
| 1853 | 480,000 | Yes | | Ballou 2002: 67 |
| 1868 | 300,000 | Yes | | Kirkpatrick 1940:33 |
| 1902 | 170,000 | Yes | | Ballou 2002: 67 |
| **Sub-Total** | **1,450,000+** | **Yes** | | |
| 1910 | 51,000+ | Yes | | Kirkpatrick 1940: 35 |
| 1914 | 146,000 | 1,107 | 147 | Ballou 2002: 77 |
| 1915 | 109,000 | 1,170 | 85 | Ballou 2002: 77 |
| 1917 | 258,000 | 1,468 | 177 | ODF 1994; Ballou 2002: 77 |
| 1918 | 184,000 | 813 | 139 | Ballou 2002: 77 |
| 1919 | 143,000 | 963 | 181 | Ballou 2002: 77 |
| **Sub-Total** | **891,000+** | **5,521** | **729** | |
| 1922 | 179,000 | 1,312 | 95 | Ballou 2002: 77 |
| 1924 | 252,000 | 1,577 | 311 | Ballou 2002: 77 |
| 1926 | 208.000 | 996 | 150 | Ballou 2002: 77 |
| 1928 | 104,000 | 913 | 154 | Ballou 2002: 77 |
| 1929 | 298,000 | 1,355 | 117 | Ballou 2002: 77 |
| **Sub-Total** | **1,041,000** | **6,153** | **827** | |
| 1931 | 188,000 | 1,536 | 85 | Ballou 2002: 77 |
| 1932 | 333,000 | 1,601 | 87 | Ballou 2002: 77 |
| 1933 | 341,000 | 943 | 239 | Ballou 2002: 77 |
| 1936 | 225,000 | 1,265 | 139 | Ballou 2002: 77 |
| 1938 | 80,000+ | 957 | 188 | Ballou 2002: 77 |
| 1939 | 309,000 | 965 | 294 | Ballou 2002: 77 |
| **Sub-Total** | **1,476,000** | **7,267** | **1,032** | |
| 1945 | 253,000 | 532 | 333 | Ballou 2002: 77 |
| 1951 | 126,000 | 850 | 104 | Ballou 2002: 77 |
| **TOTAL** | **379,000** | **1,382** | **437** | |
| 1987 | 89,000 | 976 | 605 | Ballou 2002: 77 |
| 1996 | 189,000 | 4 | 9 | Coyle 2002 |
| 2000 | 110,000 | 8 | 4 | Coyle 2002 |
| 2002 | 500,000 | | | Ballou 2002: 67 |

### 4.2.4 Fire suppression policies: 1911-1951

not a single event in more than 50 years since 1951, continuing to the present time (2003). The record, then, is a dramatic and sustained increase in numbers and events of catastrophic fire seasons for the 40 years following fire suppression, and then an apparent elimination of such events from that time to the present. "Fire suppression policies," therefore, doesn't begin to explain this history. Something that occurred shortly after World War II seems to more likely answer this abrupt change in the environment. Probably a combination of factors--e.g., public education, improved firefighting technologies, dramatic increases in road-building and clearcutting during the late 1940s and into the 1970s (see Zybach 2002a), increased reliance on fossil fuels and electricity for heating homes, and other conditions--seems to be a more likely cause of such change.

The question now is: Does the abrupt change in Coast Range catastrophic fire history signal a generally permanent departure from past wildfire events, or is it the "lull before the storm" in which fuels rebuild to hazardous levels and the area once again becomes the stage for uncontrollable wildfires of historic proportions?

## 4.3 Catastrophic Fire Patterns

The following maps show patterns of burning and logging that occurred throughout the Coast Range during historical time. GIS is used to construct subregional maps to directly compare catastrophic fire patterns with Indian burning patterns of trails and openings, as described and mapped in Chapter 3. The comparison of patterns, made possible by GIS methodology, constitutes much of the focus of this research.

### 4.3.1 North: Columbia River old-growth

The northern Oregon Coast Range has long been recognized as an area of giant trees (see Figure 4.01) and logs (Figure 4.02). Clark, for example, noted on November 9, 1805:

> At 2 oClock P M the flood tide came in accompanied with emence waves and heavy winds, floated the trees and Drift which was on the point on which we Camped [in present Washington State, across the Columbia River from modern-day Astoria] and tossed them about in such a manner as to endanger the canoes verry much, with every exertion and the Strictest attention by every individual of the party was scercely sufficient to Save our Canoes from being crushed by those monstrous trees maney of them nearly 200 feet long and from 4 to 7 feet through (Thwaites 1959a: 213).

Kimmey and Furniss (1943: 8) located almost all of their western Oregon old-growth study plots in the northern Coast Range, even though their definition of old-growth was very conservative by today's standards: "mostly over the age of 400 years, and with a usual d.b.h. ["diameter at breast height," a measure usually made in inches at 4 1/2 feet above the level surface of the ground] of about 40 to 100 inches or more (ibid: 4). Although the age of Clark's and Kimmey and Furniss' trees are unknown, the diameter measures are nearly identical. The presence of such large trees, of course, indicates the absence of stand replacement events, such as catastrophic fires, windstorms, or logging, for long periods of time. The proximity of such trees to the Columbia River increased their commercial logging value as they could be readily transported long distances by water to processing centers in California, or to international ports throughout the Pacific Ocean (Benson 1964).

Despite the presence of vast tracts of giant trees, however, wildfire, wind, insects, and diseases were major disturbance factors in northern Coast Range forests. Map 4.02, for example, is part of an extensive timber cruise of Clatsop County lands in Tsp. 4 N. (see Map 2.01). This cruise map of a 640-acre section by Wherry (Cabe et al 1913: map) is representative of the types of details mapped and tabulated during the cruise. Notes, maps, and tables are of varying quality, given the number of people involved in their creation and the general purposes for which they were assembled.

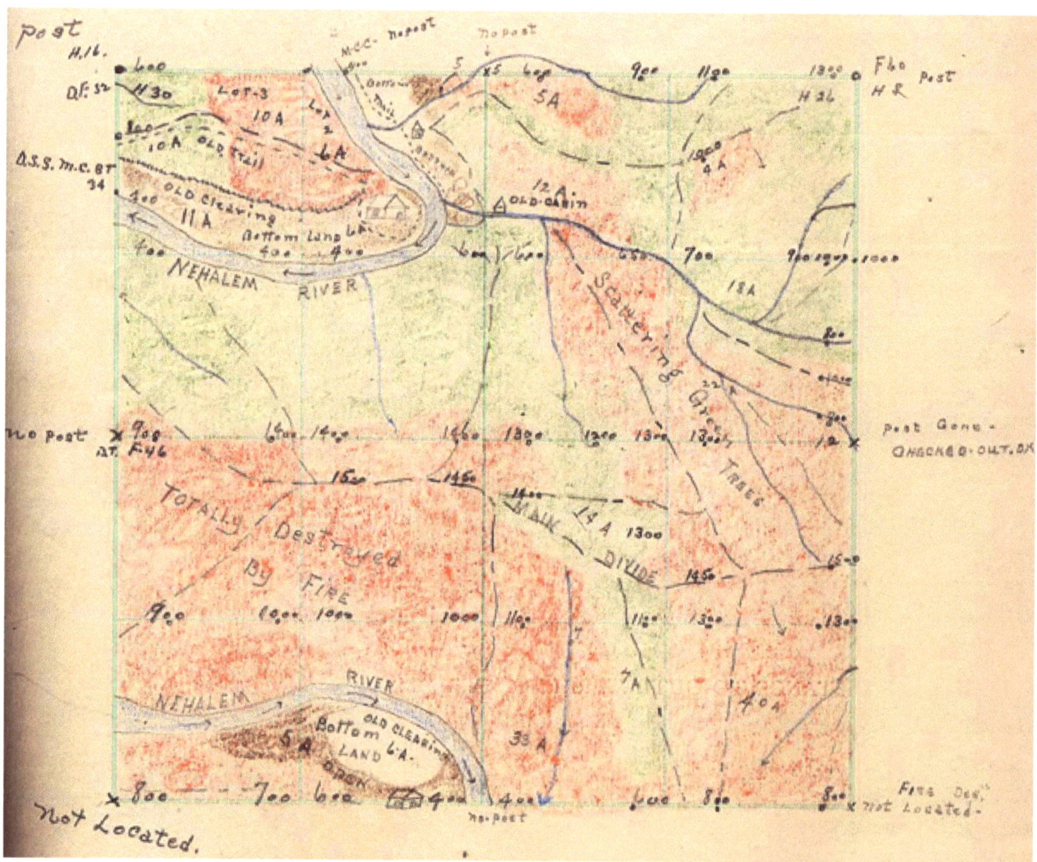


Map 4.02 Tsp. 4 N., Rng. 8 W., Sec. 35 timber cruise map, 1913.

Some of the information contained in these records include: a major windstorm (probably from the west) knocked down a significant amount of timber within 10 miles of the ocean in the 1880s; numerous fires occurred almost annually throughout the area, except within 10 miles of the ocean; a large-scale fire occurred in 1894-1895; major insect outbreaks in 1892-1893 preceded that fire and other fires in several areas; "worms" killed hemlock in much of Tsp. 4 N., generally avoiding fir; spruce and fir were also killed by "bugs" in several areas; the northern border of an extremely destructive fire from Tillamook County to the south is shown, possibly occurring during the 1902 fire year (Fick 2003: personal communication); clearing fires near the town of Elsie (Tsp. 4 N., Rng. 7 W., Sec. 6) seem to account for much of the escapement in that area. Clearings and settlements shown on the maps are likely locations of prehistoric artifacts, including Clatsop (Chinook) and Nehalem (Salish) artifacts along the coast, and Klaskani (Athapaskan) artifacts inland, particularly along the Nehalem ridges, peaks, and bottomlands. Artifacts include berry patches, grassy meadows, brakes, and balds. Onion Peak appears to be a virtual "no man's land," due to broken topography and limited access routes, although its name apparently comes from wild onions near its peak (McArthur 1982: 557).

Map 4.03 compares a hand-drawn index to the 1913 Tsp. 4 N. timber cruise (Cabe et al 1913: map) with 1850, 1890, 1920, and 1940 excerpts from the BLM "Teensma" map series (see Appendix E), and the 1900 (Gannett 1902), 1914 (Rowland and Elliott 1914: map), and 1936 (Andrews and Cowlin 1936: map) GIS patterns described in Chapter 2. The Teensma maps correlate poorly, if at all, to the historical maps and are not further considered in this analysis. The scattering of fires shown on the eastern portion of Tsp. 4 N. corresponds to similar patterns in the 1913 cruise, as do the large fire patterns on the western portion of the 1914 map. The 1936 patterns show both similar patterns of fire, and subsequent logging patterns (black polygons) that conform to the unburned portions of the 1913 cruise. Map 4.04 was constructed by comparing (see Map 4.03) and layering GIS forest and fire history data derived from GLO surveys (Christy and Alverson 2003), Gannett (1902), Rowland and Elliott (1914: map), Andrews and Cowlin (1936: map), and the patterns developed for Map 3.02. The 1720-1780 pattern was derived from a 1936 Douglas-fir size classification that also dominated the Millicoma Fire pattern at that time. There is no further evidence of forest fires from that time period. Two significant patterns emerge: 1) a large amount of forestland was left unburned during the past 250 years, and 2) there is a strong and obvious correlation between Indian burning patterns and subsequent wildfire locations (including ignition points) and wildfire boundaries.

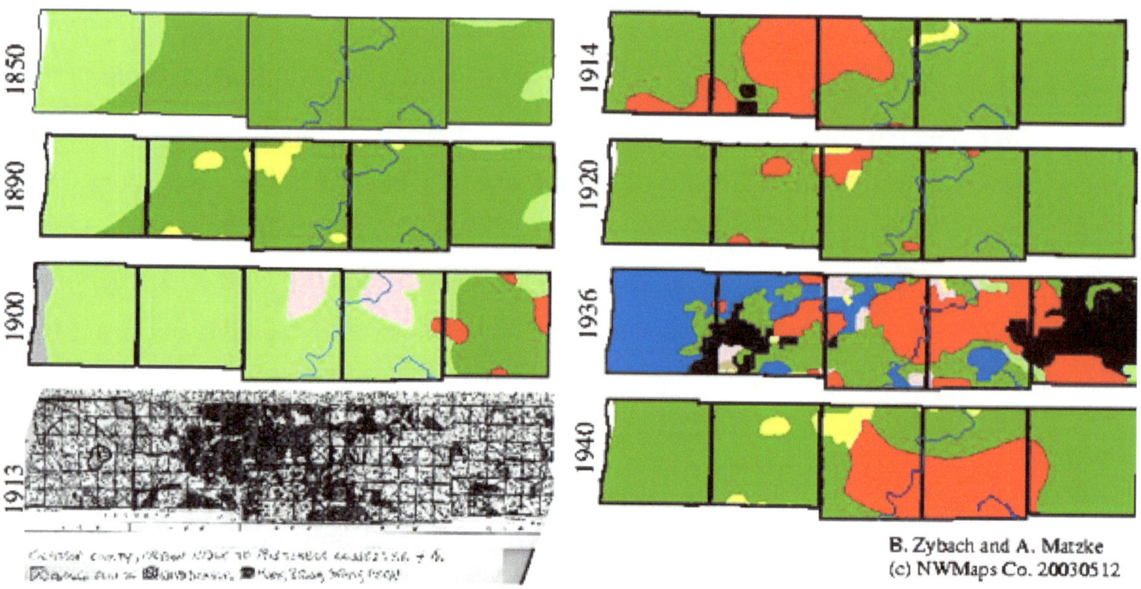


Map 4.03 Clatsop County Tsp. 4 N. fire patterns, 1868-1936.

Several likely reasons exist for the relative absence of forest fires in this subregion: three of the four boundaries are large expanses of water--the Pacific Ocean, the Columbia River, and the mainstem Willamette River from the Falls to its mouth, making fire entries from the east and north difficult and unlikely; the climate is generally foggier with more rainfall than other parts of the Coast Range (see Map 1.04); the greater precontact population was comprised of Chinookan people, who tended to spend much of their time close to riverine environments, rather than managing upland areas; a large portion of native vegetation is spruce and hemlock forest, which tends to burn less frequently than Douglas-fir forest (Agee 1993); the topography is heavily dissected compared to the other subregions of the Range, thereby making fire more difficult to carry (see Map 1.02); and much of the area was logged (see Figure 4.02) before it could burn.

There are several striking correlations between precontact Indian burning patterns and subsequent catastrophic fire patterns: 1) the majority of tideland, wetland and grassland prairies follow canoe and foot trails adjacent to the Columbia, Youngs, Clatskanie, and Nehalem rivers; 2) ignition points of most wildfires correlate closely with well-established riparian and ridgeline travel corridors and destination points; 3) the boundaries of most wildfires correlate to precontact prairies, foot trails, and canoe routes; and 4) fog belt spruce and hemlock forests show relatively minor evidence of either precontact Indian burning or subsequent forest fires.

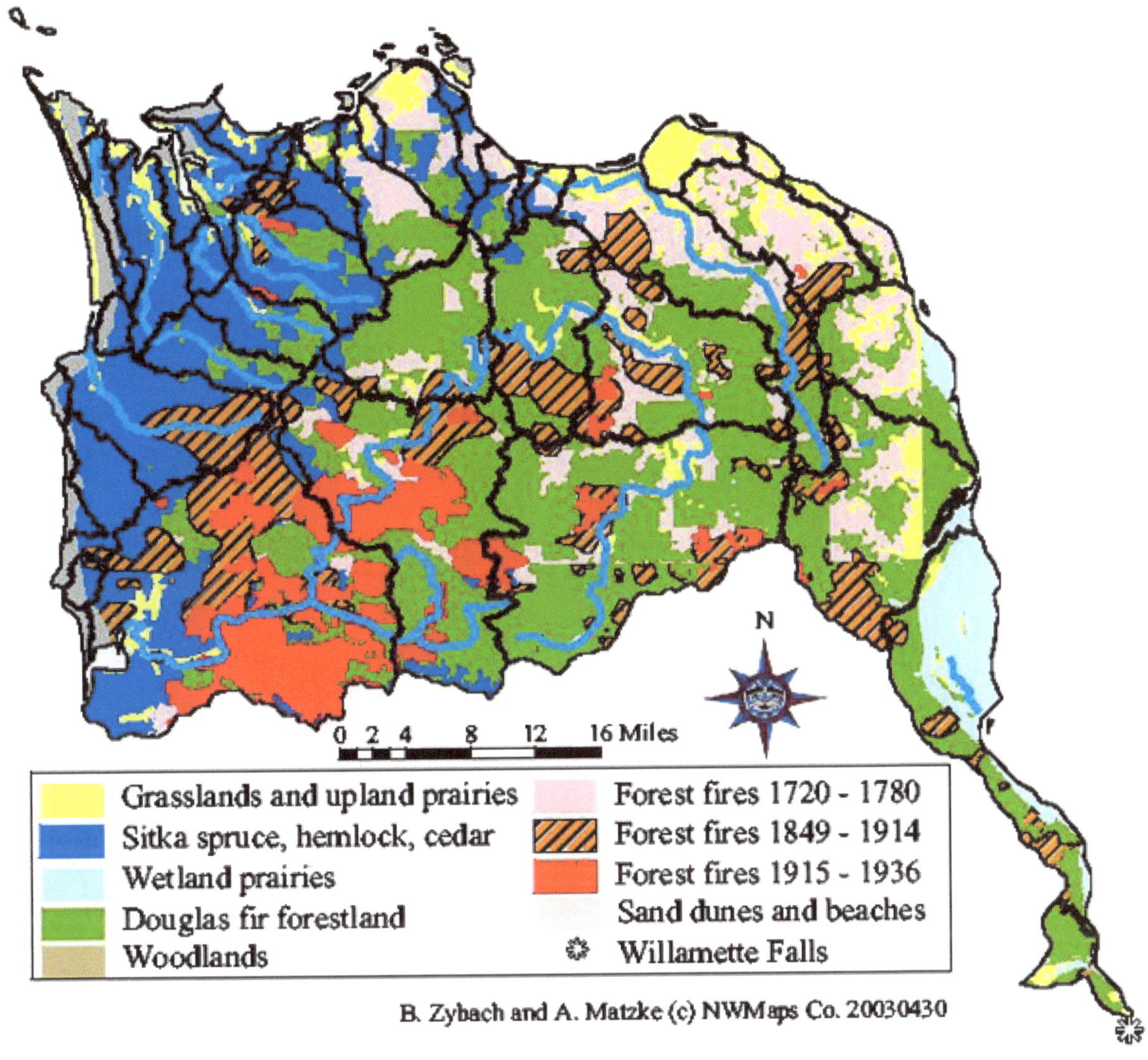


Map 4.04 Northern Oregon Coast Range forest fire patterns, 1720-1936.

An additional pattern of land use activities represented by these maps is also of interest. The 1936 pattern of logging (black polygons) correlates closely to unburned areas represented on the 1900, 1913, and 1914 patterns. Map 4.05 and Map 4.06 provide additional evidence of this relationship. These maps show logging and reforestation patterns for most of the Oregon Coast Range from ca. 1900-1935 (Lindgren 1935). The pattern of Tsp. 4 N. logging on Map 4.05 can be compared with fire and logging patterns shown on Map 4.03 to provide additional corroboration. Also note that

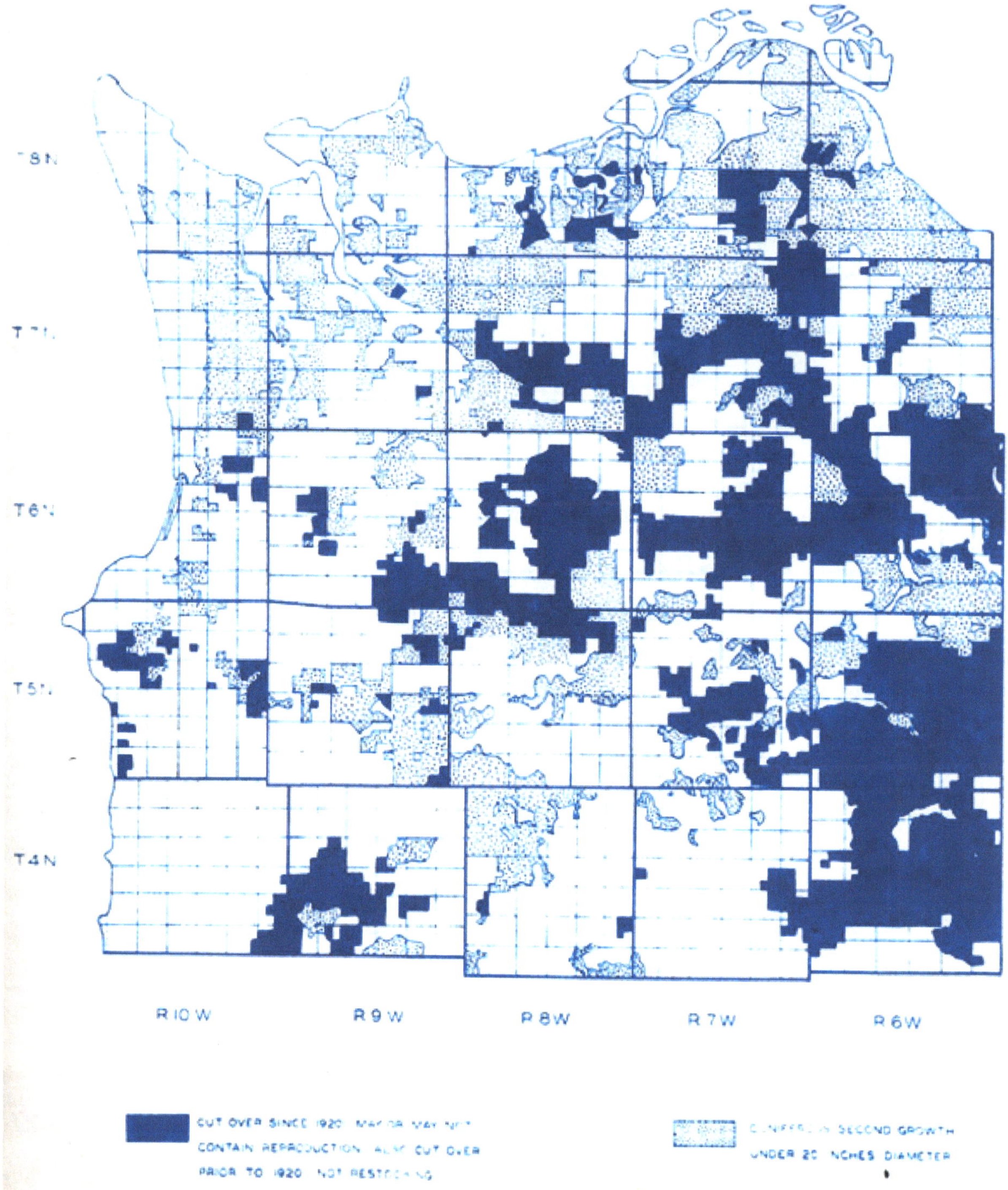


Map 4.05 Clatsop County logging patterns, ca. 1900-1935

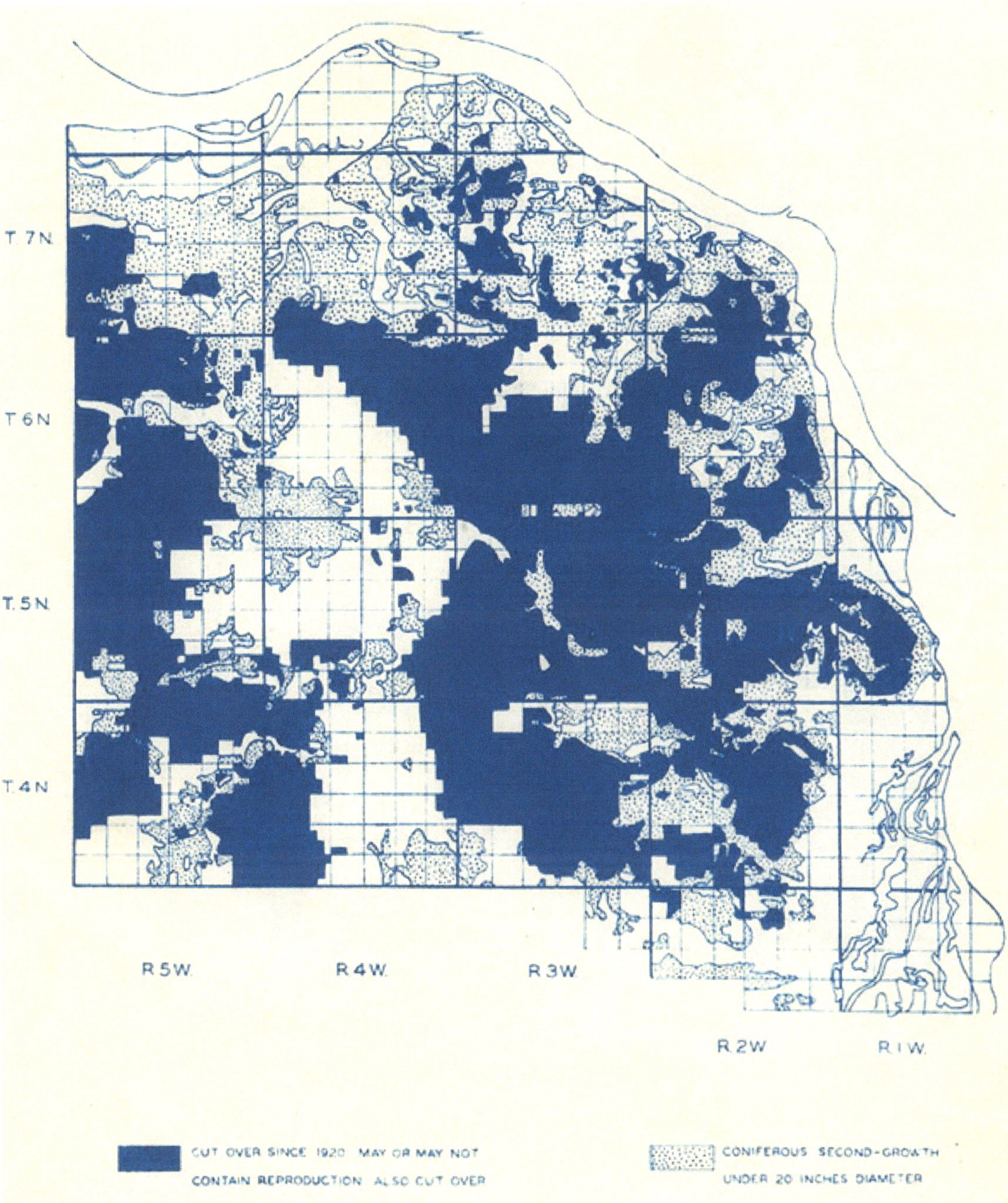


Map 4.06 Columbia County logging patterns, ca. 1900-1935.

close conformance between historical burn patterns of 1913-1914 with small diameter (less than 20 inches d.b.h.) conifer stands of 1935, adding additional credence to the assumed 1720-1780 fire pattern assumptions.

According to their source (Lindgren 1935: 8), these two maps represent 40,060 acres of pre-1921 logging and deforested burns for Clatsop County and 56,865 acres of the same classification for Columbia County (96,925 total acres), and 105,295 acres of 1921-1935 logging for Clatsop County and 117, 180 acres of the same classification for Columbia County (222,475 total acres). Total logging acreage for the northern Coast Range, therefore, is comparable to catastrophic fire event acreages in other areas of the region.

In sum, the heavily dissected landscape of the northern Coast Range, combined with heavy rainfall patterns, a large fog belt forest area, major water barriers on three sides, and lack of seasonal burning along the eastern boundary, limit the opportunity for catastrophic wildfires, and increase opportunities for trees to reach greater age and size.

### 4.3.2 East: cultural legacy

Cultural legacy is defined as the evidence of trails, savannah, prairies, fields, berry patches, brakes, balds and other environmental indications of human land uses that persist through time. This process is illustrated, among other places throughout this thesis, by the map and photographs in Appendix H. For the eastern Coast Range, the role of precontact burning practices in relation to subsequent wildfire history is obvious: precontact Kalapuyan people maintained an environment that was largely free of contiguous forestland, thereby making forest fires, by definition, impossible. Subsequent white settlers, for the most part, maintained the forest-free environment through practices of grazing, plowing, mowing, logging, planting, and urbanization. Habeck (1961: 76) observed: "By the early 1900s it had been noted that no pure pre-settlement tracts of native Willamette Valley vegetation remained." This condition, however, did not affect the basic patterns of vegetation, only the types of species present in grassland, wetland, woodland, and forestland environments. The result has been an extensively managed subregion in which catastrophic forest fires have not taken place simply because there has not been sufficient forested area to house such an event.

Soap Creek Valley (see Map 2.01) is representative of much of the eastern Coast Range in which forest expansion has occurred during historical time (see Figure 3.05). Scattered old-growth trees dating to 1647 (Starker 1939: 48) existed throughout the grasslands and, contrary to other parts of the Range, were not consumed during regular broadcast burning events. In fact, the opposite effect occurred, similar to the oak and limby Douglas-fir old-growth shown in Figure 4.01--the frequent fires eliminated most hazardous fuels from the environment, and kept competing vegetation from becoming established, thereby likely protecting these trees' existences through time and allowing them to become much older than they might in a more competitive forest environment. In other words, without sufficient ladder fuels, coarse woody debris build-up, and the development of a contiguous forest canopy, the likelihood of a stand replacement wildfire event is almost nonexistent. Map 4.07 illustrates the process of Soap Creek Valley afforestation shown in Figure 3.05 (Zybach 1999: 176-210).

As shown on this map, small pockets of older trees--barely considered old-growth at the time of initial white settlement in 1846--slowly expanded outward from their scattered locations until the early 1800s. Then a series of events occurred over time to cause a large portion of the area to become transformed from oak savannah to conifer forest during the subsequent century: disease decimated local Kalapuyan populations to just a few scattered bands by the 1830s (ibid: 190-192), probably reducing the amount of annual burning, tillage, and harvesting in the area, and allowing more tree seedlings to become established; white settlers began subdividing and fencing the land in the 1840s (ibid: 192-198), reducing disturbance in more isolated areas of the valley and further encouraging forest expansion; killing snows and the introduction of automobiles from the 1860s through the 1930s (ibid: 197-200) sporadically reduced the use of grasslands by grazing animals, thereby allowing for additional forest expansion; the removal of all farmers and grazing animals over much of the area during W.W. II (ibid: 200-202) completed the process of eliminating grazing animals that trampled and crushed new seedlings, and farmers that seasonally plowed and mowed the remaining fields and meadows. Relatively small wildfires and logging operations helped to curtail the process of prairie afforestation until the 1930s, when the nearby establishment of the Oregon Forest Nursery (ibid: 107-117) coupled with an increase in value of Douglas-fir timber resulted in the planting of forest seedlings throughout much of the remaining unforested and deforested area. Transfer of most of the lands to Oregon State College (OSU) in the early 1950s (ibid: 126-135) for purposes of agricultural and forestry research and education completed the process: what had been mostly oak savannah in 1850 had become mostly Douglas-fir forest by 1950.

Although the afforestation of Soap Creek Valley was typical of much of the eastern Coast Range, it only represented a fraction of the total area. Much of the remainder continues to remain in a

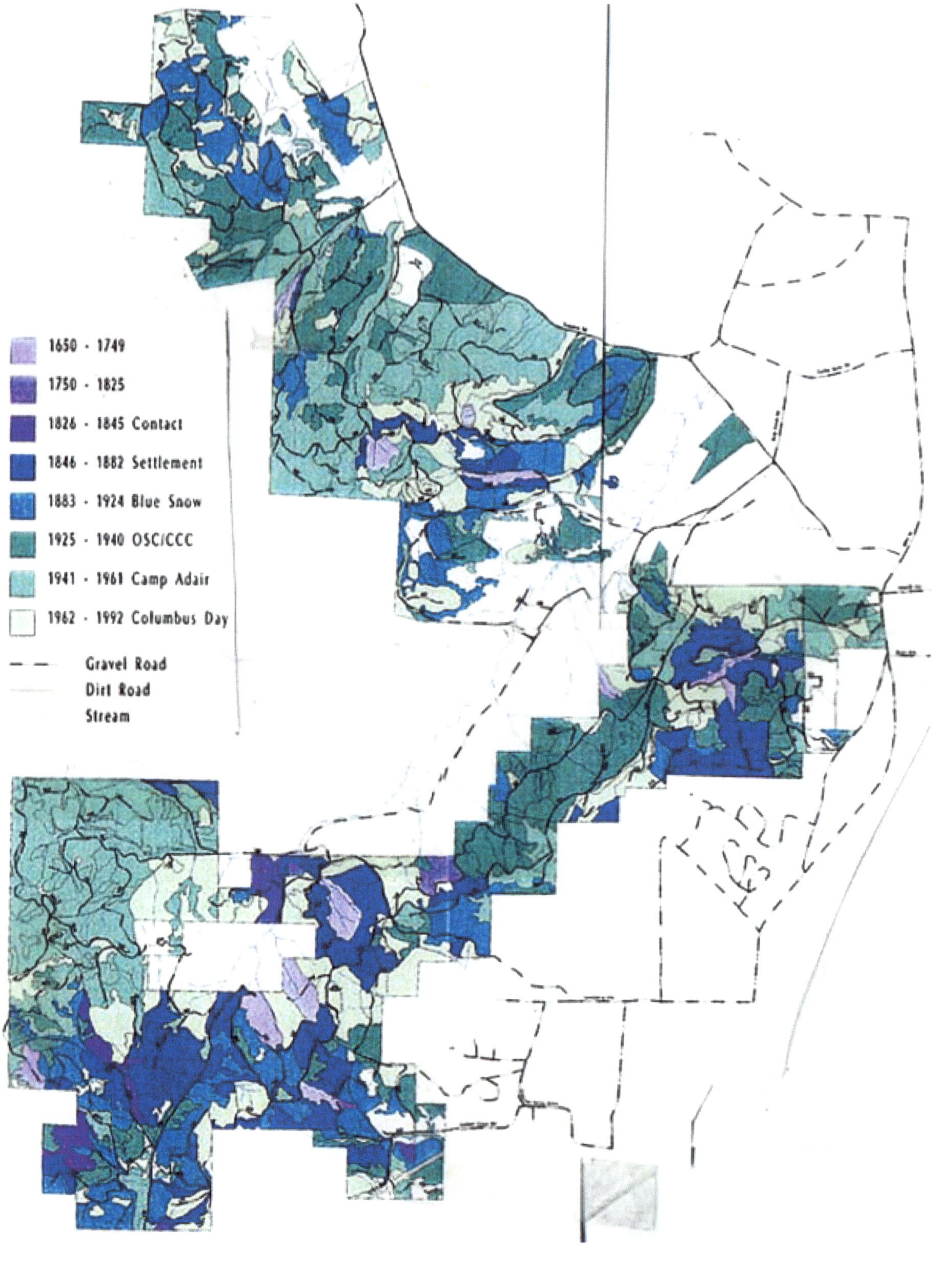


Map 4.07 Soap Creek Valley cultural forest age classes, 1650-1992.

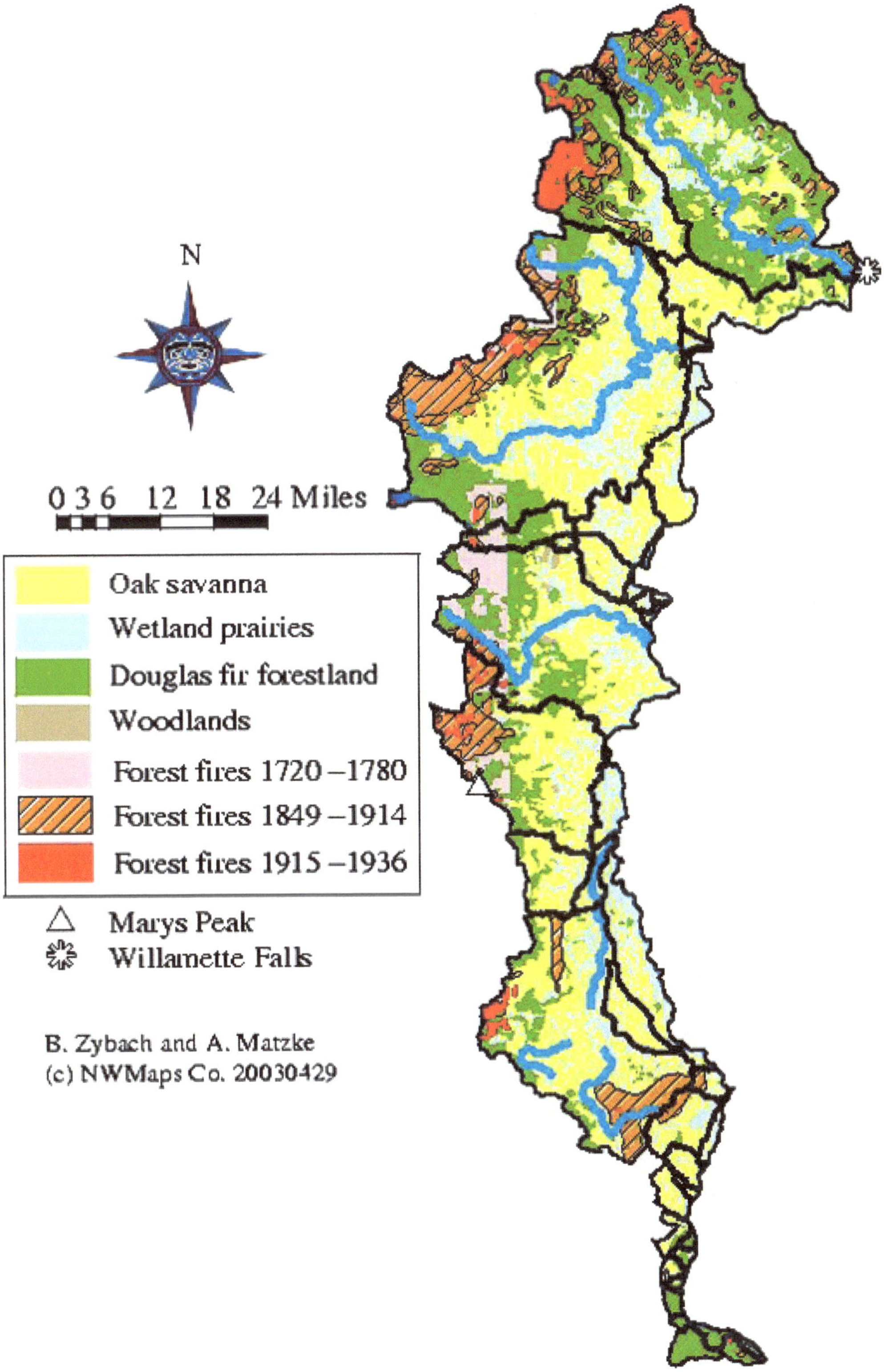


Map 4.08 Eastern Oregon Coast Range forest fire patterns, 1720-1936.

grassland condition to this time, in the forms of lawns, athletic fields, and large, industrial grass seed farms. Most of the land is owned privately, and areas that have been purposefully converted to trees are regularly harvested as pruned saplings for Christmas trees, or as second-growth logs for chipping or milling. Map 4.08, constructed in the same manner as Map 4.04, shows a primary difference between the two areas: neither subregion has been the location (except marginally) of a catastrophic fire during historical time; North, possibly because it was logged before such an event could take place; East, because it has never developed a forest of sufficient size for such an event to occur. In sum, the "cultural legacy" of Kalapuyan land management patterns is still readily apparent throughout the eastern Coast Range, although a significant shift in the variety of plant species has taken place. The lack of catastrophic fire history throughout the subregion seems entirely due to the presence and maintenance of a largely forest-free environment from precontact time to the present.

### 4.3.3 West: 6-year jinx

The Douglas-fir forests of the western Coast Range have nearly all burned at least once at one time or another during catastrophic fire events in historical time. Whereas precontact northern Coast Range forests have been mostly logged, and eastern Coast Range lands have been mostly grazed, mowed, or otherwise cultivated, the precontact forests of this subregion have mostly burned. The Yaquina, Nestucca, and Tillamook fires (see Map 4.01) all took place in the western Coast Range, extending into the southern reaches of the Nehalem River to the north, likely being ignited along the shared ridgeline brakes and forests boundary of the Willamette Valley to the east, and extending southerly to its shared wildfire history of the Siuslaw and Umpqua valleys to the south. The western boundary of these fires is generally the fog belt spruce-hemlock forests that skirt the Pacific Ocean, but sometimes is the actual ocean beaches themselves (see Appendix C).

Map 4.09 shows the extensive nature of western Coast Range catastrophic forest fires. It was constructed in the same manner as Map 4.04 and layered with the precontact burning patterns shown on Map 3.06. As with the northern Coast Range, there are several striking correlations between precontact Indian burning patterns and subsequent catastrophic fire patterns: 1) the majority of tideland, wetland and grassland prairies follow north-south canoe and foot trails adjacent to the ocean beaches and west-east canoe and foot trails that entered the interior of the

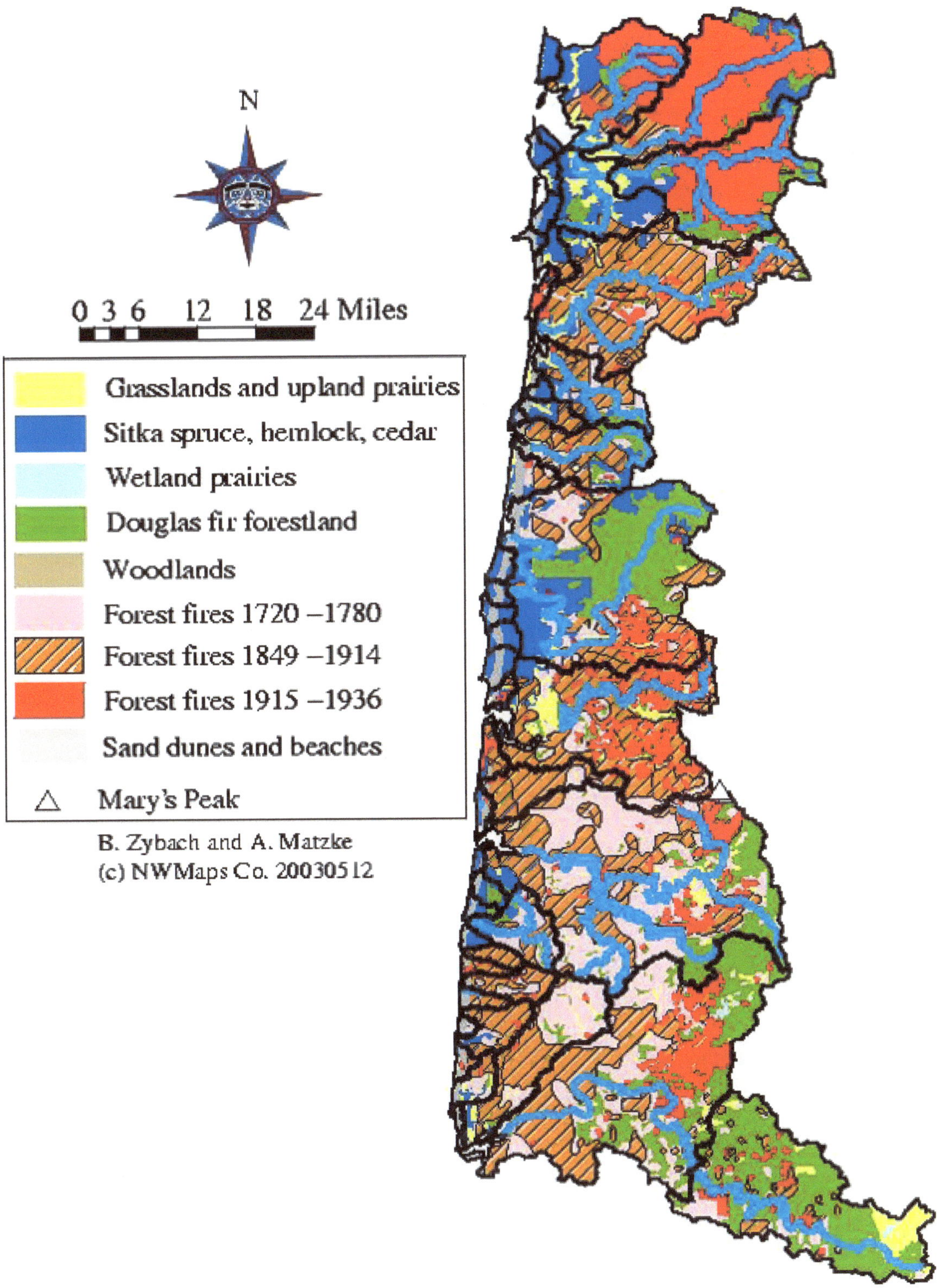


Map 4.09 Western Oregon Coast Range forest fire patterns, 1720-1936.

subregion via the major river drainages; 2) ignition points of most wildfires correlate closely with well-established riparian and ridgeline travel corridors and destination points; and 3) the boundaries of most wildfires correlate to precontact prairies, ridgeline foot trails, and canoe routes.

Because the western Coast Range contains the greater share of catastrophic fire history for the region, the following series of maps are used to better consider the potential relationships between precontact burning patterns and subsequent wildfires. Map 4.10 is excerpted from a 1929 timber cruise of Lincoln County (Pearson, Grady & Co. 1929: map) and annotated to illustrate the southern extent of the Nestucca Fire from the north, the northeast corner of the Yaquina Fire to the south, and the 1910 South Fork fire to the east. The 1966 Stott Mountain Fire is also shown. The fire and logging history of the Siletz basin is similar to that of the northern Coast Range. A large block of old-growth Sitka spruce to the west was logged along the Siletz River mainstem between 1920 and 1940; the "Siletz Gorge" to the east contained about 100,000 acres of old-growth Douglas-fir that was primarily logged during a 20-year period following W.W. II. The pattern of precontact burning is reflected in local landscape names such as Camas Prairie, Peavine Ridge, and Siletz Prairie (see Appendix B). These patterns were used as a basis for assembling Map 4.09, which clearly shows the large blocks of spruce and Douglas-fir that were left unburned, and later logged. As can be seen, the Siletz basin was one of the few areas of the western Coast Range that contained a significant amount of forestland to escape wildfire during the past 250 years. Map 4.11 is based on the "Alseya Valley" boundaries shown on Map 2.13. The same boundaries are used to compare the 1900 (Gannett 1902), 1914 (Rowland and Elliott 1914: map), and 1936 (Andrews and Cowlin 1936: map) GIS patterns described in Chapter 2. In this instance, the 1900 pattern is shown to be too general to further consider, but the 1914 and 1936 patterns are complementary to earlier GLO surveys (see Appendix D) and sufficiently accurate to layer into a regional scale map (Map 4.09), as previously determined by the earlier analysis of Map 2.15.

Map 4.12 shows the "Alseya Valley" precontact ridgeline foot-trail and riparian canoe trail network (see Figure 2.05) within context of the greater Alsea River basin. The network was mapped using 1853-1897 GLO survey notes as described in Appendix D and extended over the entire basin by use of a CLAMS 5th-field watershed GIS layer (see Table 2.03). As can be seen, there is a nearly seamless connection between the surveyed trails and the proxy measures along

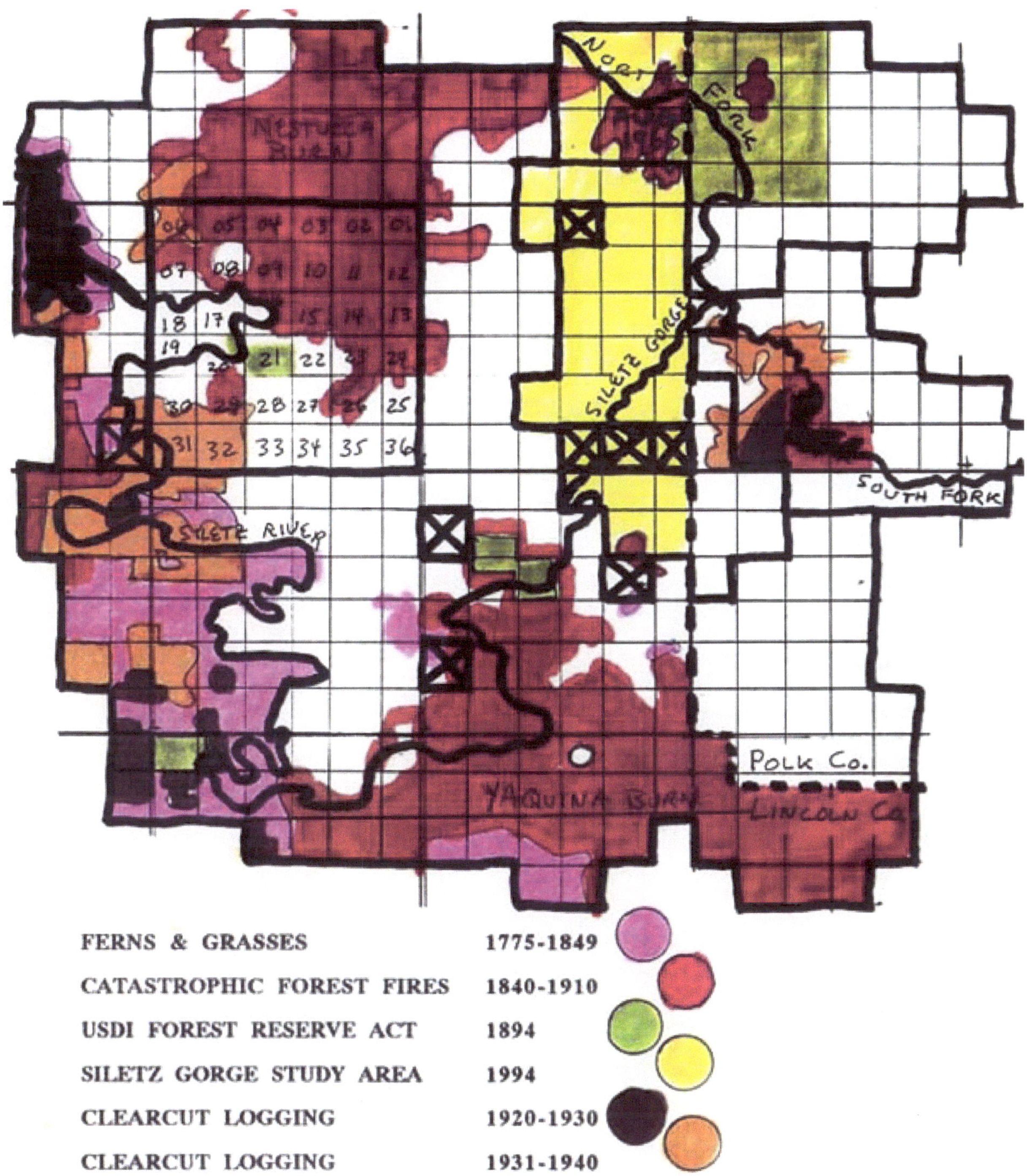


Map 4.10 Siletz River basin forest fire patterns, 1849-1966.

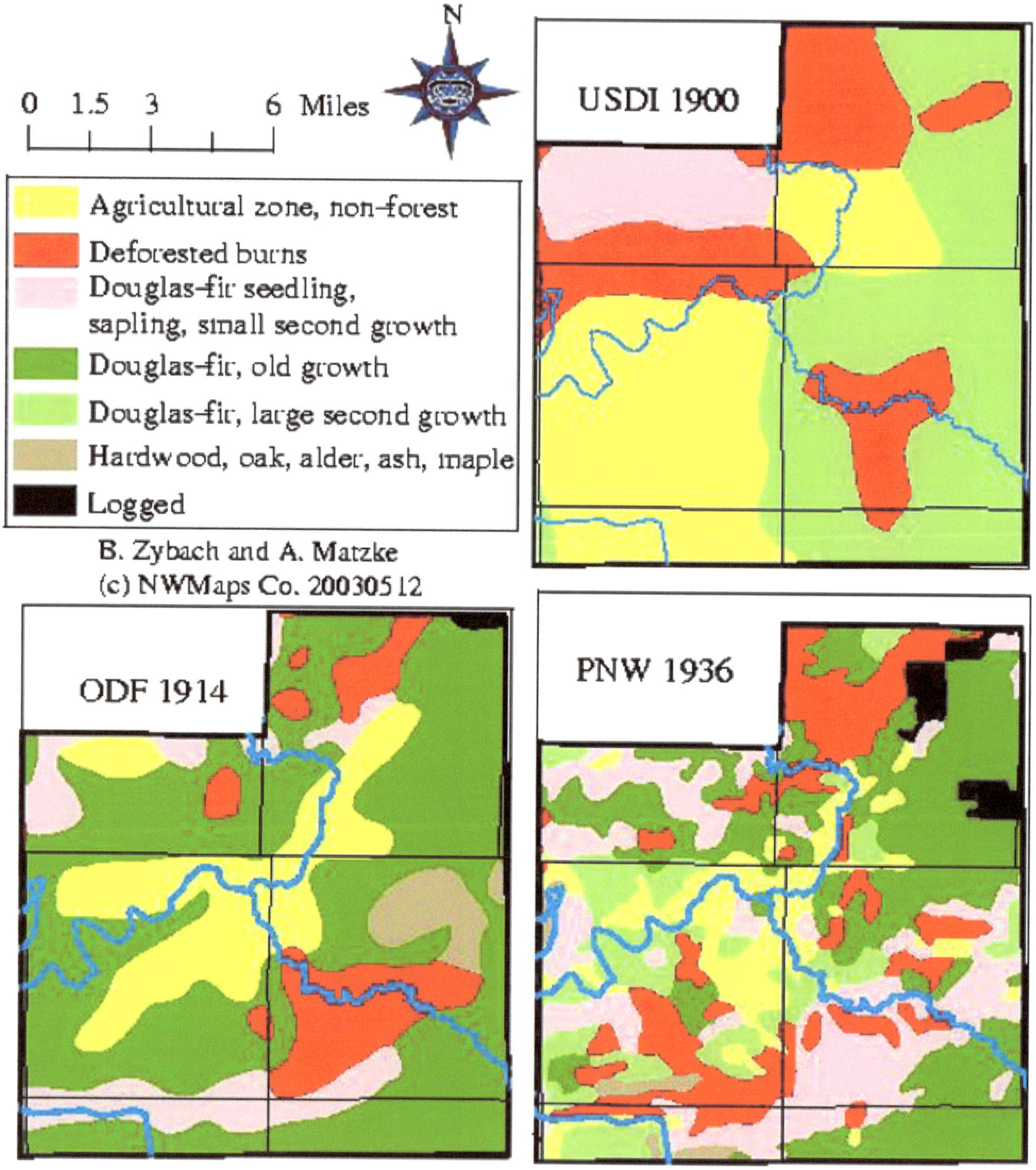


Map 4.11 Alsea Valley forest fire patterns, 1900-1936.

nearly the entire 50-mile perimeter of the Alseya Valley study area. The reasons for this conformity can be readily determined by looking at Figure 2.05 and adds significant confidence to the use of proxies to develop the entire pattern. The Klickitat horse trail of 1849-1853

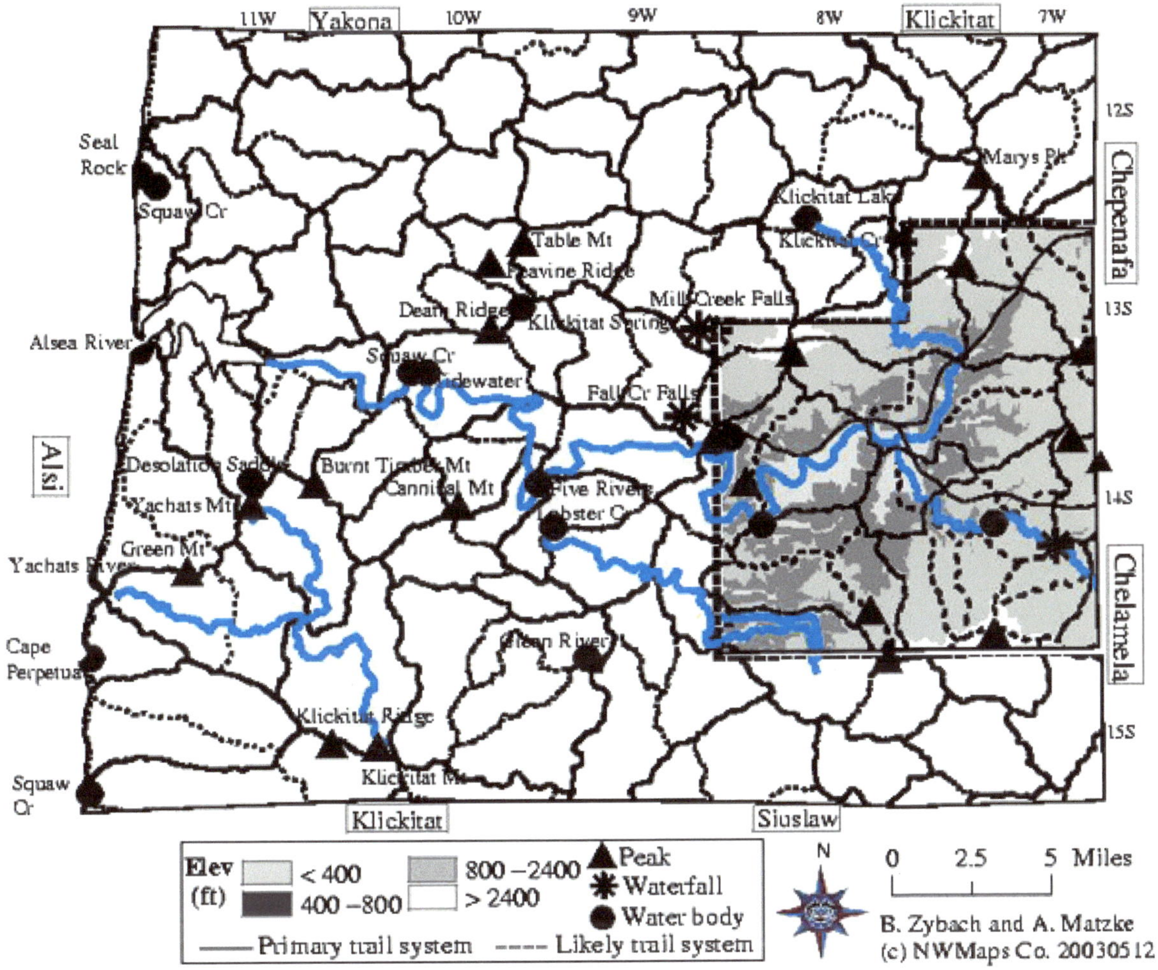


Map 4.12 Alsi tribe foot, horse, and canoe trails, 1826-1853.

That extends from Klickitat Lake near Marys Peak to Klickitat Mountain near Yachats River is also connected by this method, adding further confidence to the result.

Map 4.13 was constructed by layering the precontact trail patterns of Map 2.13 over historical wildfire and precontact vegetation patterns excerpted from Map 4.09. The result, derived from entirely different sources of information, is striking: 1) the detailed pattern of relatively small 1915-1936 fires clearly follow established routes of precontact travel; 2) fire boundaries are determined by a combination of precontact prairies and trails, sometimes to an exacting degree; 3) wildfire boundaries, over time, connect to one another major travel routes. This latter observation can be seen by the nearly due east and west series of fires across the northern portion

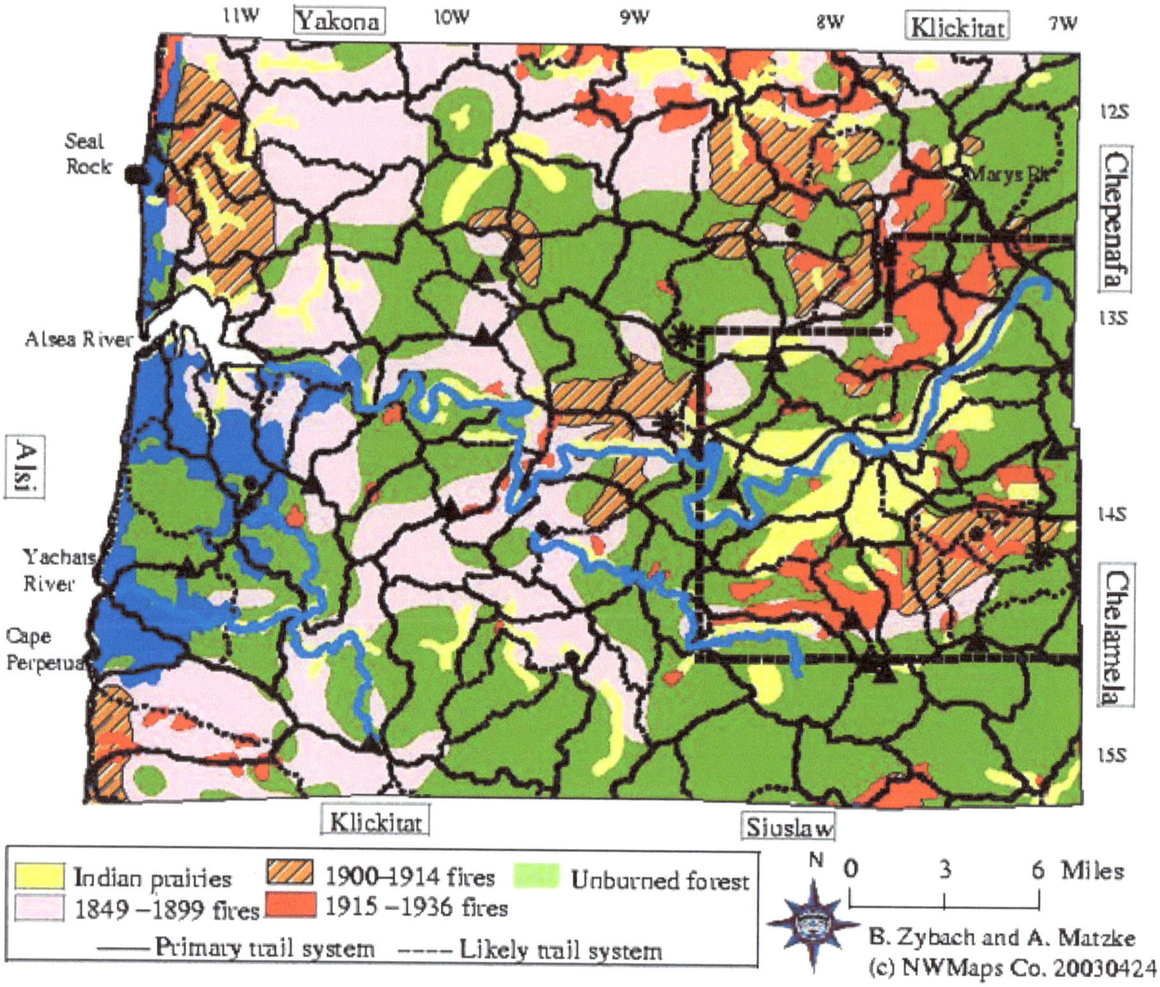


Map 4.13 Alsea River basin forest fire patterns, 1849-1936.

of the map from Seal Rocks (a well-known precontact town site or campground) to the Willamette Valley and by the diagonal series of fires from southwest to northeast that follows the Klickitat horse trail. These fire patterns are not random and clearly conform to precontact and early historical Indian burning patterns.

### 4.3.4 South: mixed

The southern Coast Range has a mixed set of landscape patterns, including patterns similar to: the heavily dissected and remote Douglas-fir forest and foot-trail pattern of the Athapaskan Klaskani to the North, along theheadwaters of the Coquille, Coos, and Umpqua rivers; the oak savannah grasslands of the Kalapuyans in the Willamette Valley, along Calapooia Creek and Umpqua Valley; the riverine "flood and firewood" pattern of tideland Clatsop and Multnomah along Coos and Coquille tidewaters; and the "lawns, corridors, and mosaic" patterns of the Yakonans, among the inland valleys and ridgeline brakes of Brewster Valley and the Millicoma and Williams rivers.

Map 4.14 shows the prairie, catastrophic forest fire, and reforestation pattern of the Coos Bay Quadrangle (see Map 2.01 and 2.06). The upper left pattern (see Map 2.07) shows the ca. 1765 Millicoma Fire (pink) and the 1868 Coos Fire (red) as they were surveyed in 1894-1895. The remaining patterns are derived from the three regional base maps (see maps 2.07, 2.08, and 2.09) and track the reforestation and logging patterns that subsequently developed from the older burns. Munger's observation for western Oregon applies equally well to the area surrounding the Coos Bay Quadrangle:

> Instead of finding and uninterrupted forest carrying 100,000 feet or more per acre reaching from the Cascades to the Pacific, the first settlers seventy-five years ago [ca. 1840] found in the valleys great areas of "prairie" land covered with grass, brakes, or brush which were burned and kept treeless by the Indians, and mountain sides upon which forest fires had destroyed the mature forest and which were then covered by a "second-growth" of Douglas fir saplings or poles (Munger 1916: 92).

Map 4.15 was constructed in the same manner as maps 4.04, 4.08, and 4.09, and combines the records of Map 4.14. The resulting patterns are similar to the findings in the other subregions of the Coast Range. The regularly burned oak savannah and grassy prairies of the eastern part of the southern Coast Range have not experienced catastrophic wildfires; fires move from east to west on late summer east winds; ignitions are by people and occur along regularly traveled routes or adjacent to settlements; and boundaries correlate to precontact subbasin-scale land use patterns. These findings are in contrast to other findings. For example, one common assumption, as stated by Spies and Franklin, is that Coast Range forests in 1900 were formed predominantly of old-growth conifers in excess of 300-years age:

> In the early part of this [20th] century most of the forested area west of the crest of the Cascade Range was covered by old-growth forests consisting of Douglas-fir, western hemlock . . . and several other large, long-lived conifer species. Most of these forests were probably more than 300 years old and many exceeded 750 years (Spies and Franklin 1988).

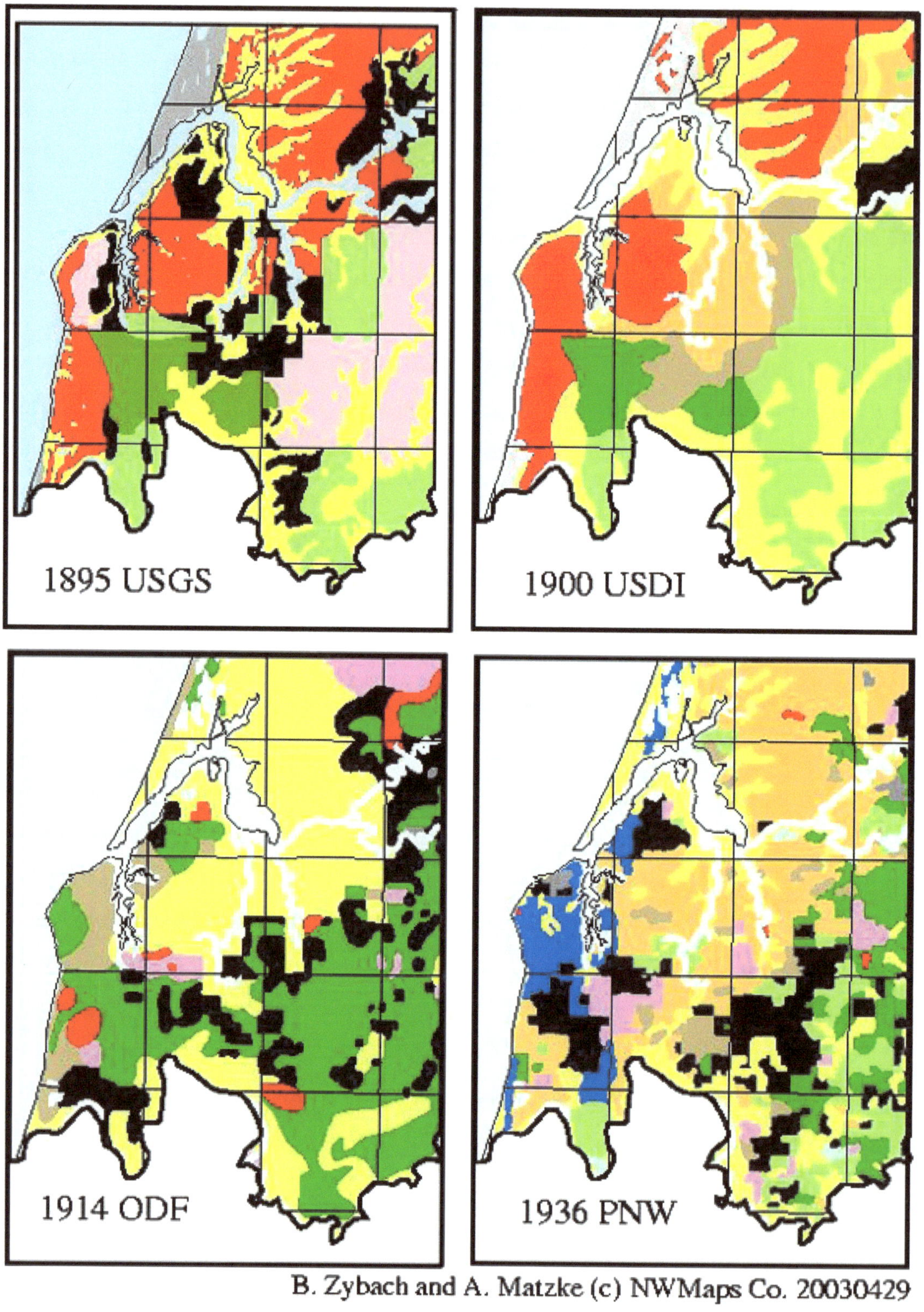


Map 4.14 Coos Bay forest fire patterns, 1895-1936.

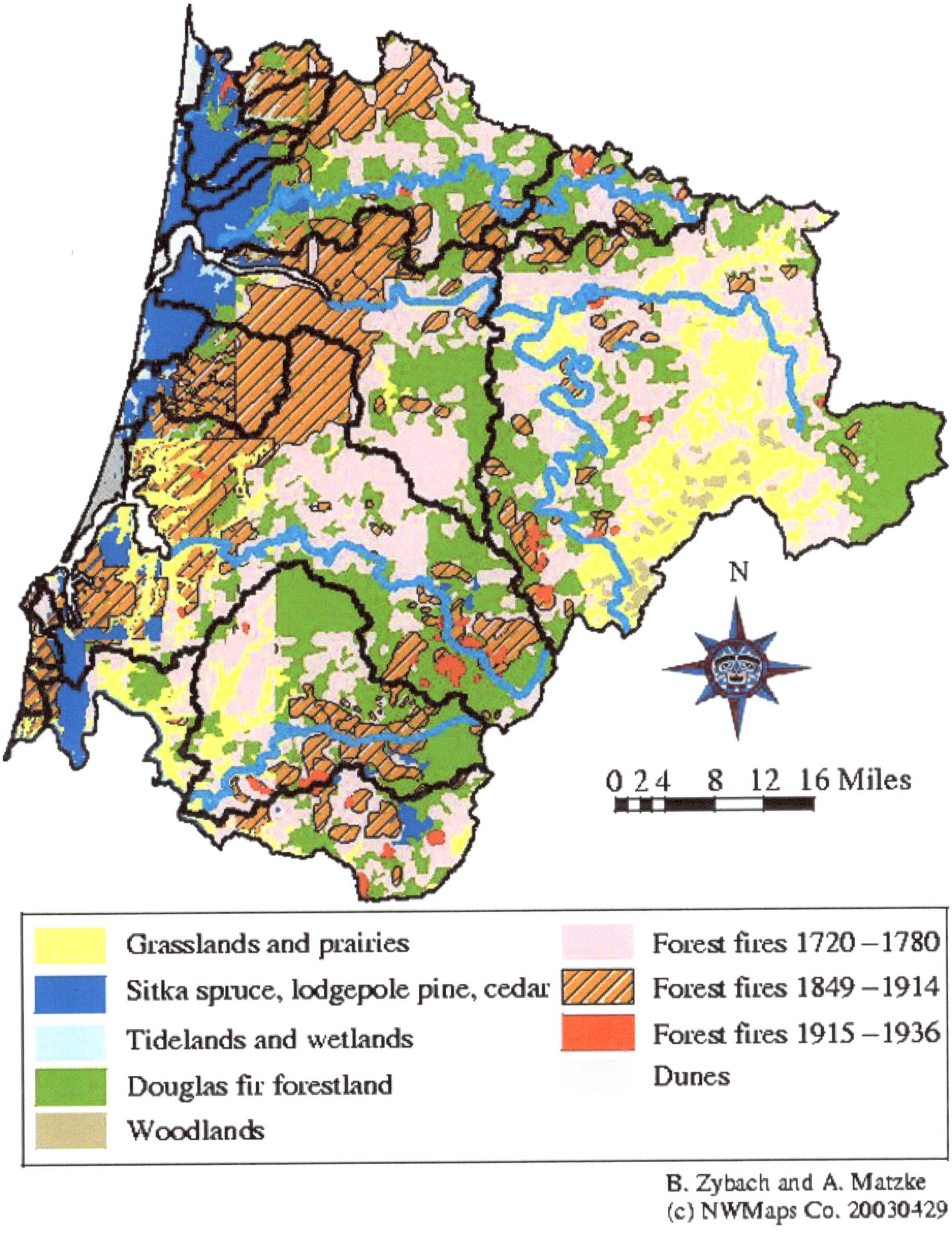


Map 4.15 Southern Oregon Coast Range forest fire patterns, 1720-1936.

The presence of the Millicoma and Coos fires before 1900 in the southern subregion make this circumstance unlikely. Fire histories of the eastern and western subregions also do not support this possibility. Only the northern Coast Range was likely comprised of "mostly" 300-year old and older trees in 1900, and North is the smallest subregion of the four (see Table 1.03).

## 4.4 Discussion and Summary: Transitional Landscape

In the 1840s, the forests and grasslands of the Oregon Coast Range began to make a sharp transition from the ownership, cultures, and management philosophies of local Indian communities to the ownership, culture, and management objectives of white European and American immigrants. This process can be called a "transitional landscape," a described by Kimmerer and Lake:

> Every ecosystem in North America has been affected in some way by a fire regime . . . manipulated by indigenous people. Much forest science, including ecological classifications of vegetation types, arose from observation of forest that were essentially in transition from conditions of indigenous fire management to post-colonial fire suppression. Our understanding of forest processes may thus be based on an anomalous, transitional landscape (Kimmerer and Lake 2001:37).

This chapter has focused on two key points: 1) the unknown factors associated with the widespread forestation of Oregon Coast Range lands in the 16th and 17th century; and 2) the strong coincidence between precontact and early historical patterns of Indian burning that had all but ended by 1848, and the subsequent patterns of catastrophic forest fires that continued until 1951. Both points are discussed in greater detail in Chapter 5. The second point is the focus of this thesis and is summarized in accordance with the types of patterns listed in Chapter 1: 1) sources of ignition; 2) timing of fires; 3) spatial boundaries of fires; 4) fire severity; 5) human safety; and 6) effects on wildlife habitat.

Indian burning and catastrophic fire pattern comparisons. The strong coincidence between precontact and early historical patterns of Indian burning and the subsequent "Great Fires" includes considerations of sources (and locations) of ignition, timing of fires, spatial boundaries of fires, fire intensities, human safety, and effects on wildlife habitat. Other patterns are associated with climate, topography, and land ownership. Catastrophic fires on the Coast Range seem to share the following characteristics: a) started by people, b) occur on the western slope of the range, c) occur mostly in August and September, d) move on east winds in a south-westerly fashion, e) cover more than 100,000 acres of forestland, and f) extinguished with fall rains.

Sources of ignition. Indian burning patterns, by definition, are caused by people. Occasional fire escapements were probably a significant part of the landscape pattern created by regular fire use. All evidence associated with catastrophic fire causes and most evidence associated with ignition sources during catastrophic fire years show people, rather than lightning or any other source, as principal causes of these events, too. Known ignition points of wildfires correlate closely with well-established precontact riparian and ridgeline travel corridors and destination points. Fires occurring after 1848, whether controlled or wild, can likewise be attributed almost entirely to whites, often for differing land management purposes than their Indian predecessors.

Timing of fires. Historical accounts of broadcast burning activities in the Coast Range occur during two fire seasons: late winter/early spring "fern burning," and late summer/early fall "field burning." All catastrophic fires are associated with the late summer/early fall "fire season," especially during the months of August and September.

Spatial boundaries. The majority of tideland, wetland and grassland prairies follow north-south canoe and foot trails adjacent to the ocean beaches and the Willamette River, and connective west-east canoe and foot trails that entered the interior of the subregion via major river drainages and ridgeline alignments. The boundaries of most wildfires correlate closely to this subbasin-scale pattern of precontact prairies, ridgeline foot trails, and canoe routes.

Fire severity. Feller (1996) describes fire severity by the amount of "forest floor consumption" that takes place during a fire. Several of the figures in this chapter demonstrate an almost complete consumption of living vegetation, both understory and overstory, resulting from Great Fire events. Severity likely increases with subsequent reburns (Isaac and Meagher 1936: 11). Indian burning patterns were based on regular and seasonal fires that consumed fuels incrementally, or covered areas with such frequency that fires seldom occurred in areas of heavy forest or fuels. Catastrophic fires, on the other hand, typically followed centuries of tree growth or burned through areas of standing snags and downed logs remaining from earlier wildfires. Differences in fire severity can be easily inferred from figures representing the localized cooking fires or seasonal "cool burns" of Indians in Chapter 3, and those illustrating the killed timber and mushroom clouds of later events in this chapter. The coincidence in these patterns is temporal: the earlier controlled burns consumed a minimum of fuel as they "flashed" through a landscape of ferns, grasses, and shrubs, while the later wildfires burned as hot and explosively through massive conifer forests as any on record.

Human safety. There are only a few cases of human deaths known to be related to Oregon Coast Range forest fires, and all are related to firefighting (Fick and Martin 1992: 89). No fatalities are on record known to be associated with precontact Indian burning practices, although fatal automobile accidents have been blamed on smoke from agricultural field burning (Venell 2003: personal communication: see Chapter 6.2.4).

Wildlife populations. There are no long-term reductions in local wildlife populations that have been associated with either Indian burning or catastrophic forest fires. Deer, elk, berry, grass, oak, and bear populations would likely decline through time as forests expanded their range (Kay 1995), and increase in numbers as forests were burned or otherwise cleared (Kay 2002: 248-250). The development and maintenance of transportation corridors, extensive oak savannahs, prairies, berry patches, filbert groves, camas fields, lawns, and balds by Indian burning practices (Table 3.03) also resulted in beneficial habitat to a number of plant (Table 3.05) and animal (Table 3.06) species. During wildfire events, these areas could also function as "refuges" for threatened wildlife species.

Climate. Graumlich (1987) analyzed tree rings to arrive at a 300-year precipitation pattern that identified specific years (1717, 1721, 1739, 1839, 1899, 1929, and 1973) and at least one decade per century (1790s, 1840s, 1860s, 1920s, and 1930s) of prolonged regional drought. The annual drought events seem to have no significant relationship to Coast Range fire history, but the prolonged droughts correlate closely with major forest fire events identified in this chapter. All of these fires, however, are known or believed to have occurred during the late summer-early fall months of August and September. Thus, there is a strong correlation between fire events and seasonal weather patterns, but people, rather than lightning or other weather-related causes, are responsible for the large majority of fires in the Coast Range (see Table 4.02)

Topography. Topography plays an important part in affecting burn patterns of Coast Range forest fires. In heavily dissected areas, such as the headwaters of the Nehalem or Siletz rivers, large areas of forest typically avoid being burned (see map 4.01). In "rolling" areas of topography, such as the Yaquina and western Coos river drainages, almost all trees over large expanses of land might be killed. "Flat" areas of the Range, such as the Willamette and Umpqua river valley floodplains, have typically been used by people for thousands of years and thus have not been subjected to forest fire at all.

Landownership patterns. Oregon Coast Range land ownerships also play a significant role in landscape patterns of purposeful burning and large scale wildfires. Precontact burning practices varied significantly between Chinookan, Kalapuyan, Athapaskan, and Yakonan tribes, for example (see Map 3.01 and Chapter 3.D), and resulting landscape patterns reflect those differences. Current forest landownership patterns, on the other hand, correlate closely to historical catastrophic fire patterns (see Map 4.01 and Map 6.01): the Millicoma fire developed a dense, even-aged stand of second-growth Douglas-fir that was owned and mostly logged by Weyerhaeuser Timber Co. (see Figure 4.03) during the latter half of the 20th century; the Yaquina and Nestucca fire boundaries of the mid-1800s (see Figure 4.05) are similar to the current boundaries of the Siuslaw National Forest; the 1868 Coos Fire (see Figure 4.06) closely matches the current boundaries of the Elliot State Forest; and the Tillamook fires of the first half of the 20th century (see Figure 4.07, Figure 4.08, and Map 4.09) evolved into today's Tillamook State Forest.

# 5. SUMMARY, DISCUSSION, AND CONCLUSIONS

*Throughout the course of this study an effort has been made to find evidence which the forester in the woods might use as a key to local forest-soil relationships--evidence that could be obtained without digging a great number of deep pits.*

--Benjamin B. Stout 1952: 27

*Men argue, nature acts.*
--Voltaire

The hypothesis of this dissertation is that western Oregon Indian burning practices of 1491 to 1848 had a direct effect on subsequent patterns of catastrophic forest fires that took place from 1849 to 1951 in the Oregon Coast Range (Chapter 1.1). These patterns are shown to vary between northern, eastern, western, and southern parts of the Coast Range (Chapter 1.6.1) due to differences in cultural and tribal traditions (see Map 3.01), topography (see Map 1.02), climate (see Table 1.01), vegetation (see Map 1.09), and distance from the ocean (see Map 4.01). This chapter summarizes the findings of this research, discusses a variety of hypotheses that might reasonably result from those findings, and provides a number of conclusions based on the "weight of evidence" (Chamberlin 1967).

The "cultural legacy" of long-term Coast Range Indian burning practices is shown to have a direct effect on subsequent spatial and temporal patterns of catastrophic forest fires within the same subregion (see Chapter 4.3). This study demonstrates a high rate of coincidence between the land management practices of precontact Indian communities of the Oregon Coast Range, and the causes (see Table 4.02), timing (tables 3.04 and 4.01), boundaries (Map 4.13), severity (see Chapter 4.1) and extent (see Map 4.01) of subsequent catastrophic forest fires in the same areas. These findings differ significantly from current assumptions that Indian burning actions affected a relatively minor portion of the landscape (Vale 2002), that Coast Range lightning fires had greater influence over precontact vegetation patterns than human fires (Whitlock and Knox 2002), and that precontact western Oregon vegetation could be characterized as a "blanket of old-growth" Douglas-fir forests (FEMAT 1993; Chase 1995: 402-407). Instead, early journalists, painters, surveyors, photographers, timber cruisers, and immigrant residents (see Chapter 2.2) found a lush, thriving mosaic of managed forests, woodlands, prairies, berry patches, camas fields, and thousands of contiguous acres of oak savannah grasslands, wildflowers, "Indian oats", and "Indian peas"; all connected by a well-established and maintained network of foot trails and canoe routes (see Chapter 3.4).

Between 1770 and 1840, a series of plagues decimated Indian families and town sites in the Coast Range in a matter of days and weeks. The diseases, including smallpox, measles, and malaria, were introduced largely by American, European, and African fur traders (se Chapter 3.1). Local communities and trade networks subsequently collapsed. In the 1840s another dramatic transition took place, in which the few thousand Indians who remained alive in the Coast Range were physically displaced by thousands of European American immigrants and hundreds of herds and flocks of domestic grazing animals. For the first time in millennia, the landscape was not being managed primarily with fire. Hundreds of new species of plants and animals--principally from Europe, but also from North and South America, Asia, and Africa--were being cultivated and released into the environment by the new immigrants. With these new people, new plants, and new animals, a century-long series of some of the largest and most spectacular forest fires in history was initiated in 1849 with the Yaquina Fire. The fire, likely within a few hours or days time, burned 300,000 acres or more forest. It was probably started on a warm day in late August or early September, and spread rapidly with the help of an east wind (see Chapter 4.1.3; Appendix C). Its boundaries and fuels had been partly determined by the Kalapuyan and Yakonan people who had lived in the area and managed the landscape in past generations (see Map 4.13).

## 5.1 Indian Burning Patterns

Precontact Oregon Coast Range Indians used fire to produce landscape patterns of trails, berry patches, root and bulb fields, nut orchards, woodlands, forests, and grasslands that varied in age and composition from time to time and place to place. These variations were due in part to demographic, cultural, topographic, vegetative, and climatic differences that existed throughout the region. Local Indian families systematically managed native plants in even-aged stands, usually dominated by a single species, throughout all river basins of the Range. Oak, filberts, camas, wapato, tarweed, yampah, strawberries, blackberries, huckleberries, brackenfern, lilies, onions, nettles, and/or other plants were raised in select areas by all known tribes, over large areas and for long periods of time. These plants were thinned, pruned, planted, tilled, peeled and otherwise affected by people during harvest and other management processes. Fire was the principal tool used to maintain large tracts of these plants. Landscape patterns that resulted from these actions were relatively stable and predictable. They produced large, reliable, and varied crops of food during all seasons of the year, and could be harvested and maintained with relatively little effort.

### 5.1.1 Cause and locations

Indian burning patterns were, by definition, purposefully created by people. Fires were started, transported, and spread for specific reasons. Cooking and heating fires were maintained constantly in and near homes, in camping spots, campgrounds, and other gathering places. These fires were concentrated within permanent settlements and communities near the mouths of rivers and streams, and at other key locations along the coastline and principal riverbanks. Seasonal cooking and heating fires were located in favored hunting and gathering spots, depending on social, food gathering, and processing activities. These fires depended on the systematic gathering, storage, and use of firewood, which was also used for other purposes, such as constructing bonfires and heating sweat lodges. Fires were also used to clear and maintain trails; rejuvenate berry patches, pea fields, root fields, and orchards; for hunting; for weed control; and to cure large fields of tarweed seeds ("Indian oats"). Patches and fields were located along trails that typically ran adjacent to rivers and streams, the coast, and along ridgelines, directly connecting communities, peaks, campgrounds, waterfalls, and other favored locations. Fires were also used to regularly burn vast areas of oak savannah and grassy prairies contained in Coast Range lands occupied by Kalapuyan people. In sum, Indian fires were caused and maintained by people in a regular and constant pattern across the landscape, in predictable locations, at predictable times of the year, and within definite topographical and political boundaries.

### 5.1.2 Seasonality

Coast Range Indian fires were used on a daily basis, at all hours of the day and night, every day of the year. Firewood gathering, trail maintenance fires, and patch burning were performed as needed and as circumstance and weather permitted. Brakes and balds were usually burned in winter or spring, following desiccation by seasonal freezing, and preceding the gathering of sprouts and digging of roots and bulbs. Filbert patches might also be burnt in the winter or spring, to encourage sprouts for weaving. Berry patches and fruit and nut orchards were burned following harvest and desiccation by late summer drought or fall senescence. Oak savannahs were broadcast burned in the summer and fall, beginning as early as July on the eastern slope of the Range, and continuing until September or October, ending with the arrival of steady fall rains. In sum, people used fire constantly at all times of the year, but certain landscape patterns were maintained largely through seasonal burning practices; typically occurring within a given area at regular one to five or ten year intervals.

### 5.1.3 Wildlife habitat

The development and maintenance of transportation corridors, extensive oak savannahs, prairies, pea fields, berry patches, filbert groves, camas fields, root fields, lawns, and balds by Indian burning practices (Table 3.03) resulted in beneficial habitats to a number of plant (Table 3.05) and animal (Table 3.06) species. During wildfire occurrences, these areas functioned as "refuges" for threatened wildlife species; even in severe, stand-replacing events, fields of grasses, berry patches, riparian meadows, and fern prairies often remained unburned, or only slightly singed. Deep pools in rivers and streams, waterfalls, tidelands, and shorelines were other important wildlife refuges during catastrophic fire events. The openings created by Indian burning practices--in what would otherwise be a nearly unbroken blanket of conifer trees throughout nearly every acre of the Coast Range--also provided important sources of food and protein for a wide variety of animal species. Bees and butterflies, songbirds, deer, elk, black bears, gray squirrels, bobcats, wolves, and California condors all benefited by food derived from unforested lands maintained by Indian burning.

### 5.1.4 Cultural legacy and fuel distribution

The term "cultural legacy" has been used in this research to mean the general and local-scale patterns of managed vegetation that have persisted from precontact times to the present. Such patterns are most apparent in the fields and grasslands of the Umpqua and Willamette valleys, but are also found in the inland riparian prairies, ridgeline trails, brakes, balds, and salmonberry patches of lesser rivers and streams. Areas kept clear of conifer forest encroachment 200 years ago continue, for the most part, to be clear of conifer forests to the present time. Likewise, areas consisting of solid, even-aged stands of shorepine, Sitka spruce, Douglas-fir, western hemlock, and true fir at the present time, are fairly good markers of the locations and extent of even-aged stands of the same species from ca. 1500 until the 1840s and later, continuing--with some notable exceptions--through to the present time. In some instances, logged spruce or Douglas-fir stands were followed by stands of red alder or bigleaf maple. Other areas were cleared for pasture or urban development, while abandoned fields and old meadows often began seeding to a more permanent cover of forest trees within a few years' time.

In general, it can be shown that: 1) oak savannahs and large, grassy prairies that were created and maintained in precontact times by regular use of broadcast burning practices, have never developed extensive forest canopies, and therefore have not experienced catastrophic forest fires during historical time; 2) extensive areas of even-aged Douglas-fir that existed in the early 1800s indicate catastrophic stand-replacement fires in precontact time, and are currently covered with even-aged stands of Douglas-fir, typically less than 50-years of age; and 3) fogbelt stands of Sitka spruce, hemlock, and cedar tended to burn less often than Douglas-fir stands to their east in precontact time, were mostly logged during the last 150 years, and are covered largely with even aged stands of similar species less than 75 years of age. In sum, current patterns of vegetation in the Oregon Coast Range (see Map 1.09) generally mirror similar patterns established by regular burning practices of precontact Indian families and communities. Differences, such as the imposition of straight lines on the landscape, sprawling urban and industrial areas, and changed grassland species, become more obvious as the scale becomes finer. These patterns also reflect precontact and historical catastrophic wildfires, which in turn indicate the dynamic spatial and temporal patterns of wildlife habitat that has characterized the Coast range for the past several centuries. Native animals have either adapted to these patterns of change or, as in the examples of condors, grizzly bears, and wolverines, been extirpated.

### 5.1.5 White settlement and transitional forests

The first white settlement in the Oregon Coast Range was Lewis and Clark's Fort Clatsop of 1805-1806, established in close proximity to towns of Clatsop, Chinook, and Kathlamet people. In 1810, Fort Astoria was established a few miles distance from the abandoned Fort Clatsop. Whereas Fort Clatsop had been built to protect a small band of men from the weather and possible hostilities, Fort Astoria was located for the purpose of international trade, in a prominent spot near the mouth of the Columbia River. In the 1820s and 1830s a few retired beaver trappers, some ministers from the eastern United States, and a handful of cattle ranchers began to occupy the lands of the Kalapuyan people in the Willamette Valley. Early town site developments in

Portland and Oregon City in the 1840s were platted on strategic town sites of the Multnomah, near the mouth of the Willamette River, and the seasonal trading center of the Clowwewalla, near Willamette Falls. By the 1850s, important white town sites had been platted near the mouth of nearly every major stream in the Coast Range; invariably these town sites were located over the remnants of precontact towns abandoned largely by the mid-1830s because of plagues and diseases.

In sum, most major precontact town sites in the Coast Range were abandoned a few years or decades in advance of white settlement of the 1840s and 1850s; subsequent settlement mirrored precontact population patterns, concentrated largely near the mouths of streams and within grasslands adjacent to shorelines accessible by canoes and steamboats. Farms and ranches were established in less densely populated areas of former oak savannahs and grassy valleys. These general demographic patterns persist to the present (see Table 3.01), although current human and domesticated animal (including pets, livestock and feral animals) population is probably far greater than at any time in the past. Although general forest vegetation and wildlife populations remain similar to precontact times, there are a number of notable changes that indicate the transitional nature of Coast Range forests that continue to adjust to abrupt changes in ownership, culture, and technology (Lake and Zybach: In Review). Some examples of measurable changes that have accompanied the 1840-1850 transition from Indian to white ownership and management include: the imposition of straight lines on the landscape; the introduction of hundreds of new species of vascular plants and dozens of new vertebrates; the expansion of conifer forests into old openings and along the perimeters of precontact prairies and oak savannahs; the extirpation of large carnivores and poisonous snakes from the environment; and the marked daily and annual reduction in firewood gathering and use.

## 5.2 Catastrophic Fire Patterns

This research considers catastrophic fire patterns in terms of size, cause of ignition, boundaries, severity, and seasonality. By definition, a catastrophic fire is 100,000 acres or greater in size. A catastrophic fire year is one in which 100,000 acres or more of forest and grassland wildfires have occurred in the Coast Range, whether as a result of a single event or as a complex of lesser events, during the course of a calendar year. Most Coast Range land subjected to forest fires is within the belt of Douglas-fir. This belt extends along the center of the entire length of the region, from the Middle Fork of the Coquille River on the south to the Columbia River on the north (see Map 1.09). Within this boundary exists some of the finest conifer growing land in the world. Vast tracts of even-aged stands of Douglas-fir mark the extent of catastrophic wildfires that occasionally burned so hot they created their own weather and killed nearly every tree for miles on end. Pinchot (1987) and Munger (1916) noted the patterns of Douglas-fir stand replacement wildfire events and subsequent even-aged regeneration in the late 1800s and early 1900s, and for the "past few centuries" before then. These patterns were carefully mapped and assigned age and size classes by trained foresters and cartographers between 1895 and 1936 (e.g., Gannett 1902; Andrews and Cowlin 1940). Aerial photographs added detail thereafter.

At this time, cities, rural real estate developments, and farms occupy the majority of eastern Coast Range grasslands, to the east of belts of higher elevation Douglas-fir. A contiguous forest has never developed in these lands during historical time, and they are shielded from large scale forest fires for this reason alone. Grasslands, fern prairies, and berry patches are common in this subregion, however, and are typically rejuvenated by fire, whether prescribed or unintended. Thus, wildfires that occur in the more heavily populated areas of the Coast Range--the brushy grasslands of the east--tend to green, flower, and fruit up almost immediately with the first rains and seasons to follow the fire (Pendergrass 1996). These fires also burn far less fuel per acre than the Douglas-fir fires to the west.

Coastal Sitka spruce and hemlock fog belt forests to the west of the Douglas-fir belt (see map 1.09) have the capability to grow more wood fiber per acre in fewer years than Douglas-fir. Despite this rapid accumulation of fuel, forest fires tend to be less frequent than in the Douglas-fir lands. The primary reasons for fewer wildfires in the spruce-hemlock forests are the cool, moist local climate that is subjected to heavy fogs and drizzly rains during any time of the year, and the regular wet, westerly winds that can readily dampen (rather than spread) fires with their moisture content. When large tracts of fog belt spruce and hemlock forests are burned it is usually the result of wildfire from the Douglas-fir forests to the east, moving toward the coast on dry, east winds.

In sum, Coast Range catastrophic forest fire patterns are heavily influenced by the type of vegetation in which they occur, and tend to be marked by even-aged stands of conifer trees. About 90% of the fuel is Douglas-fir, in the north-south Douglas-fir belt, and mostly spruce and hemlock in the fog belt forests along the coast. Forest fires tend to be relatively small in size to the east of the Douglas-fir belt because areas of conifer forest are relatively small.

### 5.2.1 Cause and locations

People have caused all of the historic Great Fires and a large majority of all documented Coast Range wildfires. Lightning is relatively rare in the region, and is often accompanied by heavy rains when it does occur. Lightning-caused fires do occasionally take place, but they are not recorded for most years, are not a regular occurrence, are not known to occur in regular places, and have never been known to reach catastrophic proportions. Because people cause wildfires on accident and on purpose, and because the Great Fires all traveled from east to west, people have tended to start the fires along roads and trails, near camp grounds, in fields, and in logging units, along the eastern boundary of the Douglas-fir belt (and mostly in August or September). Several quotations in this dissertation have made reference to known and suspected causes of these fires, and they all fit this pattern.

### 5.2.2 Climate trends and seasonality

Graumlich (1987) analyzed tree rings to arrive at a 300-year precipitation pattern for the Pacific Northwest that identified specific years (1717, 1721, 1739, 1839, 1899, 1929, and 1973) and five decades (1790s, 1840s, 1860s, 1920s, and 1930s) of prolonged regional drought. The Palmer Drought Index for "Oregon Zone 1" (Figure 1.02), beginning in 1663, shows minor droughts for the 1790s and 1840s (but not the 1860s), and the longest period in the last 300 years occurring in the 1920s and 1930s. Palmer also shows a significant drought in the 1760s that Graumlich misses. Major Coast Range fires are not recorded for the individual drought years noted by Graumlich, but coincide very well with prolonged droughts noted by either her or Palmer: the Millicoma Fire of the 1760s; the Yaquina (and maybe Nestucca) Fire of the 1840s; the Coos and Yaquina fires of 1868; and the multiple fire years of the 1920s and 1930s, including the Tillamook fires of 1933 and 1939. These fires are known or believed to have occurred in the late summer and early fall months of August and September. Further, they seem to have occurred during periods of east wind. In sum, during the past 350 years, Coast Range catastrophic fire events have most often occurred during periods of prolonged drought, in conjunction with east winds, and during the driest part of the year.

### 5.2.3 Topography and wildlife habitat

Topography plays an important role in affecting burn patterns of Coast Range forest fires. In turn, those patterns become habitat for native wildlife species. In heavily dissected areas, such as the headwaters of the Alsea, Siletz, or Nehalem rivers, large areas of forest have typically avoided being burned for centuries at a time. Habitat, in those areas, can develop "classic" old-growth attributes (FEMAT 1993). In "rolling" areas of topography, such as typify much of the Tillamook and Yaquina river basins, almost all trees over large expanses of land have been killed by fire on a more frequent basis. Habitats in these areas tend toward even-aged stands of second-growth Douglas-fir. "Flat" areas of the Range, such as the floodplains of the Umpqua and Willamette rivers, have typically been used by people for thousands of years, and so have not developed contiguous forest canopies; thus, they have not been subjected to significant forest fires during historical time. Flat and rolling areas along the coast favor spruce and hemlock trees in a moist, foggy environment that discourages wildfire in those areas at most times of the year. Coastal wildlife species are significantly different than other areas of the Range and are not always so dependent on terrestrial habitat conditions.

In sum: 1) flatter areas of the Coast Range (the most heavily populated areas of the region) tend not to have forest fires because they tend not to have Douglas-fir forests, for reasons of local climate and land use history; 2) the steepest, most heavily dissected areas of the Coast Range (typically, riverine headwaters leading from the highest peaks and ridges) tend to have numerous wildfires, but tend not to have catastrophic events, possibly because the nature of the terrain affects wind patterns and can shelter fuels in steep canyons, or shady north slopes; 3) the remainder of the Coast Range, particularly the sloping and rolling portions of the Douglas-fir forest, have been subjected to numerous and repeated catastrophic wildfires throughout historical time, and probably for thousands of years of prehistoric time as well.

### 5.2.4 Human safety

Few people have been known to be killed or injured during Coast Range forest fires, largely because they take place in areas away from most human populations. During the 1933 Tillamook Fire, for example, a CCC firefighter died (Pyne 1982: 367), when a green tree fell on him while he was sleeping in a tent. The only other fatalities documented during this research were three firefighters killed during the 1945 Tillamook Fire (Fick and Martin 1993: 89). None of the historical Coast Range Great Fires have entered cities or town sites or burned homes or vehicles while people were still in them. However, the 1936 Bandon Fire, which occurred just south of the Coast Range, demolished the town in less than five hours and killed 11 people (Pyne 1982: 187). This compares to the 20 people killed during the California brush fires of October 2003 (Nash 2003: 57), the 1902 Yacolt Fire that took the lives of 16 Lewis River Valley residents (Pyne 1982: 339; Robbins 1988: personal communication), or the infamous 1871 Peshtigo, Wisconsin fire that killed about 1500 people (Pyne 1982: 206). In sum, Coast Range forest fires have not proven to be particularly hazardous to people and are not likely to become significantly more dangerous in the near future.

### 5.2.5 Land ownership

Land ownership also plays a significant role in establishing landscape patterns of purposeful burning and large scale wildfires. Precontact burning practices varied significantly between Chinookan, Kalapuyan, Athapaskan, and Yakonan tribes, for example. White European Americans subsequently settled areas managed by these nations, in patterns that closely mirrored the burning practices of earlier generations. Also, current forest landownership patterns of the Oregon Coast Range correlate to historical catastrophic fire patterns: the boundaries of the Coos and Tillamook fires now closely approximate those of the Elliott and Tillamook State forests; the areas of the Nestucca and Yaquina fires are now mostly managed as part of the Siuslaw National Forest; and the Millicoma Fire area was subsequently purchased by Weyerhaeuser Timber Company. In sum, areas of the Coast Range that were subjected to regular Indian burning practices have been converted to cities, towns, and farms, whereas forested areas that burned less frequently have been maintained as forests by large corporations and state and federal agencies.

## 5.3 16th Century Events Hypotheses

Current scientific, resource management, and policy assumptions regarding the abundance and extent of precontact western Oregon old-growth forests in the Oregon Coast Range may be in error. Although old-growth (more than 200-year-old) conifer and hardwood trees are common throughout the Range--and probably have been common throughout the 460-year time span of this study--they existed in the forms of stands, pockets, strips, belts, groves, and individual specimens, and varied dramatically in total extent and population numbers from time to time. This is significantly different than the "blanket of old-growth" description that has been used to characterize 60% to 80% of the landscape of the Range by a number of scientists and politicians during the past two decades (e.g., Spies and Franklin 1988; Teensma et al. 1991; FEMAT 1993).

Why are older trees so uncommon, and why does so little evidence exist for earlier stands of trees; those that parented the vast tracts established in the 1500s and 1600s? How did vast tracts of Coast Range precontact trees come to be established at approximately the same time? Four hypotheses are presented in the following pages that might help to explain this phenomenon. Each idea is briefly discussed, based on the "weight of available evidence."

### 5.3.1 Catastrophic fires hypothesis

Prevailing assumptions are that a gargantuan catastrophic fire or a widespread complex of wildfires took place in the 15th century that resulted in almost total deforestation of the Oregon Coast Range, followed closely by subsequent reforestation. The reforested stands of the early 1600s then became the even-aged old-growth stands, logs, and wildfire fuels for the 19th and 20th centuries. If this hypothesis is accurate, there should have been a huge influx of carbon in lakes, swamps, and soil horizons resulting from the fires and decomposition of trees. This layer of organic material should have formed a regional "marker" (Zybach 1999: 247-249) readily identifiable by scientists engaged in soil studies, palynology, or analysis of lake sediments. To date, no such marker has been noted (Impara 1997: 265). Because there have been no events of similar magnitude during historical time, and because there is no evidence that such an event took place in precontact time, it seems unlikely that the early historical forests of the Coast Range resulted from a widespread catastrophic fire or fires in (or around) the 16th century.

### 5.3.2 Reduction in Indian burning practices hypothesis

Was widespread conifer forestation throughout the Oregon Coast Range in the 1500s and 1600s possibly triggered by a significant decline in Indian populations? Some evidence suggests large numbers of people died on the Coast Range between 1500 and 1550 (Denevan 1992). If so, a number of processes may have occurred on a grand scale similar to what happened after 1775, 1830, 1850; that is, a widespread afforestation of Indian prairies, brakes, balds and savannahs would have likely occurred. This process has been observed in the eastern US as well. For example, William Wood, who lived in Massachusetts Bay colony from 1629 to 1633, wrote:

> it being the custom of the Indians to burne the wood in November, when the grasse is withered, and leaves dryed, it consumes all the underwood, and rubbish, which otherwise would over grow the Contrey, making it unpassable, and spoyle their much affected hunting; so that by this meanes in those places where the Indians inhabit, there is scarce a bush or bramble, or any cumbersome underwood to bee seene in the more champion ground . . . In some places where the Indians died of the Plague some foureteene yeares agoe, is much underwood, as in the mid way betwixt Wessaguscus and Plimouth, because it hath not beene burned (Russell 1983: 82)

Assuming this hypothesis is correct: With the massive loss of lives, homes, and communities and the threat of white invaders, survivors would likely be less inclined to venture into isolated areas away from towns and established campgrounds. Hunting and gathering would also be greatly reduced and likely confined to known areas, and deer, elk, bear, seal, sea lion, clam, and mussel populations would likely have boomed (Kay 2002: 248-250). Many heavily tilled swamps and marshes, with a reduction or elimination of regular harvesting activities, would begin to grow trees (willow, ash, crabapple, cedar, cottonwood, etc.), as would scattered meadows, abandoned town sites, inland prairies, and the perimeters of savannahs. If oaks and/or filberts were being managed as plantations (speculative evidence), the creation of new orchards would have probably stopped. Spruce and hemlock would begin foresting coastal headlands, ridges, prairies, and valleys, as Vaughn (1923) described for the Tillamook area in the 1850s; Douglas-fir would begin populating isolated meadows, prairies, berry patches, and the perimeters of bottomland prairies and savannah, as Moravets (1932) describes; true fir would begin populating higher elevation berry patches and the perimeters of balds on the west slope (Aldrich 1973) and seeps and highlands within the Willamette Valley (Zybach 1999); and riparian areas would begin to grow brush and gallery forests in many places. The whole process would take about 20 to 60 years (e.g., Gannett 1902; Weyerhaeuser 1947; Winter et al 2002), and by 1600 or 1650, a large portion of the Coast Range would have been coated with young and thriving second growth stands. These stands would have contained few snags and little “large, woody debris," and would have been interspersed with scattered pockets and relatively small stands of parent seed trees. By 1800, many of these stands would be old-growth around 200 years of age, and by 1950 they would be 350 year-old trees very similar to "ancient" forestlands described in FEMAT (1993), but with a marked absence of charred snags, stumps, logs or other "biological legacy" (e.g.,

Impara 1997; Franklin et al 2002). Around 1700 and thereafter, large-scale "one careless match" fires, independent of changed human population numbers, would begin to enter the landscape as canopies filled in and tree mortality created fuels. A large-scale mosaic of even-aged forested tracts would begin to form. Reburns would begin to reduce the amount of snags, large woody debris, and other biological legacy in those areas as well, but there would still be more of these materials than in the parent stands, due to increased tree densities that preceded mortality.

Is it possible that the Coast Range was far more populated in the 1500s than has been estimated? If so, what evidence exists that might be proof of a higher population? The apparent surplus food stores created by wapato, camas, brackenfern, and other vegetables, as well as coastal marine animals, hunting, and anadromous fish runs, seems to point to the possibility of enough food for a much larger population than has been documented. Certainly, land and housing would have been no problem for larger numbers of people; current populations demonstrate that fact. Firewood for heat, light, and cooking was also abundant. The afforestation and forestation processes that have taken place during historical time as Indian populations and practices declined provide additional supporting evidence, especially if parallels can be made with the widespread establishment of forest trees in the 1500s. It would also help explain the consistent age of eastern Coast Range white oak with Douglas-fir, redcedar, and spruce to the west. Too, there would not be the layer of organic material that would mark widespread wildfires, as discussed earlier in this chapter, nor would there be evidence of large, older snags, downed old-growth logs, or wind throw-"stirred" soils for the same reasons. Instead, one might reasonably expect to find large tracts of abandoned croplands producing nuts, fruits, bulbs, roots, berries, stalks, and other foods, and archaeological evidence of large, thriving communities near the mouths of streams and rivers. Thousands of acres of abandoned croplands have been located, but otherwise relatively little historical or archaeological evidence has been produced indicating significantly larger populations in the 1500s than were recorded in the late 1700s. Note, however, that little effort has been made to locate such evidence, and it may yet remain to be discovered.

### 5.3.3 Climate change hypothesis

The climate change hypothesis is that an abrupt and prolonged cooling period, the so-called "Little Ice Age" of ca. 1450 to ca. 1850, created ideal weather conditions for establishing tracts of Sitka spruce, Douglas-fir, Ponderosa Pine, and other long-lived conifer species, throughout western North America (Cromack et al 2002: 180-184; Sessions 2003: 27). The warming trend that signaled the end of the Little Ice Age begins in 1850, coinciding with the first of the Great Fires. Warming and drying patterns continued over the next century, as did catastrophic fires. Current weather conditions, coupled with a possibility of Global Warming, make re-creation of historical forest conditions in regards to species mix, age, and structure, difficult or impossible (Cromack et al 2002: 216-217; Sessions et al 2003: 57). In this hypothesis, Indian burning patterns are secondary to climate and weather regarding fuel history and current planning assumptions. Atzet applies this hypothesis to the 2002 Biscuit Fire in southwest Oregon, immediately south of the Coast Range:

> Exacerbating this competitive situation is a warming climate. Much of the forest that burned in the Biscuit was established during the waning years of the Little Ice Age. Current and likely future climates are more favorable to root-sprouting shrubs than when the forests originated (Atzet 2003: personal communication).

Despite the relative proximity of the two areas (and in response to a suggestion that the Coast Range--and individual subregions of the Range--may have experienced the Little Ice Age differently than southwest Oregon) Atzet noted:

> You are correct about the spatial variability. The Coast Range does not behave as SW Oregon. Similarly high elevation has different fire regimes than the low elevations. Even the Biscuit fire can be stratified by coastal, inland, and low elevation vs. high. We have also stratified into plant association groups and see that all functions (successional, organismic, and disturbance) can have significant differences (Atzet 2003: personal communication).

Atzet's comments are consistent with Bradley and Jones (1995), who note:

> Although there is voluminous indirect evidence that climatic conditions in the past 500 years were often quite different from our contemporary experience (e.g. von Rudloff 1967; Lamb 1982; Grove 1988) the precise nature of these differences, and what caused them, remains elusive . . . In almost every glacierised mountain region of the world, glaciers grew and advanced down-valley, often to positions as extensive as at any time since the last Ice Age, more than 10,000 years ago. These changes in alpine glaciers are so characteristic of the period that it is often referred to as "the Little Ice Age". However, among those who use this term there is little consensus on when it began, or when it ended (Bradley and Jones 1995: 2).

For North America, Jones and Bradley (1995: 654) note: "No prolonged periods of cool temperatures are evident between 1640 and 1820." For the western US, they rely on Fritts and Shao's (1995) tree ring work to conclude that, from 1602 forward, all of the five western US regions "show little evidence of protracted cool periods with the coldest period occurring during the late 19th and early 20th centuries" (Jones and Bradley 1995: 654-655). In remarking on climate maps based on 1901-1970 mean values, Fritts and Shao (1995) note:

> These maps might be regarded as an estimate of the departures of the Little Ice Age climate beginning in 1602 and ending in 1900 from the 1901-1970 mean climate. The reconstructed annual temperature variations in the western United States do not show a Little ice Age cooling, but lower temperatures are clearly evident east of the Rocky Mountains (Fritts and Shao 1995: 277).

Although thousands of trees in the Oregon Coast Range, covering tens of thousands of acres, have had their rings counted, no evidence has been presented that changing climate conditions have affected tree growth patterns or migration. Graumlich (1987) looked specifically for evidence of climate change in her tree ring study, but was only able to document a few years of apparent severe drought, and no long-term trends. In addition, both Graumlich and Fritts and Shao were limited in their interpretations by having access to a limited amount of tree ring data from the Coast Range. The assumption that old-growth trees in southwest Oregon--because of their current species distribution and regeneration problems--were established under "cooler [Little ice Age] conditions," may not apply for the Coast Range to the north. The moderating effects of the Pacific Ocean and the relatively low elevation of coastal mountains have resulted in little or no glaciation of the Coast Range for thousands of years, and further indicates that the climate of the Range was not significantly different during the Little Ice Age than now. Therefore, it is unlikely that climate change is a principal factor in the establishment of old-growth forests in the 16th and 17th century in the Oregon Coast Range.

### 5.3.4 Remarkable coincidences hypothesis

A fourth possibility is that a remarkable series of coincidences occurred throughout the Coast Range in the 15th century. In this scenario insects, diseases, wildfires, droughts, earthquakes, human plagues, tidal waves and/or other events took place, coincidentally resulting in widespread deforestation followed quickly by reforestation of forestlands and afforestation of grasslands. This hypothesis--a series of interrelated catastrophes of "biblical proportions"--seems fairly unlikely and has no current proponents. This hypothesis would account for a condition (common age of historical Coast Range old-growth) that has not been widely discussed. This hypothesis, if true, should have caused an identifiable soil marker and left significant "biological legacy" (snags, stumps, logs) from earlier, destroyed stands. The answer to this hypothesis may already exist in the soil profiles, pollen and sediment cores that have been used for other research questions. A series of such events should have left ample evidence--including vivid oral traditions--that can still be sampled. The lack of such evidence does not discount this hypothesis entirely, but does help discount the possibility.

#### 5.3.5 Summary: "weight of the evidence"

Research for this thesis documented widespread forestation in the 16th century that resulted in stands and groves of old-growth Douglas-fir, oak, and Sitka spruce during historical time. Four hypotheses are discussed that might explain this condition: 1) Did a regional-scale catastrophic fire complex destroy most of the Coast Range's forests (followed by reforestation)? 2) Did a decimation of local Indian communities lead to lesser burning practices over a smaller area of the landscape (followed by reforestation of burned forests and afforestation of abandoned grasslands)? 3) Did cooling brought on by the Little Ice Age cause a unique tree seed germination condition that resulted in historical old-growth forests? 4) Did a remarkable series of coincidences, including the possibilities of insects, disease, floods, fire, wind, tidal waves, meteor strikes, or earthquakes, strike the Coast Range, resulting in widespread deforestation (followed by reforestation and afforestation)?

Massive amounts of organic matter are grown over almost every acre of Oregon Coast Range soil each year. This is because of the climate-moderating effects of the ocean, relatively low elevation, and remarkably large conifer, oak, and alder trees that grow in predictable locations. These plants leave distinct records of their existence across the landscape: trees and other vascular plants leave pollens in peat bogs, forest fires deposit charcoal in lake sediments, and dead trees leave snags, stumps, logs, and piles of rotted wood in their wake.

Region-scale catastrophic fires should have produced an identifiable carbon "marker" in soils and sediments and left significant evidence of past forests above ground. Because this evidence is apparently lacking, such an occurrence seems unlikely. An interconnected mosaic of tree-killing events is also unlikely, both because of statistical improbabilities and the same lack of evidence associated with widespread fire. It is possible that an abrupt cessation of Indian burning resulted in widespread forestation of the Range, but archaeological, anthropological, and historical evidence is also lacking in this regard. It is possible that a relatively small number of people--say a number consistent with the historical populations of the 1800s--could have systematically kept historical forested areas burned clean, but their motivation for doing so is unknown. Could forest burning have been systematically undertaken systematically during the course of extended inter-tribal or international warfare? As a method of hunting or controlling Grizzly bears? Traditional hunting practices? Any of these explanations are possible, however unlikely. In each instance, though, is the assumption that people could have kept the Coast Range largely free of conifer trees if they had wished to do so. Constant, regular fires would not have produced a signature charcoal layer in lake bottoms or soils and would have eventually eliminated any "biological legacy" remaining from earlier, forested conditions. If continuous burning patterns were stopped, whether by change in population, culture, technology, or land ownership, it would have led likely to the growth of even-aged stands of trees throughout the Range, as occurred during historical time. Evidence of forest establishment related to climate change (cooling associated with the Little Ice Age) is also lacking: first, the Little Ice Age seems to have had little, if any, effect on Coast Range climate, and second, Douglas-fir and other native Coast Range trees continue to reproduce, flourish, and grow at rates similar to those measured over the past several centuries (Zybach 1983; Tappenier et al 1997).

Until more research is completed, one can only conjecture as to what may have happened in the 1500s to spawn the vast forested tracts of subsequent centuries. If they were caused by wildfire, the fires were probably caused by people; if they resulted from afforestation, they likely grew from lands previously burned on a regular basis by people. Climate, earthquakes, insects, and diseases may have aided either or both of these processes, but were not likely the principal causes of deforestation or catalysts for forestation. In sum, the widespread establishment of conifer forests throughout the Oregon Coast Range in the 16th century was likely a result of Indian burning practices, including both wildfires and prescribed burns. Whether a significant portion of the forests resulted from a reduction in landscape burning practices, or whether such a reduction may be related to reduced human populations, is unknown.

## 5.4 Indian Burning and Catastrophic Fire Pattern Hypotheses

Is there a relationship between Oregon Coast Range Indian burning patterns and subsequent catastrophic fire patterns? If such a relationship appears possible, could other factors be involved? Could climate have been solely responsible for these patterns, as postulated by Whitlock and Knox (2002)? Or, are Indian burning practices incidental when compared to a suite of "natural" disturbances, as commonly assumed (e.g., FEMAT 1993; Tappeiner et al 1997; Franklin et al: 2002)? These questions are central to this dissertation and are the focus of the final section to this summary chapter.

The 460-year period of this study begins with the seeding and juvenile growth of vast conifer forestlands throughout the entire Coast Range, when the land was owned and occupied by local Indian families. The period ends, largely, with their destruction and harvest by wildfire and logging that took place during subsequent white ownership and occupancy. One of the interesting findings of this study is that most of the forested landscape in the study area has been even-aged, second-growth (less than 200 years) conifer for most of the study time. For two time periods, 1550 to 1700 and 1945 to the present, a majority of the Coast Range has been a mosaic of even-aged stands of mostly second-growth Douglas-fir. For a third period, 1491 to 1550, it is unknown if the majority of forestland was in a seedling, sapling, second-growth, or old-growth condition; there isn't enough information available to make an informed choice at this time. The 1550 to 1700 period is when most of the forest stands became established that developed into the fuels of the Great Fires and the old-growth logs of 19th and 20th century timber owners. From 1700 until 1848, an increasing number of 16th and 17th century stands became 200 years old. There were probably more acres of old-growth trees in the study area between 1820 and 1848 than at any other time in the past 500 years, and most certainly the greatest number of acres of old-growth for the past 400 years (1600-2000). From 1849 until 1945, of course, an increasing amount of old-growth Coast Range forest either burned, or was logged. The old-growth had become established over vast expanses of the Coast Range in a matter of decades before the arrival of whites; the new immigrants saw an entirely different landscape than had existed 200 years before. A series of catastrophic wildfires, since ended, immediately followed the simultaneous cessation of Indian burning and the arrival of whites on the landscape, and continued for a hundred years.

### 5.4.1 Climate change hypothesis

The climate change hypothesis is that forest composition and extent (fuel) is largely a function of climate, and that precontact Indian fires added little variation to the pattern of "natural" lightning fires (e.g., Whitlock and Knox 2002). Therefore, Indian fires were incidental. The Coast Range has experienced a fairly stable climate over the past 500 years, with little evidence of glaciation, tree migration, or retarded tree growth during the so-called Little Ice Age that ended around 1800 or 1850.

Regional tree ring studies were used to determine a 300-year precipitation pattern (Graumlich 1987) that identified specific years and specific decades (1790s, 1840s, 1860s, 1920s, and 1930s) of prolonged regional drought. The annual events seem to have no significant relationship to Coast Range fire history, but the prolonged droughts correlate closely with major forest fire events and catastrophic fire years identified in Chapter 4. All of the Great Fires, however, are known or believed to have occurred during the late summer-early fall months of August and September. Thus, there is a strong correlation between fire events and seasonal weather patterns; but people, rather than lightning or other weather-related causes, are responsible for the large majority of fires in the Coast Range (see Table 4.02). There is no evidence of a shift in Coast Range climate in the 1840s and 1850s, so there is no reason to assume that a prolonged drought or global warming was a significant factor in the beginning or continuation of Great Fire events after that time.

### 5.4.2 Cessation of Indian burning hypothesis

This hypothesis assumes that the abrupt cessation of Indian burning practices in the Oregon Coast Range during the 1830s and 1840s led to the buildup of ladder fuels in forested areas and rank grass and shrub lands around their perimeters. Thus, the stage was set for wildfires that could quickly and easily reach into the canopies of Douglas-fir forests in hundreds of locations, needing only a campfire or a field burn and an east wind to get started and become out of control.

Because the Great Fires began so abruptly, within a few years and decades of the cessation of Indian burning, a reasonable conclusion is that this hypothesis is correct. This hypothesis remains speculative, however, and there is little in the way of factual evidence--other than coincidence--to support this conclusion.

#### 5.4.3 White land management practices hypothesis

This hypothesis states that the Great Fires were coincidental with white settlement because new logging practices (including warming fires made with unlimited fuel) and land clearing with fire were the sources of ignition. The fires didn't happen under Indian management because Indians didn't use fire in those ways. Also, log friction was known to cause fires (the 1933 Tillamook Fire had this cause) and within a few years flammable fuels were being used for logging purposes. This argument seems unlikely, in that Indian fires were being used constantly, so the source of ignition isn't in question--just the purpose of the ignition. Did the loggers build fires in places that Indians didn't? Do rubbing logs cause that many fires? However, as with the cessation of Indian burning hypothesis, the coincidental nature of these events makes such a conclusion possible.

#### 5.4.4 Coincidence/better record-keeping hypothesis

Phillips (2003: personal communications) discusses the possibility that the Great Fires became possible due to the decadence of the stands. He supports this thought with evidence that 75 to 100-year old Douglas-fir stands are essentially "fire-proof" because they've shaded out ladder fuels (including lower limbs) and have only started to develop a significant number of dead tops and dry snags. Thus, it is just coincidental that the beginning of the Great Fires coincided with the cessation of Indian burning and settlement by whites. Cabe et al (1913), Kimmey and Furniss (1943), and Shea (1963) all add weight to this hypothesis with their observations that historical stands of old-growth trees begin dying from insect and disease infestations as they increase in age. Openings in the canopy that result from this mortality, as described by Phillips, typically produce ladder fuels in the forms of hanging mosses and understory shrubs and trees, adding credence to this possibility.

This hypothesis seems reasonable, but it is difficult to support with existing evidence. Record keeping wasn't accurate before 1850 in the Coast Range, and it is possible that there were several other catastrophic fire events in the 1700s and early 1800s that went unnoticed by this research.

Likewise, record keeping and research has been poor regarding the age and location of trees killed in the Great Fires.

### 5.4.5 Summary: "weight of the evidence"

This research has shown a strong correlation between Indian burning practices in the Oregon Coast Range and subsequent Great Fire events; however, it has not demonstrated a cause and effect relationship between the two. Table 5.01 lists similarities and differences between Indian burning and catastrophic fire patterns in the Coast Range. This table shows a high rate of coincidence between the land management practices of precontact Indian communities and the causes, timing, boundaries, and general extent of subsequent forest fires in the same area.

Table 5.01 Comparison of Indian burning and catastrophic fire patterns.

| Fire Characteristics | Indian Burning | Catastrophic Fires |
|---|---|---|
| **Causes** | People | People |
| **Location of Ignitions** | Travel corridors & destinations | Travel corridors & destinations |
| **Wind directions** | Variable | East |
| **Seasons** | Late summer/early fall--Late winter/early spring | Late summer/early fall |
| **Frequency** | Daily | Years or centuries |
| **Extent** | 100,000s of acres annually | 100,000s of acres per event |
| **Boundaries** | Ridgelines, riparian areas, forested areas, ocean | Ridgelines, riparian/unforested areas, fog belt |
| **Wildlife Habitat** | Stable, sunny, high protein mosaic of grasslands, forests & woodlands | Sudden changes in wildlife demographics & habitats |

Despite the similarities in Table 5.01, however, little evidence exists--other than timing--to suggest that the end of Indian burning directly resulted in a series of catastrophic fires. A more reasonable assumption is that a combination of factors, including the cessation of Indian burning, the increasing decadence of existing stands, and changed land management strategies introduced by white immigrants, all played a role in the occurrence of Great Fire events between 1849 and 1951.

# 6. RECOMMENDATIONS

*I think the largest single need in American forest biology is the study of man's relation to forest land. Our foresters need to understand much more than most of them do about purely human motives and aspirations with respect to the land. They ought to become genuinely knowledgeable and respectful of people's economic, social, and aesthetic institutions.*
--Hugh Raup (Stout 1981: 93)

*The landscape that people saw when they came from Europe was a landscape that was literally an expression of the culture of the Indian people . . . There is absolutely no separation between the way the landscape looked in pre-contact times, the species composition and the structure of that forest, or that prairie, and the cultural needs and expressions of the Indian people. The land is an expression of the culture, as much as the arts and crafts and ceremonies . . . We need to bring back all the plants that were used for food and for baskets and for medicine. And we need to involve the children in the process.*
--Dennis Martinez 1993: 51

The following recommendations are based on the findings of this research and personal ideas and interests. They are divided into two sections: first, recommendations for further scientific research, as suggested by the hypothesis in Chapter 1, the method of multiple hypotheses, as defined in Chapter 2, and research conclusions, as discussed in Chapter 5; and second, recommendations for the practical application of methods and findings described in chapters 2, 3, and 4. The intended audience for the first section is agency scientists, university researchers, students, teachers, and interested individuals; for the second section, the intended audience is regional landowners, natural and cultural resource managers, policymakers, and interested public.

## 6.1 Additional Research

The research in this thesis was performed in accordance with the "method of multiple working hypotheses," first described by Chamberlin in 1890 (Chamberlin 1965). The following recommendations were partly developed by this process, as was much of Chapter 5.

### 6.1.1 Carrying capacity and precontact populations

Were pre-1492 human populations in the Pacific Northwest possibly much larger than previously thought? Reliable population figures for precontact Oregon Coast Range people do not exist.

Information has been developed for early historical time, particularly by Boyd (1990; 1999a), and reasonable estimates have been developed for late precontact time that is based on this evidence (e.g., Denevan 1992; Boyd 1999a), but nothing in these estimates addresses the great amount of surplus food resources or the millions of acres of "suitable habitat" that was available to people--yet apparently went unused, based on these estimates--in precontact times.

This research has shown that a tremendous number of seacoast, riverine, and inland plants and animals were readily available as food to people in this region. The habitats that these plants and animals lived in are some of the most productive in the world. Darby (1996) estimated that a single food plant, wapato, grown in a single area, Sauvies Island, during historical time was sufficiently productive to support a population of more than 18,000 people a year. And that would exclude the need for those people to fish for salmon, sturgeon, smelt, or eels, hunt seals, elk, deer, birds, or rabbits, to gather acorns or filberts, dig camas, pick berries, gather seed, or trade for a wide variety of other foods obtained along the ocean or further inland.

Even the most generous estimates of precontact Coast Range populations do not account for the establishment and maintenance of hundreds of thousands of contiguous acres of oak savannah, tens of thousands of acres of roots, bulbs, berries, seeds, fruits, and greens systematically arranged throughout all of the river drainages, or the elaborate network of international foot trails and canoe routes that allowed people in large numbers to move quickly and efficiently from place to place, whenever need or desire might dictate.

Based on the potential volume, variety, and availability of plant and animal foods, the moderate and generally predictable climate, and long stretches of low gradient foot trails and canoe routes, it seems possible that a much larger number of people may have lived in Oregon Coast Range communities in 1550, 1491, or earlier, than indicated by current estimates (e.g., Denevan 1992; Boyd 1999a). If so, a significant number of these people may have died from disease or other causes--or, at least, abruptly left the region--sometime before 1600 or 1650; as may be partly evidenced by the establishment of landscape and regional scale, even-aged tracts of conifer (mostly Douglas-fir) forestland during that time.

The best evidence of this hypothesis would be archaeological remains around the mouths of rivers; that is, where the cities of Portland, Salem, Corvallis, Eugene, Coos Bay, Newport,

Tillamook, and Astoria now stand. Most of that evidence is likely lost for all time due to urban development. A second line of evidence would be pollen counts with a focus on human disturbance, such as performed by Russell (1980) for historical time. Large infusions of pollen from major food plants such as oak, hazel, camas, huckleberries, brackenfern, and tarweed would be strong indicators of stable, potentially large, human populations. Rapid transitions of pollen-types during the past 500 years, from food plants to conifers, for example, might indicate major changes in local human populations, or at least significant changes in local culture or technology. If large numbers of people had lived here at one time, and if those people managed a much larger area of the landscape than previously thought, this information should be of particular value to landowners and resource managers concerned with ESA wildlife regulations or old-growth habitat management issues.

### 6.1.2 Indian burning practices and wildlife habitat

Could current valued wildlife populations, including game animals, songbirds, butterflies, and certain ESA "Threatened and Endangered" (T&E) species, benefit by reintroduction of precontact habitat patterns? Wild animals flourished in these environments, both in variety and numbers, as evidenced by anthropological, archaeological and historical records. These patterns were established and maintained through thousands of years of Indian burning practices, as evidenced by current and early historical vegetation patterns (See maps 3.02, 3.05, 3.06, 3.07, and 3.08). Areas of regular burning were high in protein and fiber production, in the forms of flowers, seeds, fruits, nuts, greens, roots, and bulbs, and offered a wide range of habit types connected by well-established land trails and water routes. Native wildlife either benefited or adapted to patterns of burning and tillage over time. People shared the fruits of this form of land management with wildlife, to varying degrees. Current assumptions regarding precontact vegetation patterns guide much of the coastal land management strategies for spotted owls, marbled murrelets, and coho salmon. If these assumptions prove inaccurate, or can be refined, then targeted native wildlife species should benefit from this line of research.

### 6.1.3 Fire history and forest management practices

Research for this dissertation identified a number of threatened or obscure sources of information, including fading memories, 90-year old hand-colored timber cruise maps, collections of 60- to

120- year-old photographs, agency reports, etc. Much of this information has value to resource managers today, particularly with recent considerations of past "reference" conditions, old-growth trees, and native wildlife.

Where can this information be found? Is it worth safekeeping, duplicating, or wider distribution? If so, how should it be cited? How can someone else locate this information next time it is needed?

Resource management decisions are often said to be based--ideally--on "the best information available." Decisions vary significantly between land ownerships and regulatory constraints. Private timberland management decisions are made with different amounts and types of information than Forest Service management decisions; which can vary significantly depending on whether the decision is being made for an AMA (Adaptive Management Area), a designated Wilderness, or a Douglas-fir plantation.

Fire history research is of interest to all forest resource managers because wildfire crosses boundaries of landownership and regulatory control with indifference. Decision-making during times of wildfire management, salvage logging operations, prescribed burning, or other fire-related activities, tend to affect all immediate landowners at some level, no matter whose land houses the event or activity. Decisions are usually made with entirely different objectives based on entirely different sets of knowledge and information. Yet the outcomes--both good and bad--are often shared by all. If more pertinent information was made available to all resource managers in a manner that was efficient and allowed for improved communications among affected parties, better and cheaper decisions could likely be made faster and easier than processes in place today.

This recommendation is based on the themes of information location, evaluation, duplication, storage, and retrieval, with a focus on the fire history of the Oregon Coast Range. The questions that need to be asked are the same as stated above, for each of the types of historical data listed below:

1) Oral histories. The memories and knowledge of key individuals represent our most threatened and endangered information. It has been 52 years since the last catastrophic fire in the Oregon Coast Range and almost 65 years since foresters and ranchers quit "fern-burning." However,

grass seed farmers began burning fields regularly in 1948, and foresters began broadcast burning tree planting sites shortly thereafter. There is a body of knowledge in those enterprises that is being lost. Agency personnel and local residents with knowledge of Coast Range fire history should be contacted, interviewed, recorded, and transcribed before more of them die, become disabled, or move away. These people are also a primary source for historical records and photos, and the best remaining interpreters of many existing documents.

2) Maps, photos, and texts. One-of-a-kind colored maps, photos, old reports, and other documents should be identified and safely stored until they can be scanned. Digitized information can then be made available to the general public as it is brought into existence, unless determined sensitive and constrained for some reason by the Freedom of Information Act.

3) Database index. All materials should be inventoried and listed on a computerized database. The proposed "5 Rivers Forest History" model (Bormann et al 2003) database would be a good model to develop and integrate with existing datasets. Databases used by the Siuslaw NF, PNW Research Station, OSU Research Forests, Benton County Historical Museum, USDI BLM, the county surveyors of Benton, Lane, and Lincoln counties, and the OSU Oregon Natural Heritage Program are all similar in structure and could be modified easily at this time to develop a common dataset of historical maps, photos, reports, survey notes, contact individuals, etc. The process of constructing such a database would help locate and correct data entry errors, as well as result in a more powerful common tool. A user-friendly database with search and display capabilities would make a good common product for these agencies to develop for internal information management use and public access via the Internet.

4) Stored files. Other useful files and records regarding reforestation, road building, prescribed fire, wildfire, logging, etc., are in storage at the USDA Waldport Ranger Station office, Benton County courthouse, Oregon State Forester's office, Oregon Historical Society, and other locations. They should be systematically inventoried, evaluated, and considered for use as time and resources allow. In the interim, they should be protected against fire, water, shredding, and misplacement.

Assuming that all of the listed recommendations are followed, an incredibly rich and varied inventory of information related to wildfire history and fire use in the Oregon Coast Range would begin to emerge almost immediately. The general public and agency decision makers would be able to locate most available documents on a given topic or location almost immediately. As more historical information becomes digitized, desired photographs, maps, GIS layers, etc., could be put online and directly linked to the database. Prescribed fire and wildfire management questions of a specific nature or for a specific location could then be answered with far more detailed information than is currently available. Decisions could readily be peer-reviewed, or interested parties alerted, through the same online network of agencies, landowners, businesses, and individuals.

#### 6.1.4 Geology and soils

Could the Bretz Floods of 12,800 BP to 15,000 BP have affected the geology of western Coast Range rivers and valleys? Earlier forest history research in the eastern Coast Range has demonstrated a strong correlation between Bretz flood events from 12,800 to 15,000 years ago, to subsequent patterns of human use and development (Zybach 1999: 12-13, 77). That research was focused on the Soap Creek Valley subbasin of the Luckiamute River drainage (see Map 2.01), but presented a pattern that seemed generally accurate for much of the Willamette Valley floodplain. It also disclosed a 400-foot elevation "shelf" that existed in the study area at a slightly higher elevation than the floodplain and could be found along most neighboring hills and valleys, including those of the adjacent Marys River basin (ibid: 13). This shelf also showed strong evidence of precontact and early historical use, as indicated by the types, numbers and locations of artifacts from those times.

Research performed in Alsea Valley (Zybach 2002b; see Map 2.01), despite being in an entirely different drainage than the Willamette River, also showed a 400-foot well-used "shelf." Allen (1989: personal communication) speculated that the Bretz floods might have been so deep they spilled out over the Yamhill River-Salmon River divide and partly entered the ocean by that route, rather than entirely by way of the Columbia River. However, he didn't think they could have filled (or left) the Valley swiftly enough to carve a shelf or shape a large rock outcropping (specifically, Coffin Butte), or that they may have spilled into another westward river, such as the Alsea. Swanson (2003: personal communication) discounts the possibility that the Bretz floods entered the Pacific with such force that a "backwater" effect could have inundated local rivers (including the Alsea) and affected topography in that manner. Perhaps the well-traveled "shelf" is not common throughout the Coast Range and its existence in three nearby drainages a coincidence; or perhaps it is common through the region, but has a different source of formation than the Bretz floods. In either instance, the apparent relationship between this geological formation and local human use is interesting.

Of similar interest is the existence of areas of soil in Lincoln County in which conifer trees seem to grow much slower than in other areas (Bormann 2003: personal communication). Previous assumptions were that trees grew relatively fast throughout the Coast Range, with most growth differences within species related to distance from the ocean, elevation, poor drainage, and available rainfall (Franklin and Dyrness ca. 1988); not soil differences, as exist with serpentine soils in other areas of western Oregon. It is also possible that the underachieving trees were the result of poorly selected off-site seed sources, such as used to plant Mount Hebo in the early 1900s (Isaac 1949). This possibility could bring past methods of estimating relative tree ages or mapping past fire boundaries based on tree diameters and height, into question.

Finally, there have been recent efforts to date lake sediment cores by layers of carbon. It is assumed that major carbon infusions are indicative of forest wildfire events; or at least, prescribed forest slash fires. What little work that has been done, however, is inconclusive. Long (1996: 35), for example, looked at "carbon peaks" for two cores drilled in Little Lake. The "short core" had peaks for 1763, 1910, 1967, and 1982. The 1763 date correlates very well with the Millicoma Fire. The 1910 date correlates to the fires (including local--see Rust 1992) of that year. However, the important years of 1849, 1868, 1902, and 1933 are missed entirely. One reason for this may be the location of Little Lake: to the east of the Great Fire events. Charcoal and cinders would have been traveling away from the Lake and toward the ocean on an east wind. The author also suggests that the methodology itself might be the cause for limited or poor conformance to historical records, when he notes:

> An examination for peaks at short core raw charcoal accumulation rates revealed four peaks: 1982, 1967, 1910, and 1763. However, the only documented wildfire occurring over the last 230 yr in the watershed, Lake Creek Fire of 1929, did not register as a charcoal peak in accumulation rates . . . The 1982 peak may be associated with a known Bureau of Land Management debris burn that same year. The 1967 peak, as well as the general increase in charcoal accumulation rates since 1950, may be the result of undocumented local or extra-local fires (Long 1996: 35).

A similar test--but for a lake in western Tillamook or Lincoln counties, where historical fires would be expected to leave distinct charcoal peaks--might be worth trying, to see if there is value in this method.

### 6.1.5 Tree rings and climate change

Did a Little Ice Age climate change affect the Oregon Coast Range? There is recent concern that chilling temperatures associated with the Little Ice Age were responsible for the ca. 1500 to 1850 establishment of forests in the western US, and that current plans should take that possibility into account. The importance of this concern is based on research showing that only a few degrees increase in temperature can determine whether a tree seed germinates or not. Hermann (1989: personal communication), for example, has developed evidence that in western Oregon even a two or three degree increase in temperature might stop natural reproduction of Douglas-fir. Cromack et al (2000), cites evidence that such a climate change has affected yellow pine reproduction in an area of Colorado, and Sessions et al (2003) voice concern a similar reproductive effect may have taken place in southwest Oregon.

In the past 20 years public concern has shifted from Nuclear Winter to Global Warming. Either type of climate change, of course, would affect the forest composition and fire history of the Oregon Coast Range. And, at some point—perhaps now--the climate is beginning to change again, whether towards warmer or colder weather. Available evidence seems to show that the Coast Range climate has been remarkably stable over the past 500 years, and doesn't indicate a Little Ice Age climate change. If true, this is likely due to the temperature moderating presence of the Pacific Ocean and low elevations. Pollen data seems to show the climate for the Coast Range has been fairly stable for more than two thousand years, but this type of information is too coarse to measure seasonal changes. Tree rings are a good source of proxy climate measures and show seasonality. Significant tree ring data has been gathered for the Coast Range, but little has been analyzed for climate change. It is possible that existing sets are sufficient to analyze for this purpose. Many of the trees cored over the past 50 or 60 years, however, have been logged, and much data has also been retrieved from freshly cut stumps. Because climate change is of national concern, a systematic coring of living trees in the Coast Range might provide an excellent dataset for analyzing past conditions and monitoring current changes.

### 6.1.6 Long-term experimental fire reintroduction

The second part of this chapter recommends large-scale reintroduction of Indian-type burning practices. However, little is known about the methods needed to make or manage such a change, or what the long-term effects on wildfire, plants, and animals will be. Arguments are given that reintroduced burning practices will greatly reduce wildfire damage, improve wildlife habitat, and benefit local economies, but these are untested assumptions. For example, Pendergrass has done the greatest amount of work regarding the reintroduction of Indian-type burning in eastern Coast Range wetland prairies, but only concludes that burning can "possibly reduce" the visual dominance of woody species over time (Pendergrass et al 1998) because her experiment only lasted two replications. Other efforts are being made to reintroduce Indian-type burning in conjunction with oak preservation by the City of Corvallis, USDI Salem BLM, and USDA Middle Fork Willamette NF, but these projects are in their beginning stages and do not seem to have an experimental component.

A series of long-term experiments should be established to measure the social, economic, and ecological costs and benefits of Indian-type burning practices in the Oregon Coast Range. The Forest Service, BLM, or the State of Oregon could make land available. Information could be provided to cooperating agencies and private landowners for local decision-making purposes. Opportunities and problems associated with reintroduction of fire are different than opportunities and problems associated with maintaining regular burning schedules. What short and long term effects these practices will have on local plant and animal populations is unknown, but can be measured with a well designed monitoring strategy. How will exotic plants and animals respond to regular burning? Should they be controlled or encouraged? How? Research questions, many of practical management value, can be addressed from a wide range of disciplines and perspectives with an experiment of this nature. Based on her own research, Pendergrass has concluded:

> If the management goal is to maintain the species balance that evolved with Native American burning practices, burning must be reintroduced. A long-term commitment to the reintroduction of fire as a management tool will be required to maintain native prairies in the [Willamette] valley (Pendergrass et al 1998: 309).

## 6.2 Reintroduction of Indian-type Burning Practices

The previous section of this chapter recommends long-term experimental studies of Indian-type burning, to test the objectives and methods needed for reintroduction. Enough knowledge exists at this time to make immediate reintroduction of these management practices both possible and desirable. The four key management steps (removal of excess fuels, control of undesired exotics, regular burning schedules, and smoke management) for reintroduction are fairly well understood and can only be refined through experimentation and technological change. The principal reasons for doing so are less well understood and not generally recognized. Anthropologists have listed more than 70 reasons for broadcast burning by Indians; however, many of those reasons might be used to argue against, rather than for, reintroduction. Several of the objectives (e.g., hunting, removing enemy cover, communication) are no longer practical, necessary, or even legal. Other objectives, such as weed control, site preparation, and fuel management, are in common with resource managers of today. A principal difference is that Indians had only fire and hand labor for meeting those objectives, whereas modern resource managers have many additional tools, including heavy equipment, chain saws, grazing animals, and herbicides. Common objectives, once achieved almost exclusively via fire, can now be achieved through a variety of means.

Indian burning practices were daily and seasonal events, and involved hundreds of thousands of acres of land a year. A key characteristic of Indian burning patterns is the absence of straight lines. Both fire boundaries and travel routes were established geologically, with river mainstems, ocean shorelines, and ridgelines guiding most human traffic to and away from homes or camps. Major travel routes appear to have been linked on a subbasin-scale, and most subbasins seem to have been bounded by local ridgeline and riparian trails. "Burning units" seem to have developed at a similar scale (usually several thousands of acres each), typically containing numerous patches and fields, and miles of trails. Any recommendation made to reintroduce Indian burning practices, therefore, must consider the great amount of land needed to reintroduce such practices on a landscape scale, and must also consider the least amounts of land needed to reintroduce such practices for educational, aesthetic, or cultural purposes.

Map 6.01 identifies Oregon Coast Range landowners most capable of reintroducing Indian-type burning on a landscape scale. Large Coast Range landowners include the US Forest Service, USDI BLM, and the Oregon Department of Forestry. As an agency, the US Fish and Wildlife

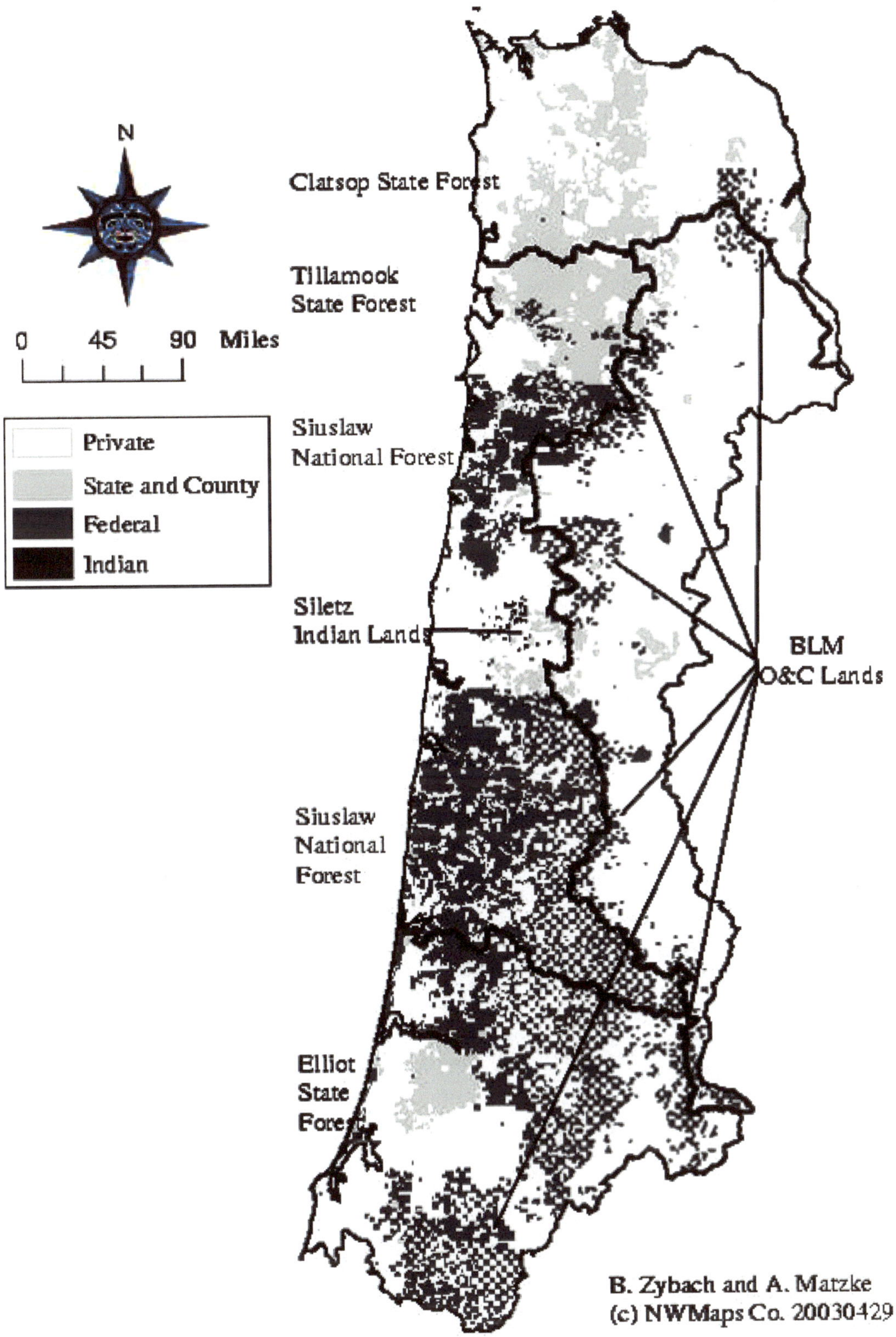


Map 6.01 Land ownership patterns of the Oregon Coast Range, 2003.

Service has control over many other federal lands, depending on the status of T&E species in an area. Other large landowners include private timberland owners in forested areas of the Range, and grass seed growers in the eastern grasslands. Both types of landowners are familiar with seasonal broadcast burning methods. Siletz, Coquille, Coos, Siuslawan, Kalapuyan, and Grand Ronde tribal members have demonstrated a cultural interest in reintroducing past land management activities, and in producing certain foods, medicines, and basketry materials via burning (Phillips 2003: personal communication; Kentta 2003: personal communication; Lake 2003: personal communication).

### 6.2.1 Aesthetics

Beauty is in the eye of the beholder. Many people enjoy a landscape covered with grasses and wildflowers; others enjoy the appearance and ambiance of deep shady forests; still others prefer a variegated landscape featuring both flowers and trees. In reading the literature of early journalists in Oregon, particularly those traveling the Willamette Valley, and the eastern Coast Range, two impressions seem to recur regularly: the first is that the appearance of the landscape--except after being recently burned--is incredibly beautiful; and the second is that the arrangement of plants on the landscape appears to have been done "as if by the hands of man." Two good examples of these observations are by Wilkes in 1841 and Neall in 1845:

> They found the country beautiful, and the land was rich. Their route lay over hills and through prairies. The hills were wooded, with large pines and a thick undergrowth of rose-bushes, Rubus, Dogwood, and Hazel. The prairies were covered with variegated flowers, and abounded in Nuttalia, Columbines, Larkspurs, and bulbous-rooted plants, which added to the beauty, as well as to the novelty of the scenery . . . The country in the southern part of the Willamette Valley, stretches out into wild prairie-land, gradually rising in the distance into low, undulating hills, which are destitute of trees, except scattered oaks; these look more like orchards of fruit trees, planted by the hand of man, than groves of natural growth (Wilkes 1845b: 218, 222).

> [Salt Creek Valley was] some two miles wide, perfectly level, and for all the world looking like some immense field of growing hay, not a speck of brush or undergrowth, and the rolling hills on either side divided here and there by small ravines emptying into the broad valley. These ravines, frequently fringed by a scattering growth of tall fir trees, formed a landscape of natural beauty most charming to behold. It was hard to realize that this was nature's own doing, unaided and untouched by the hand of man (Neall 1977: 41).

The point of these observations, of course, is that the environment Wilkes and Neall were viewing had been created and maintained by people, even if those people were no longer living and able to claim credit, and that the environment appeared beautiful, even by European standards. That is not to say that landscape aesthetics can be generally described and are common to all cultures, only that the regular burning actions of western Oregon Indians produced effects that were pleasing and appreciated by a large number of people from all kinds of backgrounds. Although Indian-type burning practices are not apt to be reintroduced entirely for aesthetic reasons, an expected outcome of the process would be a pleasant and attractive appearing landscape.

Map 6.02 compares subbasin-scale landscape patterns for the 90,000-acre "Alseya Valley" study area. The top pattern is a generalized and hypothetical landscape, as managed by the Alsi with prescribed fire in late precontact time (see Map 2.13). The middle pattern is based on the 1936 PNW Forest classification map (Andrews and Cowlin 1940). This pattern reflects 85 years of white settlement, including farming, ranching, logging, and sawmilling. The bottom pattern is a theoretical construct, based on what the landscape might look like if it were managed in accordance to the Northwest Forest Plan (FEMAT 1993). Although much of the area is owned by the USDA Siuslaw NF and the USDI BLM and is actually being managed in this manner (Bormann 2003: personal communication), most federal land is in forested areas, rather than along the river, where most historical and precontact pattern variation occurs. Each of these patterns represents an aesthetic choice: 1) Alsi management, 2) early white settlement, or 3) old-growth forest. Each of these choices requires a commitment to specific objectives over time and will result in significant differences in appearance and human, plant, and animal population numbers and locations. A key difference between the patterns is that the top two represent the works of people, whereas the FEMAT (1993) pattern represents an absence of people, much as stated in the 1964 Wilderness Act:

> A wilderness, in contrast with those areas where man and his own works dominate the landscape, is hereby recognized as an area where the earth and its community of life are untrammeled by man, where man himself is a visitor who does not remain. US Public Law 88-577 (September 3, 1964).

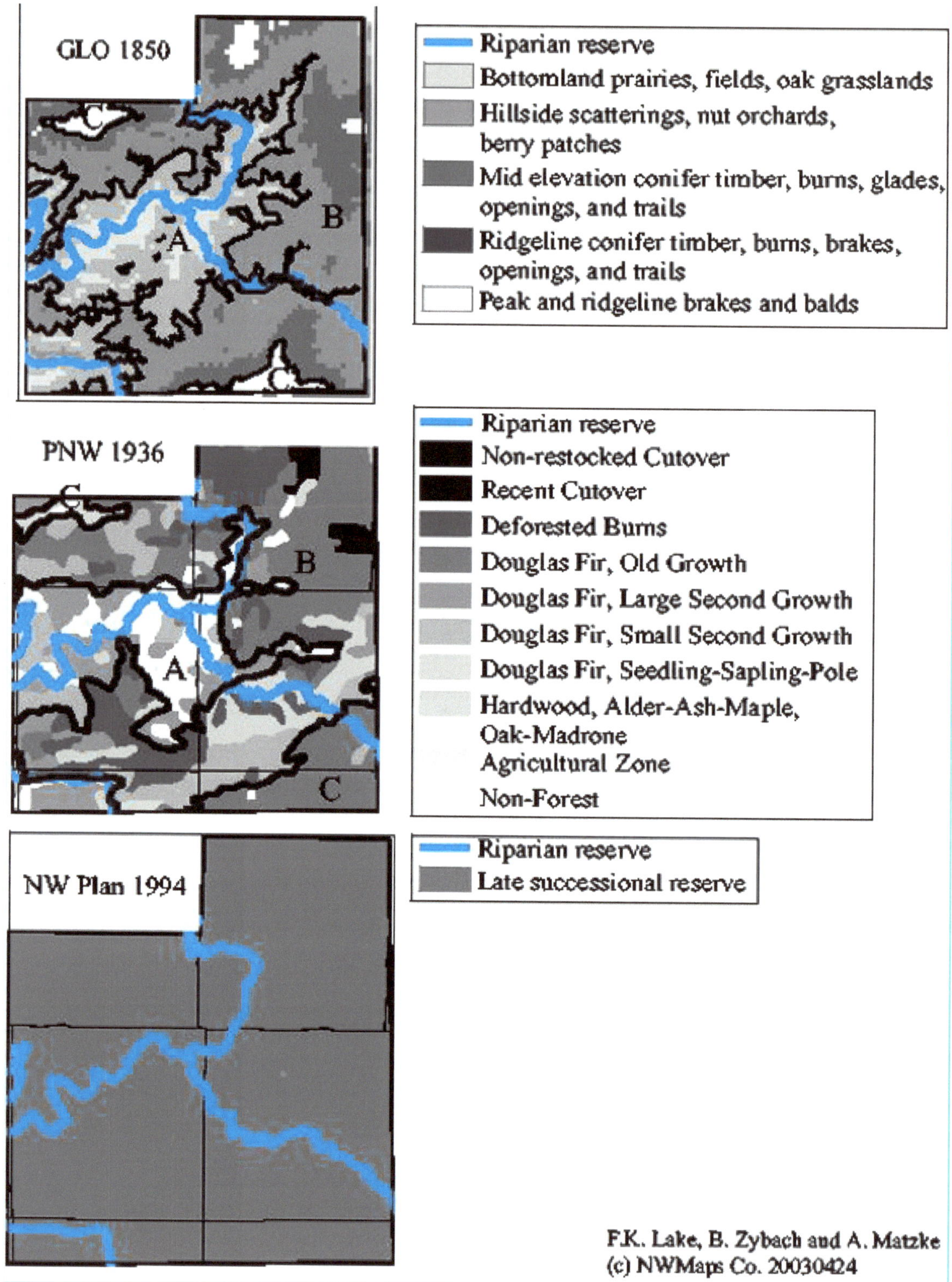


Map 6.02 Alsea Valley: GLO ca. 1800, PNW 1936, FEMAT 1993 patterns.

### 6.2.2 Cultural landscapes and events

The reintroduction of Indian-type burning practices does not necessarily mean that a landscape will be transformed to a representative pattern of the past. Burning would have to be performed systematically and purposefully to achieve that type of specific result. This research has shown that different burning patterns were associated with different groups of people in western Oregon during precontact time: Kalapuyans maintained vast tracts of oak savannah with annual broadcast burns; Chinookans did little more than gather firewood and maintain small prairies around their river-front communities of plank houses; while Yakonans maintained a complex network of root fields, balds, berry patches, and nut orchards connected by foot trails and canoe traffic. None of these people are still living, and their descendants are few, scattered, and have little current knowledge of the landscape management practices of their distant ancestors. There is little living memory or developed knowledge of the landscape patterns of plants and animals that typified the cultural heritage and homelands of Oregon Coast Range families in precontact time.

The question that might reasonably be asked is: If so little remains of the Coast Range people and cultures who created and maintained landscape patterns in precontact and early historical time, what could the purpose be in re-creating those patterns? There are at least four good reasons: 1) respect for the historical and cultural past, and the families that tended these lands in past times, 2) many plants and animals on federal and state T&E listings depended on, or were resilient to, generations of Indian burning, 3) the opportunity for annual and seasonal public service events and spectacles, 4) proven buffers and resiliency to catastrophic fire events.

Winkler and Bailey (2002: 2) argue that precontact landscapes are worth managing for their cultural value and, on federal lands, are required to be considered for this purpose. They cite the Secretary of the Interior's Standards that were revised in 1992 to consider "Rural Historic Landscapes" from an "ethnographic" perspective, as a directive to "protect" such landscapes. Kay (2002: 259) goes further and, although he is referring specifically to the 1964 Wilderness Act, infers there may be a moral or ethical obligation for considering such management:

> After years of study and internal debate, I have come to the conclusion that "wilderness" must be purged from our legal system and the American psyche. Not only am I opposed to the "creation" of any more officially designated wilderness, but all existing wilderness areas should be de-authorized, and the

> Wilderness Act repealed, because it is racist legislation. By permitting this deception to continue, not only do we ignore the genocide of the past, but we allow it to color our ongoing treatment of America's original owners (Kay 2002: 259).

Martinez (1993: 51), as cited at the beginning of this chapter, is more moderate than Kay, but promotes such landscape management for both cultural and educational values.
A second advantage of reintroducing precontact cultural landscapes is the opportunity for community events and spectacles. When ridgeline brakes or the balds of the highest peaks, such as Marys Peak, Prairie Peak, Grass Mountain, and Saddle Mountain, were burned, the fire would have been visible for miles. If the burn were done at night, the appearance would have been more pronounced and perhaps served other functions--such as hunting, religious celebrations, or seasonal markers--as well. These areas were probably burned about every one to three years, so these events would likely be both seasonal and annual. Modern equivalents might be Yule logs, 4th of July fireworks, and homecoming bonfires. When Jordan argues for reintroduction of Indian-type burning as the best method of retaining old prairie lands in the Midwest, he speaks in terms of "communal restoration rituals" in which communities might get together for large celebratory burns that "provide them with new and vital links to the historic landscape" and as a "basis for a new relationship between the immigrant peoples and the indigenous people who originally in habited the area" (Jordan 1993: 26).

An additional advantage for reintroducing precontact cultural landscape patterns into the environment is that these patterns have provided safe harbor for people and wildlife during catastrophic fires in the past (Robbins 1988: personal communication). Morris, for example, when reporting on the 1933 Tillamook Fire, noted:

> The southwest front was burning below the Trask House on the Trask River with great fury about 10:00 P. M. according to an old rancher who remained at this home on the river and watched the fire pass. The northwestern front moved even more rapidly. The flames were in the tree tops at the north-fork junction of the Wilson River at noon as reported by twenty-five fire fighters who stayed in an old homestead clearing there (Morris 1935: 16)

That these fires have maintained or regenerated similar patterns of tree, shrubs, grass, ferns, and bulbs almost immediately following wildfire events is shown by the even aged nature of Coast Range vegetation, dating to precontact times, and by the cultural legacy of cultural landscape patterns that has persisted beyond numerous wildfire events and land management processes to the present time. Part of that legacy is also represented by the numbers of people and wildlife that survived the Great Fires by safely watching from prairies, pastures, and town sites originally created and maintained by local, precontact families and communities.

### 6.2.3 Wildfire management

This dissertation examines the relationship between Indian burning practices and catastrophic forest wildfires. Because the practices and events happened, for the most part, in different time periods, it is difficult to consider the exact effects the reintroduction of Indian-type burning practices might have on future wildfire events. However, because there is evidence of catastrophic wildfires in precontact time, and because large areas of relict precontact landscape patterns have persisted through historical wildfire events, it is possible to make reasonable estimates of the consequences of reintroduced burning (Wade et al 1998).

A strategy of reintroducing burning with an objective of reducing frequency, damage, or management costs of subsequent wildfires should likely focus on two aspects of Indian-type burning: firewood gathering and situational patch burning. Most of the Coast Range land that was broadcast burned in precontact time has long since been converted to cities, towns, farms, and ranches. Most broadcast burning since World War II has been done by grass seed growers and industrial timberland owners. The scale of the majority of these burns is generally local, rather than landscape, but cumulatively, over a few seasons or year's time, they have affected a significant portion of the region's visible landscape.

Although people do not generally use wood heat in western Oregon any more, it is possible to efficiently gather firewood and kindling size fuels for cogeneration facilities, pulp, or mulch products. Just as Indians probably gathered fuels from around their homes and communities first, and quickly salvaged the products of nearby wind damage or snow breakage, modern home owners and communities can use similar tactics to remain free of fuel build-ups around homes and other structures. Patch burning of nuisance weeds or ladder fuels near the same localities can further reduce fire hazards.

### 6.2.4 Site preparation

Broadcast burning has long been used to prepare sites for reforestation and grass seed production in the Oregon Coast Range. These practices are similar to Indian-type burning in that they take place during the fall burning season and that they involve burning a broad surface of land for a variety of objectives. The principal difference is that Indians didn't broadcast burn forestlands (although wildfire did), and the timing of fires had to do more with plant combustion than with smoke management.

The value of broadcast burning to Douglas-fir regeneration and commercial grass seed growing is well known and recognized, but the practice has been greatly reduced in the past 20 years for two principal reasons: people have become increasingly concerned with air quality issues and even occasional intrusions of smoke into populated areas can be expected to generate negative publicity and possible lawsuits (Associated Press 2003: D10), and a fatal August 3, 1988 automobile accident that killed seven people and injured more than 35 others, was blamed on smoke from an Albany-area field burning. Subsequent legislation greatly reduced the amount of field burning that can take place as a direct result of the accident, and Willamette Valley grass seed growers now burn several hundreds of thousands of fewer acres annually than they did before 1988 (Venell 2003: personal communication).

Because the Willamette Valley, including the eastern Coast Range, is so heavily populated, it may be that occasional smoke intrusions and smoke visibility are issues that will tend to limit the amount of broadcast burning done in the future. Because forested areas are further from populated areas and because issues of wildfire control and T&E species habitat are better understood and are more popular topics than grass seed economics, it may be that broadcast burning can be returned to those environments at past, or increased, levels with less resistance or controversy.

Other methods of preparing sites for planting or seeding include scarification, plowing, herbicides, or hand labor. Fire isn't a critical tool for this purpose. Burning is often cheaper, doesn't disturb soil beneath the duff layer, and is generally the most desirable preparation for establishing or maintaining native plant populations (Zybach 1982; Walstad and Seidel 1990).

(However, site preparation alone is probably not a good justification for reintroduction of Indian-type burning practices.)

### 6.2.5 Weed and pest control

Two objectives of Indian broadcast burning were weed and pest control (Williams 2000: 45). Grass prairies were burned to reduce invasive species, such as Douglas-fir (Thompson 1991: 33), and living areas were occasionally burned to control fleas and other vermin (Stewart 2002). The same reasons are currently used to justify broadcast burning of grass seed fields, which cleanses the area of bacteria and insects harmful to crops (Venell 2003: personal communication). In Douglas-fir forest environments, regular burns have rarely been used for the specific purpose of weed or pest control, but these objectives are viewed as additional rationale for using fire to prepare sites. The principal objective of broadcast burning for site preparation is to remove logging debris, flammable materials, and other impediments to planting from an area (Walstad and Seidel 1990), and to expose mineral soil before seedlings are planted. Secondary objectives are to temporarily eliminate or control weeds that might compete with seedlings for light and moisture, and to remove browse that might attract animals that also eat young trees.

When patch burning was performed to rejuvenate a berry patch, clear a brake for fiddlehead and root harvest, or induce sprouting in certain shrubs or trees, secondary objectives likely included weed and pest control. It is expected that, should Indian-type burning be reintroduced in the Coast Range for reasons of wildfire management or wildlife habitat maintenance, weed control and pest outbreak costs may decrease as lands are regularly burned and monitored.

### 6.2.6 Traditional crops

The reintroduction of Indian-type burning practices would result in greater production of native crops, including firewood, berries, nuts, bulbs, and wildflowers. The plants that produce these crops have historical and cultural value because of their place in the landscape. The foods, medicines, dyes, construction materials, weaving fibers and other products made from native crops, also have historical and cultural values attached to their harvest, preparation, appearance, taste, texture, and aroma.

A certain amount of actual Indian burning takes place every year in the western US with a focus on producing traditional crops and products (Kentta 2003: personal communications; Lake 2003: personal communications). Often, burns are just a few acres in size, and are for a specific purpose, such as hazel sprout production, berry patch rejuvenation, or bulb field maintenance. This limited amount of burning for specialty crops is sufficient to meet current needs of education and special occasions, such as pow-wows, conferences, and family gatherings. Other native crops, such as firewood, wild blackberries, chittam, mushrooms, and decorative ferns and shrubs, have continuing market value (e.g., Kirkpatrick: ca. 1940), but have not been specifically managed with fire. Several of these commercial crops would also be enhanced with the reintroduction of regular burning and trail management practices. Other crops still, such as chokecherries, brackenfern roots, cat-tail stalks, or nettle fiber, have little if any demand at the present time, are not endangered, and are not being used for commercial or cultural purposes.

In sum, although there are reasonable historical and cultural reasons for using fire to produce native food and fiber crops, there doesn't appear to be any compelling market demand to do so. However, with the growing popularity of pow-wows and other cultural events in recent years, and a growing public awareness and appreciation for native foods and flowers, it is possible that market demands will increase or be created in the foreseeable future. The reintroduction of Indian-type burning practices for other reasons, such as wildfire control or T&E species habitat, would likely increase the amount and availability of native food and fiber crops and possibly increase markets for them as well. If berries, fern shoots, acorns, and tiger lilies continue to go unused by people, however, numerous wildlife species, such as birds, deer, and boomers, should benefit.

#### 6.2.7 Wildlife habitat: old-growth and grasslands

The reintroduction of Indian-type burning into the landscape would help protect old-growth forest stands and rejuvenate grasslands and shrub species used by wildlife. Pendergrass, for example, states that regular burning is "required" to maintain native wetland prairies in eastern Coast Range (Pendergrass et al: 1998). Regular burning can improve habitat conditions and long-term survival potential for T&E Coast Range species, as well. Schultz and Crone, for example, propose regular prairie burning to enhance Fender's Blue Butterfly (Schultz and Crone 1998). The same process can be used on other isolated prairies and meadows, such as the rare plant assemblages on

Mary Peak and Saddle Mountain (Detling 1954), or to create wildlife refuges during wildfire events (Krech 1999: 112). Robbins (1988: personal communication).

Old-growth forest stands can be protected by maintaining ladder fuel-free zones around their perimeters, as done during precontact times (although li8kely for different purposes). Ridgeline brakes and riparian meadows were often immediately adjacent to stands of old-growth trees in early historical time, continuing to the present. Major stands existed in the northern Coast Range, the headwaters of the Siletz, the headwaters of the Alsea, and the headwaters of the Coos, Coquille, and Umpqua rivers in the southern Coast Range. These stands often abutted prairies, meadows, and brakes, which kept fires from crowning into the larger trees during prescribed fire or wildfire events.

## 6.3 Discussion and Summary

This research, partly by design, has raised a number of questions that can be addressed by an interdisciplinary research approach, or through one of several different disciplines, including anthropology, geology, paleoecology, geography, forestry, history, wildlife sciences, and others. Some of the questions are almost purely academic: Were Coast Range Indian populations much greater in 1491 than previously estimated? Did the Bretz floods affect the geography of coastal Oregon rivers? Other recommended research focuses on the management of resources, including fire, wildlife habitat, climate monitoring, and information. All research costs money to conduct, and that will be the true test of these recommendations--whether any of them are ever funded or not.

### 6.3.1 Research: long-term experimentation

Probably the most important recommendation, from a management perspective, is the one for long-term Indian-type burning experiments, to be placed on public lands. Such experiments would provide good direction to agencies and landowners considering regular burning practices on their own lands, and they would provide critical information to others already engaged in such practices. There appear to be several good and compelling reasons to reintroduce Indian-type burning practices in the Oregon Coast Range:

1) much of the land is federal or state owned, and regulatory controls having to do with cultural landscapes and wildlife habitat might require prescribed burning in order to be in compliance with existing laws;

2) precontact cultural landscapes seem to be an effective buffer against many of the adverse effects of catastrophic wildfires, and may form a strategic tool for the management of future wildfire events;

3) the process would result in long-term local employment and development of refined vegetation and fire management skills that would add benefit to rural economies and be useful for curtailing or managing local wildfires as they occur;

4) increased burning regularity would produce an aesthetically pleasing environment, with good educational, recreational, historical, and cultural values, including seasonal and annual opportunities for community events and spectacles;

Compelling arguments also exist against the reintroduction of regular burning practices: it might cost too much; there is too great a risk of fire escapement; smoke is hazardous to human health; and the appearance of smoke is unsettling and unsightly. The following list of those concerns includes some discussion and response.

### 6.3.2 Reintroduction strategy

Remove all firewood and kindling sized wood material to a height of eight or ten feet around all structures. Coast Range Indians seemed to have their communities and major campgrounds in open, grassy areas. These areas were probably burned at least once a year. Kindling and firewood were probably gathered for a wide range from these places. Because many people today like to live in structures near large trees, and because they don't regularly remove tinder, kindling, and firewood from the areas around their structures, the likelihood of burning homes and possible fatalities is increasing through time.

Specific research is needed in the area of methods, for both the reintroduction of burning and the maintenance of burning schedules. To reintroduce burning, all excess fuels should be removed from the site first. Regularly burned areas tend to develop low fuel loads and burn quickly. This is sometimes called "cool burning." Douglas-fir is the principal invasive species in old prairies, berry patches, meadows, brakes, and balds in the Coast Range. Young Douglas-fir are pitchy and burn hot. Dead limbs and trees make excellent firewood, burn hot, and hold a flame or burn as coals for long periods. These types of fuels probably need to be removed before fire is reintroduced. Once they are removed, regular burning will see that they don't return.

Regular burning can encourage certain types of exotic plants, such as digitalis or pasture grass, to the possible detriment of native species. Changed burning cycles, herbicides, time, or hand labor might resolve the problem. Long-term experimental plots would provide an excellent opportunity to examine such problems. Also, Indian-type burning practices include the types of fuels that are gathered, that are burned in patches, and that are broadcast burned; the time of year in which the burn takes place; and the boundaries of the burn.

### 6.3.3 Reintroduction economics

Because Indian-type burning patterns are landscape scale in size, any program committed to their reintroduction--even on an experimental basis--would be expensive. One answer to this concern is that the most common invasive weed found throughout Coast Range prairies, meadows, brakes, balds, and berry patches is Douglas-fir. Other invasive species include Sitka spruce, red alder, and hemlock. Most of these species can be removed at a good profit, more than off-setting any costs needed to further prepare the land for reintroduced burning practices. The principal cost comes into play when the land must be burned again and again for decades, without showing any profitable timber growth the entire time. However, regular burning will produce other products, such as pulp, mulch, or firewood, and likely increase the production or profitability of secondary native forest crops, such as fern, salal, berries, and mushrooms. There is also a likely savings in the cost of insuring and managing future wildfire events; with a regularly maintained system of trails and grasslands, and an existing deployment of locally trained and equipped fire fighters, firefighting capabilities should be improved dramatically, and wildfire insurance premiums should be reduced accordingly. With strategic planning, the implementation of Indian-type burning practices could be profitable in the short term. The long-term picture is fuzzier, because the cost of maintaining trained fire and vegetation management crews would continue, even as hundreds of acres of timberland are removed from further production. An additional cost for private landowners is property taxes. Dramatic reductions in wildfire management costs and in fire damage to private property, however, might further account for crew maintenance costs; reduced losses to adjacent stands of timber might also be taken into account over time, so that the process may remain profitable.

#### 6.3.4 Reintroduction employment.

The reintroduction of Indian-type burning practices would result in immediate and sustained employment for numerous individuals in rural communities. These individuals would gain expertise in managing fire and maintaining vegetation. Communities would likely experience reduced future fire control costs due to local access, expertise, and equipment. Worker familiarity with local landscapes and fuels would like result in improved control strategies and operational policies. Damage would likely be minimized, but there would likely be occasional injuries and m sizes of burns are additional lines of inquiry.

#### 6.3.5 Risk of escapement

There have been several examples in the past few years where prescribed burns have become wildfires and caused extensive damage and loss of life. One of the main concerns to all landowners, but especially private landowners, is the liability attached to the individual, business, or agency responsible for wildfire damage resulting from a prescribed burn. This is a legitimate concern, and the news contains many examples of individuals and organizations suing over smoke intrusions from prescribed burns. However, numerous examples exist where landowners and communities have joined forces to reduce landowner risk and liability. For example, the California legislature legalized "Controlled Burning" in 1945 and appropriated $100,000 for trial tests and supervision. Within a few years' time, the California Division of Forestry was coordinating newly-formed county Range Improvement Associations of local landowners to conduct large scale broadcast burns on ranchlands. Most burning was done in late summer, there were few escapes, and adjacent landowners signed mutual "non-responsibility agreements" before burns could take place (Leonard 1954: 15). Similar actions could be taken at this time in western Oregon, perhaps with the use of existing watershed councils instead of forming new organizations.

### 6.3.6 Smoke

The concerns with smoke are two-fold: first, it poses a very real health hazard, especially when inhaled over long periods of time, or in toxic amounts; second, many people find the appearance of smoke disturbing or unsightly. Large, dark mushroom clouds, or thick smoke clouds blocking sunlight are two types of problem smoke conditions that can result from successful broadcast burning projects. People complain about their appearance and take legal and political actions to reduce their occurrence. When smoke goes out of control and affects breathing or visibility, and especially in instances where it is associated with fatalities such as the Albany accident, public opposition becomes much stronger and can affect burning practices and policies for long periods of time.

Prescribed fires are planned months and years in advance, and then burned under as ideal circumstances as possible. Under normal circumstances, smoke will go high into the air and disperse without people having to breathe it or without obstructing traffic. Unfortunately, the higher it goes, the more people can see it. Citizens with smoke related health problems could be easily mapped and inventoried so they would be notified in advance of local burns. In cases of smoke intrusion, they could be transported to another location or given such local assistance as needed, until the smoke had cleared. Burns could be scheduled on as few days' possible, for economic, ecological, and social reasons, and special occasions, events, holidays, and weekends could probably be avoided in most instances. That would reduce the likelihood of many people being exposed to smoke most of the time, particularly vacationers and weekend travelers. People hired to do burning and fight fires would likely want to take off those days as well.

Good planning, then, should take care of most concerns related to smoke about health and safety, but only some concerns about its appearance. This might be more of a public education issue than just a smoke management problem. Until recently, smoke has been a daily part of Coast Range history. Before gas, oil, coal, and electricity, everyone cooked and heated with wood. The smoke from home fires, campfires, and piles of burning leaves and brush became the beautiful sunsets of fall. In the late 1800s and early 1900s, fern burners burned spring pastureland for their goats, cows, pigs, horses, and other animals. After World War II, people began burning fields in the course of grass seed production, began burning logging units in order to plant tree seedlings, and began using wigwam burners to burn sawmill waste; and many families, even in cities, still cooked and heated with wood and burned piles of leaves and garden wastes in the fall. By the 1980s, wigwam burners had been banned, most families, even in rural areas, had stopped heating and cooking with wood, leaf burning, field burning, and slash burning were drawing negative public attention, and the Clean Air Act had been adopted. Even visible automobile emissions and cigarette smoke were becoming regulated. By this time, for the first time, a generation has lived to adulthood in an environment that has been largely smoke free, and even the appearance of smoke has been given a negative connotation. In order to be successful, the reintroduction of Indian-type burning practices will depend on public acceptance of smoke in the environment on a regular basis.

# REFERENCES

Citations are noted within the body of the text by the inclusion of page numbers for specific text quotations or references; with the words "personal communication" for spoken words, notes, letters, and emails; and with the word "map" for individual maps, atlases, or GIS layers.

Agee, James K. 1993. Fire Ecology of Pacific Northwest Forests. Island Press, Washington, DC: 493 pp.

Aikens, C. Melvin 1975. Archaeological Studies in the Willamette Valley, Oregon. University of Oregon Anthropological Papers, No. 8. Eugene, OR: 552 pp.

Aikens, C. Melvin 1993. Archaeology of Oregon. USDI Bureau of Land Management, Portland, OR: 134 pp.

Alban, David H. 1978. Influence on Soil Properties of Prescribed Burning Under Mature Red Pine. USDA Forest Service Research Paper NC-139, North Central Forest Experiment Station: 8 pp.

Albert 1848. Oregon Spectator, Oregon City, OR. September 7: pg. 1.

Aldrich, Frank T. 1973. A Chorological Analysis of the Grass Balds in the Oregon Coast Range. PhD dissertation, Oregon State University, Corvallis, OR: 156 pp.

Allen, John E. 1984. The Magnificent Gateway: A Layman's Guide to the Geology of the Columbia River Gorge. Timber Press, Forest Grove, OR: 144 pp.

Allen, John E. and Marjorie Burns, with Sam C. Sargent 1986. Cataclysms on the Columbia. Timber Press, Portland, OR: 211 pp.

Allen, John E. 1989. Personal communications. Oregon State Geologist, Department of Geology, Portland State University.

Allison, Ira S. 1946. "Early Man in Oregon: Pluvial Lakes and Pumice," The Scientific Monthly. Vol. LXII: 63-65."

Allison, Ira S. 1953. Geology of the Albany Quadrangle, Oregon. Bulletin No. 37, Department of Geology and Mineral Industries, State of Oregon. Portland, OR: 18 pp.

Alsea High School Students 1997. "Alsi Indians," IN: Kingfisher Magazine # 13. Linn-Benton Educational Service District, Albany, OR: 30-51.
Internet URL: www.ORWW.org/Reports/S/007J/0912/Kingfisher/ (October 15, 2003).

Anderson, E. William and Thomas E. Bedell 1987. Northwest Common-Name Check List of Plants. Agricultural Experiment Station, Oregon State University, Corvallis, OR: 45 pp.

Anderson, M. Kat 1996. "Tending the Wilderness," Restoration and Management Notes. Vol. 14, No. 2: 154-166.

Anderson, M. Kat 1997. "Indigenous Cultivation of Geophytes in California," Journal of Ethnobiology. Vol. 17, No. 2: 149-169.

Andrews, H. J. and Cowlin, Robert W. 1936. Forest Type Map, State of Oregon. Compiled by the Forest Survey Staff of the USDA Pacific Northwest Forest Experiment Station, Portland, OR. 4 sheets, scale 1 inch = 4 miles. (Andrews and Cowlin 1940).

Andrews, H. J. and Robert W. Cowlin 1940. Forest Resources of the Douglas-Fir Region. USDA Miscellaneous Publication No. 389, US Government Printing Office, Washington, DC: 169 pp. (Andrews and Cowlin 1936: map).

Associated Press 1997. "UW Professor Discovers Grove of Ancient Trees: Some of the Trees Have Been Dated Over 800 Years Old," Corvallis Gazette-Times. Corvallis, OR. (November 11): pg. A5.

Associated Press 2003. "Forest Service is Cited for Planned Burn's Smoke," The Oregonian. Portland, OR (July 7): D10.

Atzet, Thomas 2003. Personal communications. Forest ecologist, USDA Siskiyou National Forest.

Bagley, J. H. 1915. Benton County, Oregon 1915 Cruise, 2 Vol. Hand-colored timber cruise maps with tables, text, and black and white elevation maps for the timbered areas of Benton County, Oregon. Location in the County Clerk's Office, Corvallis, OR (Benton County Commissioners 1914).

Bailey, Tim and Jane Kertis 2002. Jim's Creek Savanna: The Potential for Restoration of an Oregon White Oak and Ponderosa Pine Savanna. Unpublished report, USDA Willamette NF, Oakridge, OR: 8 pp.

Ballou, Brian 2002. "A Chronology of Wildland Fire Protection in Oregon," IN: Fire in Oregon's Forests: Risks, Effects, and Treatment Options, Stephen A. Fitzgerald (ed.), Oregon Forest Resources Institute (OFRI), Portland, OR: 67-77.

Ballou, Brian 2003. Personal communications. Information Officer, Central Oregon District, Oregon Department of Forestry.

Balster, Clifford A. and Rowen B. Parsons 1969. "Late Pleistocene Stratigraphy, Southern Willamette Valley, Oregon," Northwest Science. Vol. 43, No. 3: 116-129.

Barry, J. Neilson 1926. "Broughton on the Columbia in 1792," Oregon Historical Quarterly. Vol. 27, No. 4: 397-411.

Baum, Willa K. 1985. Transcribing and Editing Oral History. American Association for State and Local History, Nashville, TN: 127 pp.

Benchmark Maps 1998. Oregon Road and Recreation Atlas. Benchmark Maps, Medford, OR: 111 pp.

Benda, Lee E. 1990. "The Influence of Debris Flows on Channels and Valley Floors in the Oregon Coast Range, USA," Earth Surface Processes and Landforms. Vol. 15: 457-466.

Benner, Patricia and James Sedell 1997. "Upper Willamette River Landscape: A Historic Perspective," IN: River Quality: Dynamics and Restoration. Antonius Laenen and David A. Dunnette (eds.), Lewis Publishers, New York, NY: 23-47.

Bennett, C. N. and V. B. Stanbury 1937. Reference Data Showing Aerial Surveys in Oregon. Works Projects Administration, Salem, OR: 16 pp.

Bennett, Stanley M. 1948. An Interpretation of Annual Rings of Douglas Fir. "Senior paper," School of Forestry, Oregon State College, Corvallis, OR: 38 pp.

Benton County Commissioners 1914. "Contract for Cruise of Timber Lands," Benton County Court Journal, Vol. 10. Corvallis, OR: 73-89 (Bagley 1915: map).

Berkley, Evelyn L. 2000. Temporal and Spatial Variability of Fire Occurrence in Western Oregon, A.D. 1200 to Present. MS thesis, University of Oregon, Eugene, OR: 110 pp.

Berreman, Joel V. 1937. Tribal Distribution in Oregon. Memoirs of the American Anthropological Association, No. 47, American Anthropological Association, Menasha, WI: 67 pp.

Blodgett, James 1984. Personal communications. Life-long resident of Blodgetts Valley, Benton County, and family historian.

Bonnicksen, Thomas M. 2000. America's Ancient Forests: From the Ice Age to the Age of Discovery. John Wiley & Sons, Inc., New York, NY: 594 pp.

Bormann et al. 2003. Personal communications. Principal Investigator, Five Rivers Landscape Management Study, Pacific Northwest Research Station.

Botkin, Daniel B. 1992. Discordant Harmonies: A New Ecology for the Twenty-first Century. Oxford University Press. New York, NY: 241 pp.

Botkin, Daniel B. 1996. Our Natural History: The Lessons of Lewis and Clark. The Berkeley Publishing Group, New York, NY: 300 pp.

Bourdo, Eric A. 1956. "A Review of the General Land Office Survey and of its Use in Quantitative Studies of Former Forests," Ecology. Vol. 37, No. 4: 754-768.

Bowen, Lola 1990. Family and Friends of Lobster Valley. Self-published. Copy available in Benton County Historical Museum Library, Philomath, OR: 95 pp.

Boyce, John S. and J. W. Bruce Wagg 1953. Conk Rot of Old-growth Douglas-fir in Western Oregon. Bulletin 4, Oregon Forest Products Laboratory and Oregon State Forestry Department, Salem, OR: 95 pp.

Boyd, Robert 1986. "Strategies of Indian Burning in the Willamette Valley," Canadian Journal of Anthropology. Vol. 5: 65-86.
Boyd, Robert 1999a. The Coming of the Spirit of Pestilence. University of Washington Press, Seattle WA: 403 pp.

Boyd, Robert (ed.) 1999b. Indians, Fire and the Land in the Pacific Northwest. Oregon State University Press, Corvallis, OR: 313 pp.

Boyd, Robert 1999c. "Strategies of Indian Burning in The Willamette Valley," IN: Indians, Fire and the Land in the Pacific Northwest, Robert Boyd (ed.), Oregon State University Press, Corvallis, OR: 94-138.

Boyd, Robert 1999d. "Conclusion: Ecological Lessons from Northwest Native Americans," IN: Indians, Fire and the Land in the Pacific Northwest, Robert Boyd (ed.), Oregon State University Press, Corvallis, OR: 292-297.

Bradley, Raymond S. and Philip D. Jones 1995. "Climate Since A.D. 1500: Introduction," IN: Climate Since A.D. 1500. Raymond S. Bradley and Philip D. Jones (eds.), Routledge, New York, NY: 1-16.

Braman, Tom 1987. "Indian Trail to Coast Teaches History, Ecology," Benton Bulletin. Philomath, OR (January 10): 1, 5.

Braun, E. Lucy 1950. Deciduous Forests of Eastern North America. The Blakiston Company, Philadelphia, PA: 596 pp.

Brown, James 2002. "Wildfire in Oregon: What are We Doing About It?" Presentation IN: Fire in Oregon's Forests: Assessing the Risks, Effects and Treatment Options. Oregon Forest Resources Institute Conference, Bend, OR: October 23, 2002.

Buan, Carolyn M. and Richard Lewis 1991. The First Oregonians. Oregon Council for the Humanities, Portland, OR: 128 pp.

Buckman, Robert E. 1964. "Effects of Prescribed Burning on Hazel in Minnesota," Ecology. Vol. 45, No. 3: 626-629.

Buckman, Robert E. 2003. Personal communications. Former Director of USDA Pacific Northwest Forest and Range Experiment Station.

Burke, Constance J. 1979. Historic Fires in the Central Western Cascades, Oregon. MS thesis, Oregon State University, Corvallis, OR: 130 pp.

Burnham, Howard J. 1952. "The Annual Society Address: For the Land's Sake," Oregon Historical Quarterly. Vol. 53, No. 4: 223-234.

Byram, R. Scott 2002. Brush Fences and Basket Traps: The Archaeology and Ethnohistory of Tidewater Weir Fishing on the Oregon Coast. PhD dissertation, University of Oregon, Eugene, OR: 359 pp.

Cabe, J. [?], Chas. F. Corgan [?], Lewis Cunningham, George Duncan, N. C. Ingram, W. F. McKinnon, J. C. Murray, F. W. Olinger, M. L. Stoddard, L. M. Tuttle, Roy Wherry 1913. Timber Cruise of Tsp. 4 N., Clatsop County, Oregon. Series of sectional hand colored timber cruise maps compiled during the summer of 1913 by a crew of timber cruisers operating separate from one another, but with the same set of general instructions. Fick & Martin Tillamook Forest Archive, Oregon Department of Forestry, Forest Grove, OR.

Carey, Charles H. 1971. General History of Oregon. Binfords & Mort, Publishers, Portland, OR: 916 pp.

Chamberlin, Thomas C. 1965. "The Method of Multiple Working Hypotheses," Science. (May 7): 754-759.

Chase, Alston 1995. In a Dark Wood. Houghton Mifflin Company, New York, NY: 535 pp.

Chen, Shu-Huei 1997. Characterization of Fire Effects on Forest Ecosystems in the Tillamook Forest, Oregon. MS thesis, Oregon State University, Corvallis, OR: 105 pp.

Christy, John A., Edward R. Alverson, M. P. Dougherty, S. C. Kolar, C. W. Kolar, and S. M. Hawes 2003. Historic Vegetation of the Willamette Valley, Oregon, 1851-1910. ArcView coverage, Version 6.0, Oregon Natural Heritage Information Center, Oregon State University, Corvallis, OR: 5 pp.

Christy, John 2002. Personal communication. Wetlands Ecologist, OSU Oregon Natural Heritage Program.

Christy, John A. and Edward R. Alverson 2003. Synopsis of Methods Used to Map Historic Vegetation in Oregon, Based on General Land Office Survey Notes. Oregon Natural Heritage Information Center, Oregon State University, Corvallis, OR: 5 pp.

Clar, C. Raymond 1959. California Government and Forestry: From Spanish Days until the Creation of the Department of Natural Resources in 1927. Division of Forestry, Department of Natural resources, State of California, Sacramento, CA: 623 pp.

Clyman, James 1984. Journal of a Mountain Man. Mountain Press Publishing Company, Missoula, MT: 295 pp.

Coblentz, Bruce E. 1980. "Production of Oregon White Oak Acorns in the Willamette Valley, Oregon," Wildland Society Bulletin. Vol. 8, No. 4: 348-350.

Collier, Charles 1891-1892. "Exteriors and Subdivisions of T. 13 S., R. 7 W.: Surveying Contract No. 572, dated April 18, 1891," IN: County Surveyor's Record, Benton County, Oregon n.d. Government Field Notes, Vol. 3. Corvallis, OR: 332-407.

Collier, Charles M. 1893. "South Boundary and Subdivisions of T. 14 S., R. 7 W.: Surveyors Contract No. 592, dated April 14, 1892," IN: County Surveyor's Record, Benton County, Oregon n.d. Government Field Notes, Vol. 3. Corvallis, OR: 168-230.

Collins, Lloyd R. 1951. The Cultural Position of the Kalapuya in the Pacific Northwest. MS thesis, University of Oregon, Eugene, OR: 138 pp. w/appendices.

Connolly, Thomas J. 1991. The Standley Site (35DO182): Investigations into the Prehistory of Camas Valley, Southwest Oregon. University of Oregon Anthropological Papers No. 43, Department of Anthropology and Oregon State Museum of Anthropology, University of Oregon, Eugene, OR: 243 pp.

Cook, Sherburne F. 1955. The Epidemic of 1830-1833 in California and Oregon. University of California Publications in American Archaeology and Ethnology, University of California Press, Berkeley, CA. Vol. 43, No. 3: 303-326.

Corvallis Gazette 1902. "The Fire in Alsea," Corvallis Gazette, September, 23: 2.

Corning, Howard M. 1973. Willamette Landings: Ghost Towns of the River. Oregon Historical Society, Portland, OR: 224 pp.

Coultin, Kevin G., Philip B. Williams and Patricia A. Benner 1996. An Environmental History of the Tillamook Bay Estuary and Watershed. Tillamook Bay National Estuary Project Technical Report 09-96. Garibaldi, OR: 68 pp. w/39 figures.

Cox, Ross 1957. The Columbia River. University of Oklahoma Press, Norman, OK: 398 pp.

Coyle, Jim 2002. Class "C" and Larger Forest Fires. Unpublished report, Oregon Department of Forestry, Salem, OR: 7 pp.

Cressman, Luther S. 1946. "Early Man In Oregon: Stratigraphic Evidence," The Scientific Monthly. Vol. LXII: 43-51.

Croker, Jr., Thomas C. 1987. Longleaf Pine: A History of Man and a Forest. USDA Forest Service, Atlanta, GA: 37 pp.

Cromack, Jr., Kermit, Johanna D. Lansberg, Richard L. Everett, Ronald Zeleny, Christian P. Giardina, Eva K. Strand, Tom D. Anderson, Robert Averill, Rose Smyrski 2000. "Assessing the Impacts of Severe Fire on Forest Recovery," Journal of Sustainable Forestry. Food Products Press, The Haworth Press, Inc. Vol. II, No. 1/2: 177-228.

Cromack, Jr., Kermit 2003. Personal communications. Soil scientist, Department of Forest Science, Oregon State University.

Crombie, Noelle 2003. "Molalla Man Feeling Tired, Sore After Being Struck By Lightning," The Oregonian. Portland, OR (April 15): C1.

Cronemiller, Lynn F. 1961. "Logging the Tillamook Burn," American Forests. (May): 29-31, 51-52.

Crosby, Jr., Alfred W. 1972. The Columbian Exchange: Biological Consequences of 1492. Greenwood Press, Westport, CN: 268 pp.

Dague, Charles I. 1934 (July). "The Weather of the Great Tillamook, Oregon, Fire of August 1933," Monthly Weather Review, Volume 62. USDA Weather Bureau, Government Printing Office, Washington, DC: 227-231.

Daly, Christopher and George Taylor 2002. PRISM Climate Mapping Program. www.ocs.orst.edu/prism/prism_new.html: November 4, 2002.

Darby, Melissa C. 1996. Wapato for the People: An Ecological Approach to Understanding the Native American Use of *Sagittaria Latifolia* on the Lower Columbia River. MA thesis, Portland State University, Portland, OR: 136 pp.

Davies, Kenneth G. (ed.) 1961. Peter Skene Ogden's Snake Country Journal 1826-27. Publications of the Hudson's Bay Record Society, London, England: 255 pp. w/2 maps.

Denevan, William M. 1992. The Native Population of the Americas in 1492. University of Wisconsin Press, WI: 340 pp.

Detling, LeRoy E. 1954. "Significant Features of the Flora of Saddle Mountain, Clatsop County, Oregon," Northwest Science. Vol. 28, No. 2: 52-60.

Deur, Douglas 2000. A Domesticated Landscape: Native American Plant Cultivation on the Northwest Coast of North America. PhD dissertation, Louisiana State University, LA: 361 pp.

de Velde, Paul and Henriette R. de Velde 1980. "South Lincoln County and Its Early Settlers: Charles L. Litchfield," IN: Lincoln County Lore. Lincoln County Historical Society, Newport, OR: 16-24.

DeVoto, Bernard 1953. The Journals of Lewis and Clark. Houghton Mifflin Company, Boston, MA: 504 pp.

Diaz, Nancy, Jane Kertis, and Dave Peter 1993. Quantitative Assessment of Current and Historic Landscape Structure (REAP #7). Unpublished report. USDA Forest Service, Pacific Northwest Region, Regional/Subregional Ecological Assessment, Portland, OR: 25 pp.

Dick, J. M. 1873. "Extract from Standard Parallel South between Tsps. 13 and 14 S., R. 9 W." Surveying Contract dated July 20, 1873: IN: County Surveyor's Record, Benton County, Oregon n.d. Government Field Notes, Vol. 3: 237-254.

Dickson, Evelyn M. 1946. Food Plants of Western Oregon Indians. MA thesis, Stanford University, CA: 218 pp.

Douglas, David D. 1904. "A Sketch of a Journey to the North-Western Parts of the Continent of North America, During the Years 1824, 5, 6, and 7," Oregon Historical Quarterly. Vol. 4, No. 3: 215-271; No. 4: 325-369.

Douglas, David D. 1905. "A Sketch of a Journey to the North-Western Parts of the Continent of North America, During the Years 1824, 5, 6, and 7," Oregon Historical Quarterly. Vol. 5, No. 1: 76-97.

Douthit, Nathan 1986. A Guide to Oregon South Coast History. River West Books, Coos Bay, OR: 157.

Downey, Tom, Darin Rilatos, Annette Sondenaa, and Bob Zybach 1993. The Siletz Eels: Oral History Interviews with Siletz Tribal Elders and Neighboring Residents Regarding the Decline in Siletz River Lamprey Populations. Preliminary report presented to the Tribal Council of the Confederated Tribes of Siletz Indians by the Oregon State University (OSU) Chapter of American Indians in Science & Engineering Society (AISES) and OSU College of Oceanic & Atmospheric Sciences' Native Americans in Marine Science (NAMS) Program, in Siletz, OR: 115 pp.

Drew, Linda G. (ed.) 1975. Tree-Ring Chronologies of Western America. Chronology Series 1, Laboratory of Tree-Ring Research, University of Arizona, Tucson, AZ: 45 pp.

Drucker, Philip 1965. "Contributions to Alsea Ethnography," IN: University of California Publications in American Archaeology and Ethnology, Volume XXXV, 1934-1943. A. l. Kroeber, R. H. Lowie, T. D. McCown, R. L. Olson (eds.), Kraus Reprint Corporation, New York, NY: 81-101.

Dunaway, David K. and Willa K. Baum (eds.), 1996. Oral History: An Interdisciplinary Anthology. Alta Mira Press, Walnut Creek, CA: 432 pp.

Dunaway, David K. 1996. "The Interdisciplinarity of Oral History," IN: Oral History, An Interdisciplinary Anthology. David K. Dunaway and Willa K. Baum (eds.), Alta Mira Press, Walnut Creek, CA: 7-22.

Elliott, Thomas C. 1917. "The Log of H. M. S. 'Chatham,'" Oregon Historical Quarterly. Vol. 18, No. 4: 231-243.

Elliott, Thomas C. 1928a. "Captain Robert Gray's First Visit to Oregon: Haswell's Log of Sloop Washington," Oregon Historical Quarterly. Vol. 29, No. 2: 162-188.

Elliott, Thomas C. 1928b. "Captain James Cook's Approach to Oregon," Oregon Historical Quarterly. Vol. 29, No. 2: 265-277.

Elliott, Thomas C. 1928c. "John Meares' Approach to Oregon," Oregon Historical Quarterly. Vol. 29, No. 2: 278-287.

Fagan, David D. 1885. History of Benton County, Oregon: Including Its Geology, Topography, Soil and Productions (etc.). A. G. Walling Publishing Company, Portland, OR: 531 pp.

Feller, M. C. 1998. "The Influence of Fire Severity, Not Fire Intensity, on Understory Vegetation Biomass in British Columbia," IN: Proceedings of the 13th Fire and Forest Meteorology Conference, Lorne, Australia 1996. IAWF, USA: 335-348.

FEMAT 1993. "Forest Ecosystem Management: An Ecological, Economic, and Social Assessment," IN Appendix A: Draft Supplemental Environmental Impact Statement on Management of Habitat for Late-Successional and Old-Growth Forest Related Species Within the Range of the Northern Spotted Owl. Presented to President Clinton by the Forest Ecosystem

Management Assessment Team (FEMAT), and used as the basis for the "Northwest Forest Management Plan," Washington, DC: texts w/appendices.

Fick, Lawrence R. and George E. Martin 1992. The Tillamook Burn: Rehabilitation and Reforestation. Oregon Department of Forestry, Forest Grove, OR: 320 pp.

Fick, Lawrence R. 2003. Personal communications. Tillamook State Forest historian, Forest Grove District Office, Oregon Department of Forestry.

Finnegan, Ruth 1996. "A Note on Oral Tradition and Historical Evidence," IN: Oral History, An Interdisciplinary Anthology. David K. Dunaway and Willa K. Baum (eds.), Alta Mira Press, Walnut Creek, CA: 126-134.

Frachtenberg, Leo J. 1913. Coos Texts. Columbia University Contributions to Anthropology, Washington, DC: 304 pp.

Frachtenberg, Leo J. 1914. Lower Umpqua Texts and Notes on the Kusan Dialects. Columbia University Press, New York, NY: 156 pp.

Frachtenberg, Leo J. 1920. Alsea Texts and Myths. Smithsonian Institution Bureau of American Ethnology, Bulletin 67, Government Printing Office, Washington, DC: 304 pp.

Franchere, Gabriel and Hoyt C. Franchere 1967. Adventure at Astoria, 1810-1814. University of Oklahoma Press, Norman, OK: 190 pp.

Franklin, Jerry F. and C. T. Dyrness ca. 1988. Natural Vegetation of Oregon and Washington. Oregon State University Press, Corvallis, OR: 452 pp.

Franklin, Jerry F., Thomas A. Spies, Robert Van Pelt, Andrew B. Carey, Dale A. Thornburgh, Dean Rae Berg, David B. Lindenmayer, Mark E. Harmon, William S. Keeton, David C. Shaw, Ken Bible, Joquan Chen 2002. "Disturbance and Structural Development of Natural Forest Ecosystems with Silvicultural Implications, Using Douglas-fir Forests as an Example," Forest Ecology and Management. 155 (2002): 399-423.

Franklin, Jerry F. 1981. "Vegetation of the Douglas-Fir Region," IN: Forest Soils of the Douglas-Fir Region. Paul E. Heilman, Harry W. Anderson, and David M. Baumgartner (eds.), Washington State University Extension Service, Pullman, WA: 92-112.

Freeman, James E. 1852. "Field Notes of the Survey of Subdivisions and Meanders of T. 10 S., R. 2 (4) W.: Surveying Contact No. 16, IN: County Surveyor's Record, Benton County, Oregon, Government Field Notes, Vol. 3. Corvallis, OR: 112-151.

French, David 1999. “Aboriginal Control of Huckleberry Yield in the Northwest," IN: Indians, Fire and the Land in the Pacific Northwest. Robert Boyd (ed.), Oregon State University Press, Corvallis, OR: 31-35.

Friedel, Dorothy E., Lynn Peterson, Patricia R. McDowell and Thomas L. Connolly 1989. Alluvial Stratigraphy and Human Prehistory of the Veneta Area, Long Tom River Valley, Oregon: The Final Report of the Country Fair/Veneta Archaeological Project. Report to the

National Park Service, US Department of the Interior, and Oregon State Historic Preservation Office. Oregon State Museum of Anthropology and Department of Geology, University of Oregon, Eugene, OR: 105 pp.

Fritts, H. C. and X. M. Shao 1995. "Mapping Climate Using Tree-Rings from Western North America," IN: Climate Since A.D. 1500. R. S. Bradley and P. D. Jones (eds.), Routledge, New York, NY: 269-295.

Fritz, Jean, Katherine Paterson, Patricia McKissack, Frederick McKissack, Margaret Mahy, and Jamake Highwater 1992. The World in 1492. Henry Holt and Company, New York, NY: 168 pp.

Gannett, Henry 1902. The Forests of Oregon. USDI US Geological Survey, Professional Paper No. 4, Series H, Forestry, 1, US Government Printing Office, Washington, DC: 36 pp. (Thompson and Johnson 1900: map)

Gesner, Alonzo 1891a. "Exterior Lines and Subdivisions of T. 14 S., R. 8 W." Contract #557, dated November 13, 1890, IN: County Surveyor's Record, Benton County, Oregon, Government Field Notes, Vol. 3: 238-257, 279-313.

Gesner, Alonzo 1891b. "Portion of West Boundary of T. 15 S., R. 8 W." Contract #557, dated November 13, 1890, IN: County Surveyor's Record, Benton County, Oregon, Government Field Notes, Vol. 3: 415-443.

Gibbs, George and Edward A. Starling 1851. Sketch of the Wallamette Valley. Prepared for US Board of Commissioners appointed to treat with the Indians of Oregon. Several copies of various scales located in the Oregon State University Valley Library Map Room, Corvallis, OR: 1 sheet.

Gilkey, Helen M. and La Rea J. Dennis 2001. Handbook of Northwestern Plants. Oregon State University Press, Corvallis, OR: 494 pp.

Gilsen, Le 1989. Luckiamute Basin Survey. Oregon State Historic Preservation Office, Salem, OR: 179 pp.

Goodchild, Michael F. 1987. "A Spatial Analytical Perspective on Geographical Information Systems," International Journal of Geographical Information Systems. Vol. 1, No. 4: 327-334.

Gordon, Harvey and Josiah W. Preston 1854. "West Boundary of T. 13 S., R. 6 W.: Surveying Contract No. 40, dated Feb. 22, 1854," IN: County Surveyor's Record, Benton County, Oregon, Government Field Notes, Vol. 3. Corvallis, OR: 271-275.

Gottesfeld, Leslie M. Johnson 1994. "Aboriginal Burning for Vegetation Management in Northwest British Columbia," Human Ecology. Vol. 22, No. 2: 171-188.

Grant, C. Frank 1990. The Grants Of Grant Valley. C. Frank Grant, Publisher, Corvallis, OR: 174 pp.

Grantham, J. B. 1953. The Relationship of Lumber Recovery to Log Quality in 29 Old-Growth Douglas-fir Trees of the Oregon Coast Range. Report No. G1, Oregon Forest Products Laboratory, Corvallis, OR: 18 pp.

Graumlich, Lisa J. 1987. "Precipitation Variation in the Pacific Northwest (1675-1975) as Reconstructed from Tree Rings," Annals of the Association of American Geographers. Vol. 77, No. 1: 19-29.

Grefe, Ray, Robert Cowlin, and Dahl Kirkpatrick 1943. Tillamook Burn Study: June - 1943. Fick & Martin Tillamook Forest Archive. Unpublished report with "Draft and Incomplete Copy . . ." typed on cover page. Authors' names handwritten in pencil on sticky tab attached to same page; otherwise un-noted. USDA Forest Service Region Six, Portland, OR: 81 pp. w/maps and appendices.

Grigg, Laurie D. and Cathy Whitlock 1998. "Late-Glacial Vegetation and Climate Change in Western Oregon," Quaternary Research. Vol. 49: 287-298.

Habeck, James R. 1961. "The Original Vegetation of the Mid-Willamette Valley, Oregon," Northwest Science. Vol. 35, No. 2: 65-77.

Hagenstein, William D. 2003. Personal communications. Retired Executive Vice President of the Industrial Forestry Association.

Haines, Donald A. 1982. "Horizontal Roll Vortices and Crown Fires," Journal of Applied Meteorology. Vol. 21, No. 6: 751-763.

Hall, Roberta L. 1992. "Language and Cultural Affiliations of Natives Residing Near the Mouth of the Coquille River Before 1851," Journal of Anthropological Research, Vol. 48: 165-184.

Hall, Roberta L. 1995. People of the Coquille Estuary. Words & Pictures Unlimited, Corvallis, OR: 223 pp.

Hall, Roberta L. 2001. Nah-so-mah Village, Viewed Through its Fauna. Department of Anthropology, Oregon State University, Corvallis, OR: 59 pp.

Hansen, Henry P. 1941. "Paleoecology of a Peat Deposit in West Central Oregon," American Journal Of Botany: 206-212.

Hansen, Henry P. 1942. "A Pollen Study of Lake Sediments in the Lower Willamette Valley of Western Oregon," Bulletin of the Torrey Botanical Club. Vol. 69, No. 4: 262-280.

Hansen, Henry P. 1944. "Further Pollen Studies of Peat Bogs on the Pacific Coast of Oregon and Washington," Bulletin of the Torrey Botanical Club. Vol. 71, No. 6: 627-636.

Hansen, Henry P. 1947. Postglacial Forest Succession, Climate, and Chronology in the Pacific Northwest. Transactions of the American Philosophical Society, Philadelphia, PA. Vol. 37, Part 1: 130 pp.

Hansen, Henry P. 1949. "Pollen Content of Moss Polsters in Relation to Forest Composition," The American Midland Naturalist. Vol. 42, No. 2: 473-479.

Hansen, Henry P. 1961. Cycles and Geochronology. California Academy of Sciences Occasional

Papers, No. 31, San Francisco, CA: 24 pp.

Hansen, Henry P. 1967. "Chronology of Postglacial Pollen Profiles in the Pacific Northwest (U.S.A.)," Review of Paleobotany and Palynology, Amsterdam, Netherlands: 103-105.

Hansen, Henry P. 1989. Personal communications. Retired Dean Emeritus of the Graduate School and Professor Emeritus of Palynology, Oregon State University.

Haskin, Leslie L. 1934. Wild Flowers of the Pacific Coast. Metropolitan Press, Portland, OR: 407 pp.

Haskin, Leslie L. (ed.) 1948. Lincoln County Lore: From the Journal of Lieut. Theodore Talbot, U.S.A. On His Journey Through Lincoln County and Along the Oregon Coast in 1849. Reprinted from the files of Newport News, Publication No. 1, Lincoln County Historical Society, Newport, OR: 14 pp.

Hathorn, Dennis 1854a. "Field Notes of the Surveys of Donation Land Claims in Township 11 South, Range 5 West: Surveying Contract No. 19, dated October 8, 1853," IN: County Surveyor's Record, Benton County, Oregon, Vol. 4: Donation Land Claim Index. Corvallis, OR: 75-122.

Hathorn, Dennis 1854b. "Field Notes of the Surveys of Donation Land Claims in Township 10 South, Range 4 West and Township 10 South, Range 5 West: Surveying Contract No. 30, dated May 10, 1854," IN: County Surveyor's Record, Benton County, Oregon, Vol. 4: Donation Land Claim Index. Corvallis, OR: 30-54.

Hathorn, Dennis 1856a. "Portion of West Boundary and Subdivisions of Fractional T. 13 S., R. 7 W.: Surveying Contract No. 63, dated March 22, 1856," IN: County Surveyor's Record, Benton County, Oregon, Government Field Notes, Vol. 3. Corvallis, OR: 323b-332.

Hathorn, Dennis 1856b. "Portion of Subdivisions of Fractional T. 13 S., R. 8 W.: Surveying Contract No. 63, dated March 22, 1856," IN: County Surveyor's Record, Benton County, Oregon, Government Field Notes, Vol. 3. Corvallis, OR: 411-413.

Hathorn, Dennis 1856c. "Third Standard Parallel South through R. 7 W. and Range 8 West, West Boundary of T. 14 S., R. 7 W., and Subdivisions of Fractional T. 14 S., R. 7 W.: Surveying Contract No. 63, dated March 22, 1856," IN: County Surveyor's Record, Benton County, Oregon, Government Field Notes, Vol. 3. Corvallis, OR: 151-164, 233-237.

Hawes, S. M., J. A. Hiebler, E. M. Nielsen, C. W. Alton and John A. Christy 2003. Historic Vegetation of the Pacific Coast, Oregon, 1851-1910. ArcView coverage, Version 1.0. Oregon Natural Heritage Information Center, Oregon State University, Corvallis, OR: 5 pp.

Heilman, Paul E., Harry W. Anderson, and David M. Baumgartner (eds.) 1981. Forest Soils of the Douglas-fir Region, Washington State University Cooperative Extension Service, Pullman, WA: 298 pp.

Heinrichs, Jay 1983. "Tillamook," Journal of Forestry, Vol. 81, No. 7: 442-446.

Heinselman, Miron L. 1981. "Fire and Succession in the Conifer Forests of Northern North America," IN: Forest Succession: Concepts and Application, C. Darrell West, Herman H. Shugart, and Daniel D, Botkin (eds.), Springer-Verlag, New York, NY: 374-405.

Heinselman, Miron L. 1996. The Boundary Waters Wilderness Ecosystem. University of Minnesota Press, Minneapolis, MN: 334 pp.

Henderson, Jan A., David H. Peter, Robin D. Lesher, and David C. Shaw 1989. Forested Plant Associations of the Olympic Peninsula. USDA Forest Service Pacific Northwest Region, R6 ECOL Technical Paper 001-88: 502 pp.

Henderson, Jan A. 1990. Trends in Amount of Old-Growth Forest for the Last 1000 Years in Western Oregon and Washington. Report prepared in response to a request from Congressman Robert F. Smith to US Forest Service Chief F. Dale Robertson dated September 28, 1989: 7 pp.

Hennessey, Jr., Joseph P. and Charles D. Craig 1976. The Field Burning Climatology of the Willamette Valley. Technical Report No. 76-1, Department of Atmospheric Sciences, Oregon State University, Corvallis, OR: 73 pp.

Hermann, Richard K. 1976. "Man And Forests: A Prodigal Relation," In: Forests And Future Resource Conflicts. Department of Printing, Oregon State University, Corvallis, OR: 29-51.

Hermann, Richard K. 1985. The Genus Pseudotsuga: Ancestral History And Past Distribution. Special Publication 2b, Oregon State University Forest Research Laboratory, Corvallis, OR: 32 pp.

Hermann, Richard K. 1988. Personal communications. Professor emeritus, Department of Forest Science, Oregon State University.

Heusser, Calvin J. 1960. Late-Pleistocene Environments Of North Pacific North America. American Geographical Society, New York, NY: 308 pp.

Hitchcock, C. Leo and Arthur Cronquist 1973. Flora of the Pacific Northwest. University of Washington Press, Seattle, WA: 730 pp.

Hodges, M. Constance 1979. Lords of Themselves: A History of East Lincoln County, Oregon. Delcon Historical Publications, Harlan, OR: 292 pp.

Hoffman, Alice 1996. "Reliability and Validity in Oral History," IN: Oral History, An Interdisciplinary Anthology, David K. Dunaway and Willa K. Baum (eds.), Alta Mira Press, Walnut Creek, CA: 87-93.

Hyde, George W. 1852a. "Field Notes of the Survey of Second Standard Parallel South through Range 5 W.: Surveying Contract No. 14, dated March 25, 1852," IN: County Surveyor's Record, Benton County, Oregon, Government Field Notes, Vol. 3. Corvallis, OR: 377-388.

Hyde, George W. 1852b. "Field Notes of the Survey of Subdivisions of T. 11 S., R. 5 W.: Surveying Contract No. 19, dated May 27, 1852," IN: County Surveyor's Record, Benton County, Oregon, Government Field Notes, Vol. 3. Corvallis, OR: 389-427.

Impara, Peter C. 1997. Spatial and Temporal Patterns of Fire in the Forests of the Central Oregon Coast Range. PhD dissertation, Oregon State University, Corvallis, OR: 287 pp.

Isaac, Leo A. and G. S. Meagher 1936. Natural Reproduction on the Tillamook Burn Two Years After the Fire. Fick & Martin Tillamook Forest Archive. Pacific Northwest Forest Experiment Station, Portland, OR: 20 pp.

Isaac, Leo A. and G. S. Meagher 1938. Natural Reproduction on the Tillamook Burn Four Years After the Fire. Fick & Martin Tillamook Forest Archive. Pacific Northwest Forest Experiment Station, Portland, OR: 16 pp.

Isaac, Leo A. 1949. Better Douglas Fir Forests from Better Seed. University of Washington Press, Seattle, WA: 65 pp.

Ivy, Don B. and R. Scott Byram (eds.) 2002. Changing Landscapes: "Sustaining Traditions." Proceedings of the 5th and 6th Annual Coquille Cultural Preservation Conferences, 2001 and 2002, Coquille Indian Tribe, North Bend, OR: 190 pp.

Jacobs, Melville 1945. Kalapuyan Texts. University of Washington Publications in Anthropology, No. 11, Seattle, WA: 394 pp.

Johannessen, Carl L., William A. Davenport, Artimus Millett and Steven McWilliams 1971. "The Vegetation Of The Willamette Valley," Annals Of The Association Of American Geographers, Vol. 61, No. 2: 286-305.

Jones, Philip D. and Raymond S. Bradley 1995. "Climatic Variations Over the Last 500 Years," IN: Climate Since A.D. 1500, Bradley, Raymond S., and Philip D. Jones (eds.), Routledge, New York, NY: 649-665.

Jordan III, William R. 1993. "Rituals of Restoration," The Humanist. Vol. 53, No. 6: 23-26.

Juday, Glenn P. 1976. The Location, Composition, and Structure of Old-Growth Forests of the Oregon Coast Range. Phd. dissertation, Oregon State University, Corvallis, OR: 206 pp.

Kane, Paul 1925. Wanderings of an Artist Among the Indians of North America, from Canada to Vancouver's Island and Oregon through the Hudson's Bay Company's Territory and Back Again. The Radisson Society of Canada, Ltd., Toronto, Canada: 325 pp.

Kay, Charles E. 1995. "Aboriginal Overkill and Native Burning: Implications for Modern Ecosystem Management," Western Journal of Applied Forestry. Vol. 10 (4): 121- 126.

Kay, Charles E. 2002: "Afterword: False Gods, Ecological Myths, and Biological Reality," IN: Wilderness and Political Ecology: Aboriginal Influences and the Original State of Nature, Charles E. Kay and Randy T. Simmons (eds.), The University of Utah Press, Salt Lake City, UT: 238--261.

Keeley, Jon E. 2002. "Native American Impacts on Fire Regimes of the California Coastal Ranges," Journal of Biogeography. Vol. 29: 303-320.

Kent, William E. 1977. The Siletz Indian Reservation, 1855-1900. MS thesis (1973), Portland State University, Portland, OR. Lincoln County Historical Society, Publication No. 15, Newport, OR: 53 pp.

Kentta, Robert 2003. Personal communication. Cultural Resources Director for the Confederated Tribes of Siletz Indians.

Kimmerer, Robin W. and Lake, Frank K. 2001. "The Role of Indigenous Burning in Land Management," Journal of Forestry. Vol. 99, No. 11: 36-41.

Kimmerer, Robin W. 2003. Gathering Moss: A Natural and Cultural History of Mosses. Oregon State University Press, Corvallis, OR: 168.

Kimmey, J. W. and R. L. Furniss 1943. Deterioration of Fire-Killed Douglas-Fir. USDA Technical Bulletin No. 851, US Government Printing Office, Washington, DC: 55pp.

Kimmins, J. P. 1987. Forest Ecology. MacMillan Publishing Company, New York, NY: 531 pp.

King, A. R. 1964. "Characteristics of a Fire-Induced Tornado," Australian Meteorological Magazine. No. 44: 1-9.

Kirkpatrick, Dahl ca. 1940. A History of the Siuslaw National Forest, Oregon, as of December 31, 1939. USDA Forest Service, US Government Printing Office, Washington, DC: 43 pp. w/appended 1950-52 annotations.

Krauss, Michael E. 1990. "Kwalhioqua and Clatskanie," IN: Handbook of North American Indians, Volume 7: Northwest Coast (edited by Wayne Suttles). William C. Sturtevant, series ed., Smithsonian Institute, Washington, DC: 530-532.

Krech, Shepard, III, 1999. The Ecological Indian: Myth and History. W. W. Norton & Company, New York, NY: 318 pp.

Krewson, E. A. 1955. Tioga's Pigs. Binfords and Mort, Publishers, Portland, OR: 140 pp.

Lake, Frank K. and Bob Zybach (In Review). "Historical Indian Burning Practices, Oregon Coast Range 1770 to 1848." Paper submitted to Western Journal of Applied Forestry, November, 2003.

Lake, Frank K. 2002. "The Role of Indigenous Burning for Land Management" (oral presentation), IN: Traditional Use of Fire and the National Fire Plan Conference Proceedings, Grand Ronde and Sweet Home, OR. Prepared by National Federation of Federal Employees, Local 457 for the National Network of Forest Practitioners: 5-6.

Lake, Frank K. 2003. Personal communications. Graduate student in Environmental Sciences, Pacific Traditional Ecological Knowledge Program, Oregon State University.

LaLande, Jeff and Reg Pullen 1999. "Burning for a 'Fine and Beautiful Open Country,'" IN: Indians, Fire and the Land in the Pacific Northwest, Robert Boyd (ed.), Oregon State University Press, Corvallis, OR: 255-276.

Latham, Don 1991. Lightning Flashes From a Prescribed Fire-Induced Cloud," Journal of Geophysical Research. Vol. 96, No. D9: 17,151-17,157.

Lee, Daniel and Joseph Frost 1968. Ten Years in Oregon. Ye Galleon Press, Fairfield, WA: 344 pp.

Lee, Jennifer A. and Royal G. Jackson 1984. Thurman J. Starker: Educator, Timberman, Civic Leader. Horner Museum Oral History Program, Monograph No. 3, Horner Museum, Oregon State University, Corvallis, OR: 84 pp.

Leiberg, John B. 1900. "Cascade Range Forest Reserve, Oregon, from Township 28 South to Township 37 South, Inclusive; Together with the Ashland Forest Reserve and Adjacent Forest Regions from Township 28 South to Township 41 South, Inclusive, and from Range 2 West to Range 14 East, Willamette Meridian, Inclusive," IN: Twenty-First Annual Report of the United States Geological Survey, Part V.--Forest Reserves. US Department of the Interior, Washington, DC: 209-498.

Leopold, Estella B. and Robert Boyd 1999. "An Ecological History of Old Prairie Areas in Southwestern Washington," IN: Indians, Fire and the Land in the Pacific Northwest. Robert Boyd (ed.), Oregon State University Press, Corvallis, OR: 139-163.

Lewis, Henry T. and Theresa A. Ferguson 1999. "Yards, Corridors, and Mosaics: How to Burn a Boreal Forest," IN: Indians, Fire and the Land in the Pacific Northwest, Robert Boyd (ed.), Oregon State University Press, Corvallis, OR: pp. 164-184.

Lindgren, Harry A. 1935. Report on the Feasibility of Grazing Livestock on the Cutover Timber Lands of Columbia and Clatsop Counties, Oregon. Oregon State College Extension Service, Corvallis, OR: 19 pp.

Long, Colin J., Cathy Whitlock, Patrick J. Bartlein, and Sarah H. Millspaugh 1998. "A 9000-Year Fire History from the Oregon Coast Range, Based on a High-Resolution Charcoal Study," Canadian Journal of Forest Research. Vol. 28: 774-787.

Long, Colin J. 1996. Fire History of the Central Coast Range, Oregon: A ca. 9000 Year Record from Little Lake. MA thesis, University of Oregon, Eugene, OR: 147 pp.

Longworth, Basil N. 1972. Memorandum on Thoughts, Reflections, and Transactions as Described by Basil Nelson Longsworth on his Journey from Washington Township, Guernsey County, Ohio to Oregon in the Summer of 1853. Le Galleon Press, Fairfield, WA: 43 pp.

Longwood, Franklin R. 1940. A Land use History of Benton County, Oregon. MF thesis, Oregon State College, Corvallis, OR: 139 pp.

Losey, Robert J. 2000. "Oral Tradition of Earthquakes and Tsunamis on the Central Cascadia Coast: Variations of Accounts and Relations to Historically Observed Patterns across the Northwest Coast," Changing Landscapes: "Telling Our Stories." Proceedings of the 4th Annual Coquille Cultural Preservation Conference, 2000, Jason Younker, Mark A. Tveskov, and David G. Lewis (eds.), The Coquille Indian Tribe, North Bend, OR: 3-15.

Lucia, Ellis ca. 1963. The Big Blow: The Story of the Pacific Northwest's Columbus Day Storm. News-Times Publishing, Portland, OR: 65 pp.

MacCleery, Douglas W. 1992. American Forests: A History of Resiliency and Recovery. USDA Forest Service (FS-540), Forest History Society, Durham, ND: 59 pp.

Mackey, Harold 1974. The Kalapuyans: A Sourcebook on the Indians of the Willamette Valley. Mission Mill Museum Association, Inc. Salem, OR: 165 pp.

Maloney, Alice B. (ed.) 1943. "Fur Brigade to the Bonaventura: John Work's California Expedition of 1832-33 for the Hudson's Bay Company," California Historical Society Quarterly. Vol. XXII: 193-222, 323-348.

Mann, Charles C. 2002. "1491," The Atlantic Monthly. Vol. 289, No. 3: 41-53.

Mardis, Joseph 2002. Personal communication. Assistant County Surveyor, Benton County, Oregon.

Martin, R. J. and E. Corbin, ca. 1931. Climatic Summary of the United States, Section 3--Western Oregon. USDA Weather Bureau, Washington, DC: 48 pp.

Martin, George E. and Lawrence R. Fick 2002. A Road in the Wilderness: The Salem to Astoria Military Road. Oregon Department of Forestry, Forest Grove, OR: 156 pp.

Martin, George E. 2003. Personal communications. Tillamook State Forest historian, Forest Grove District Office, Oregon Department of Forestry.

Martin, Robert E. 1990. "Goals, Methods, and Elements of Prescribed Burning," IN: Natural and Prescribed Fire in Pacific Northwest Forests, John D. Walstad, Steven R. Radosevich, and David V. Sandberg (eds.), Oregon State University Press, Corvallis, OR: 55-66.

Martinez, Dennis 1993. "Land & Culture," Winds of Change. Vol. 8, No. 1: 50-53.

McArthur, Lewis L. 1982. Oregon Geographic Names, 5th Edition. Western Imprints, Oregon Historical Society, Portland, OR: 839 pp.

Meany, Jr., Edmond S. 1935. The History of the Lumber Industry in the Pacific Northwest to 1917. PhD dissertation, Harvard University, MA: 412 pp.

Mercer, George 1865. "South Boundary of T. 12 S., R. 7 W.: Surveying Contract No. 112, dated July 20, 1865," IN: County Surveyor's Record, Benton County, Oregon, Government Field Notes, Vol. 3. Corvallis, OR: 159-160.

Mercer, George 1878. "South Boundary of Fractional T. 14 S., R. 7 W," Surveyors Contract No. 277, dated Nov. 30, 1877," IN: County Surveyor's Record, Benton County, Oregon, Government Field Notes, Vol. 3: Corvallis, OR: 164-168.

Metsker, Charles A. 1929a. Township 10 S., Range 4 W.W.M., Polk County Ore. & Benton County Ore. Metsker Map Co., Tacoma, WA: 1 sheet.

Metsker, Charles A. 1929b. Township 10 S., Range 5 W.W.M., Polk County Ore. & Benton County Ore. Metsker Map Co., Tacoma, WA: 1 sheet.

Metsker, Charles A. 1929c. Township 11 S., Range 5 W.W.M., Benton County Ore. Metsker Map Co., Tacoma, WA: 1 sheet.

Miller, Jay and William R. Seaburg 1990. "Athapaskans of Southwestern Oregon," IN: Handbook of North American Indians, Volume 7: Northwest Coast (edited by Wayne Suttles, (ed.), William C. Sturtevant (series ed.), Smithsonian Institute, Washington, DC: 580-588.

Minor, Rick, Kathryn A. Toepel, and Ruth Greenspan 1987. Archaeological Investigations at Yaquina Head, Central Oregon Coast. USDI BLM, Oregon State office, Portland, OR: 170 pp.

Minor, Rick 1989. Archaeology of the North Yaquina Head Shell Middens, Central Oregon Coast. USDI BLM, Oregon State Office, Portland, OR: 86 pp.

Minore, Don 1972. The Wild Huckleberries of Oregon and Washington--A Dwindling Resource. USDA Forest Research Paper - 143, Pacific Northwest Forest and Range Experiment Station, Portland, OR: 14 pp.

Minto, John 1900. "The Number and Condition of the Native Race in Oregon When First Seen by White Men," Oregon Historical Quarterly, Vol. 1, No. 3: 296-315.

Moore, John M. 1851. Instructions to Surveyor of Public Lands in Oregon, 1851. General Land Office, Washington, DC: 86 pp.

Moore, John M. 1855. Instructions to the Surveyors General of Public Lands of the United States, for Those Surveying Districts Established In and Since the Year 1850. A. O. P. Nicholson, Public Printer, Washington, DC: 92 pp.

Moravets, F. L. 1932. "Second Growth Douglas Fir Follows Cessation of Indian Fires," Service Bulletin. USDA Forest Service, Vol. 16, No. 20: 3.

Morgan, Dale L. 1964. Jedediah Smith and the Opening of the West. University of Nebraska Press, Lincoln, NB: 458 pp.

Morris, William G. 1934a. Lightning Storms and Fires on the National Forests of Oregon and Washington. USDA Pacific Northwest Forest Experiment Station, Portland, OR: 27 pp.

Morris, William G. 1934b. "Forest Fires in Western Oregon and Western Washington," Oregon Historical Quarterly. Vol. 35: 313-339.

Morris, William G. 1935. The Details of the Tillamook Fire from Its Origin to the Salvage of the Dead Timber ("State Forester's Office Version"). Unpublished report with some apparent dating errors. Copy in possession of author. USDA Pacific Northwest Forest and Range Experiment Station, Portland, OR: 26 pp.

Munger, Thornton T. 1916. "The Productive Capacity of the Douglas Fir Lands of Western

Oregon and Washington," The University of California Journal of Agriculture. Vol. 4, No. 3: 92-93.

Munger, Thornton T. 1940. "The Cycle From Douglas Fir to Hemlock," Ecology. Vol. 24, No. 4: 451-459.

Munger, Thornton T. 1944. "Out of the Ashes of the Nestucca," American Forests. (July): 342-345, 366-368.

Nabut, Emily. "Last Official Indian Ceremonial Ever Held in North Lincoln," IN: Pioneer History of North Lincoln County, Oregon, Earl M. Nelson (ed.), The Telephone Register Publishing Company, McMinville, OR. Vol. 1: 105.

Nash, J. Madeline 2003. "A State in Flames: How a Combustible Mix of Drought, Development and Fierce Desert Winds Sparked One of California's Worst Natural Disasters," Time. Vol. 162, No. 19: 54-61.

Nash, Wallis 1976. Oregon: There and Back in 1877. Oregon State University Press, Corvallis, OR: 289 pp.

Neall, James 1977. A Down-Easter in the Far West: The Reminiscence of James Neall in Oregon & California 1845-50. The Oregon Book Society, Ashland, OR: 133 pp.

Nevins, Allan 1996. "Oral History: How and Why It was Born," IN: Oral History, An Interdisciplinary Anthology, David K. Dunaway and Willa K. Baum (eds.), Alta Mira Press, Walnut Creek, CA: 29-38.

Newton, Michael 1970. A Wooden History Book of Philomath, Oregon: Grange Hall Road Big Oak. Unpublished manuscript and cross section of oak tree limb, in possession of M. Newton, Philomath OR: 1 pg.

Nordhoff, Charles 1874. "The Columbia River and Puget Sound," Harper's New Monthly Magazine. Vol. XLVIII.

Norse, Elliott A. 1990. Ancient Forests of the Pacific Northwest. The Wilderness Society Island Press, Covelo, CA: 327 pp.

Norton, Helen H., Robert Boyd, and Eugene Hunn 1999. "The Klikitat Trail of South-central Washington: A Reconstruction of Seasonally Used Resource Sites," IN: Indians, Fire, and the Land in the Pacific Northwest. Robert Boyd (ed.), Oregon State University Press, Corvallis, OR: 65-93.

Norton, Helen H., Hunn, E. S., Martinsen, C. S. and Keely, P. B. 1984. "Vegetable Food Products of the Foraging Economies of the Pacific Northwest," Ecology of Food and Nutrition. Vol. 14: 219-228.

Norton, Helen H. 1979a. "Evidence for Bracken Fern as a Food for Aboriginal Peoples of Western Washington," Economic Botany. Vol. 35, No. 4: 384-396.

Norton, Helen H. 1979b. The Association Between Anthropogenic Prairies and Important Food Plants in Western Washington. Northwest Anthropological Research Notes 13: 175-200.

Oliphant, J. Orin 1932. "Winter Losses of Cattle in the Oregon Country, 1847-1890," Washington Historical Quarterly. Vol. XXIII, No. 1: 3-17.

Oliver, Gordon 2002. "The Dalles Grass Fire Jumps Over the Columbia," The Oregonian, Portland, OR (date?): B4.

ODF (Oregon Department of Forestry) 1933. "Fire Loss Runs to Staggering Totals: Owners and Public Officials Study Serious Problems of Salvage of Merchantable Values," The Forest Log. Vol. 4, No. 5: 1, 4.

ODF (Oregon Department of Forestry) 1941. "Unseasonal Fires Keep Crews Busy: Mid-March Dry Weather Contributes to Condition Causing Spread of Fern Fires in Forests," The Forest Log. (April).

ODF (Oregon Department of Forestry) 1945a. "Northwest Oregon Scorched by Blaze: Army and Civilian Fire Fighters Make Strenuous Effort to Control Fire in Old Tillamook Burn," The Forest Log. Vol. 15, No. 1: 1, 5.

ODF (Oregon Department of forestry) 1945b. "Tillamook Planting Starts December 1: Ambitious Program of Reforestation Proposed Which Will Take in Tillamook Fire Area," The Forest Log. Vol. 15, No. 2: 1, 5.

ODF (Oregon Department of Forestry) 1951a. "Fourth Tillamook Disaster Averted: Fire covers 7500 Acres Before Being Stopped by Fire Fighters and Favorable Weather," The Fortest Log. Vol. 20, No. 10: 1, 3.

ODF (Oregon Deparment of Forestry) 1951b. "Fire rages Thru Oregon Forests: Smoldering Fires at Sardine Creek and Tilamook Jump Lines and Consume 40,000 Acres," The Forest Log. (October): 3.

ODF (Oregon Department of Forestry) 1994. "A Snapshot of Fire Trends," The Forest Log, Salem, OR: 7 (November).

ODF (Oregon Department of Forestry) 1997. Tillamook Burn to Tillamook State Forest. Oregon Department of Forestry, Salem, OR: 23 pp.

Orr, Elizabeth L., William N. Orr, and Ewart M. Baldwin 1992. Geology of Oregon (Fourth Edition). Kendall/Hunt Publishing Company, Dubuque, IA: 254 pp.

Palmer, Joel 1983. Journal of Travels Over the Rocky Mountains. Ye Galleon Press, Fairfield, WA: 186 pp.

Pearson, Grady & Co. 1929 [1936?]. Timber Type Map. Lincoln County. Produced for S. P. & S. Railway. Original map located in Oregon Historical Society (OHS) map collection in Portland, OR. (Note: *Photographed in 1991 with color slide film for COPE (Zybach 1992) files. As of July 16, 1992, the map was unable to be relocated by OHS map librarian, Elizabeth Winroth*).

Pendergrass, Kathy L., P. M. Miller, and J. Boone Kauffman 1998. "Prescribed Fire and the Response of Woody Species in Willamette Valley Wetland Prairies," Restoration Ecology. Vol. 6, No. 3: 303-311.

Pendergrass, Kathy L. 1996. Vegetation and Response to Fire of Native Willamette Valley Wetland Prairies. MS thesis, Oregon State University, Corvallis, OR: 241 pp.

Peterson, James 1994. "Voices in the Forest: An Interview with Bob Zybach," Evergreen. March/April: 7-22.

Peterson, James 2000. "Blow Up," Evergreen. Winter: 7-9.

Phillips, Jerry 1988. Coos County Forest Fire History, 1750-1868. Hand drawn map of the 1868 Coos Fire and the "1770" Millicoma Fire, sketched over an ODF Coos Fire District map. In possession of the author.

Phillips, Jerry 2003. Personal communications. Retired Manager of the Elliott State Forest.

Phillips, Patricia Whereat 2003. Personal communications. Coos linguist.

Phinney, Mark 2000. WPA Historical Records Survey, Benton County, Oregon: Interviews. Danell Aukerman, (ed.), Benton County Genealogical Society, Philomath, OR: 457 pp.

Pinchot, Gifford 1987. Breaking New Ground. Island Press, Washington, DC: 522 pp.

Pipes, Nellie 1934. "The Journal of John H. Frost, 1840-43," Oregon Historical Quarterly. Vol. 35, No. 1: 50-73; No. 2: 139-167; No. 3: 235-262; No. 4: 348-375.

Pittmon 1997. Pittmon's Recreational Atlas of Oregon. Indexed atlas of Oregon roads and landmarks, arranged by county. Oregon Blue Print Co., Portland, OR: 178 pp.

Poage, Nathan J. 2000. Structure and Development of Old-Growth Douglas-fir in Central Western Oregon. PhD dissertation, Oregon State University, Corvallis, OR: 164 pp.

Pullen, Reg 1996. Overview of the Environment of Native Inhabitants of Southwestern Oregon, Late Prehistoric Era. Report prepared for USDA Forest Service and DOI Bureau of Land Management by Pullen Consulting, Bandon, OR: 198 pp.

Pyne, Stephen J. 1982. Fire in America: A Cultural History of Wildland and Rural Fire. Princeton University Press, Princeton, NJ: 654 pp.

Pyne, Stephen J. 1984. Introduction to Wildland Fire: Fire Management in the United States. John Wiley & Sons, New York, NY: 455 pp.

Raup, Hugh M. and Reynold E. Carlson 1941. The History of Land Use in the Harvard Forest. Bulletin No. 20, Harvard Forest, Petersham, MA: 64 pp.

Raup, Hugh M. 1966. "The View From John Sanderson's Farm: A Perspective for the Use of the

Land," Forest History, Vol. 10, No. 1: 2-11.

Reckendorf, Frank F. and Robert B. Parsons 1966. "Soil Development Over a Hearth In Willamette Valley, Oregon," Northwest Science. Vol. 40: 46-55.

Redmond, Kelly T. 1985. An Inventory of Climate Data for the State of Oregon. Report SCP-3, Office of the State Climatologist, Climatic Research Institute, Oregon State University, Corvallis, OR: 160 pp.

Redmond, Kelly T. and George Taylor 1997. "Climate of the Coastal Temperate Rain Forest," IN: The Rain Forests of Home. Peter K. Schoonmaker, Peter K., Bettina von Hagen and Edward C. Wolf (eds.), Island Press, Washington, DC: 25-41.

Ripple, William J., K. T. Hershey, and Robert G. Anthony 2000. "Historical Forest Patterns of Oregon's Central Coast Range," Biological Conservation. Vol. 93: 127-133.

Robbins, Benjamin 1988. Personal communication. Survived 1902 Yacolt Fire, as a child with his family, by fleeing to Speelyai ["Coyote"] Prairie during the event, which killed 16 neighbors in the Lewis River basin.

Robock, Alan 1988. "Enhancement of Surface Cooling Due to Forest Fire Smoke," Science. Vol. 242: 911-913.

Ross, Alexander 1923. Adventures of the First Settlers on the Oregon or Columbia River. The Lakeside Press, R. R. Donnelley & Sons Company, Chicago, IL: 388 pp.

Ross, Alexander 1986. Adventures of the First Settlers on the Oregon or Columbia River, 1810-1813. University of Nebraska Press. Lincoln, NE: 319 pp.

Ross, Charles R. 1971. Ferns to Know in Oregon. Extension Bulletin 785, Oregon State University Cooperative Extension Service, Corvallis, OR: 17 pp.

Rouse, Cary 1986. Fire Effects in Northeastern Forests: Oak. USDA North Central Forest Experiment Station, General technical Report NC-105, St. Paul MN: 7 pp.

Rowland, Theodore and F. A. Elliott 1914. Map of the State of Oregon. Copyright by F.A. Elliott. Published by the State Board of Forestry, Salem, OR. Scale, 1 in.: to 6 mi. Size, 54 in. x 74 in. (Peterson 1994: 12)

Ruby, Robert H. and John A. Brown 1986. A Guide to the Indian Tribes of the Pacific Northwest. University of Oklahoma Press, Norman OK: 289 pp.

Ruby, Robert H. and John A. Brown 1988. The Chinook Indians: Traders of the Lower Columbia. University of Oklahoma Press, Norman, Oklahoma: 349 pp.

Russell, Emily W. B. 1980. "Vegetational Change in Northern New Jersey from Precolonization to the Present: A Palynological Interpretation," Bulletin of the Torrey Botanical Club. Vol. 107, No. 3: 432-446.

Russell, Emily W. B. 1983. Indian-set Fires in the Forests of the Northeastern United States," Ecology. Vol. 64, No. 1: 78-88.

Rust, Elma 1984. The Pioneers of Lake Creek Valley. Self published. Copy in the Valley Library, Corvallis, OR: 307 pp.

Ruth, Robert H. and A. S. Harris 1979. Management of Western Hemlock-Sitka Spruce Forests for Timber Production. General Technical Report PNW-88, Pacific Northwest Forest and Range Experiment Station, USDA Forest Service, Portland, OR: 196 pp.

Ruth, Robert H. 1954. Cascade Head Climatological Data, 1936 to 1952. USDA Pacific Northwest Forest and Range Experiment Station, Portland, OR: 29 pp.

Rycraft, Squire L. ca. 1922. Series of transcribed newspaper interviews published shortly after Rycraft's 94th birthday on September 30, including articles from October 3, 1922 Corvallis Gazette-Times, Corvallis, OR, and October 3, 1922 Oregon Daily Journal, Portland, OR. On file with Benton County Historical Museum, Philomath, OR: 7 pp.

Sargent, Charles S. 1880. Report on the Forests of North America (Exclusive of Mexico). USDI Census Office, US Government Printing Office, Washington, DC: 580 pp.

Sauter, John and Bruce Johnson 1974. Tillamook Indians of the Oregon Coast. Binfords & Mort, Publishers, Portland, OR: 196 pp.

Schmidt, R. L. 1970. "A History of Pre-settlement Fires on Vancouver Island as Determined from Douglas-fir Ages," IN: 107-108.

Schobinger, Juan 1994. The First Americans. William B. Eerdmans Publishing Company, Grand Rapids, MI: 195 pp.

Schulte, Lisa A. and David J. Mladenoff 2001. "The Original US Public Land Survey Records: Their Use and Limitations in Reconstructing Presettlement Vegetation," IN: Journal of Forestry, Vol. 99, October: 5-10.

Schultz, Cheryl B. and Elizabeth E. Crone 1998. "Burning Prairie to Restore Butterfly Habitat: A Modeling Approach to Management Tradeoffs for the Fender's Blue," Restoration Ecology. Vol. 6, No. 3: 244-252.

Scott, Leslie M. 1923. "John Work's Journey from Fort Vancouver to Umpqua River, and Return, in 1834," Oregon Historical Quarterly. Vol. 24, No. 3: 238-268.

Scott, Leslie M. 1928. "Indian Diseases as Aids to Pacific Northwest Settlement," Oregon Historical Quarterly. Vol. 29, No. 2: 144-161.

Seaburg, William R. 1982. Guide to Pacific Northwest Native American Materials in the Melville Jacobs Collection and in Other Archival Collections in the University of Washington Libraries. University of Washington Libraries, University of Washington, Seattle, WA: 113 pp.

Seaburg, William R. and Jay Miller 1990. "Tillamooks," IN: Handbook of North American

Indians, Volume 7: Northwest Coast. Wayne Suttles (ed.), William C. Sturtevant (series ed.), Smithsonian Institute, Washington, DC: 560-567.

Sessions, John, Robert Buckman, Mike Newton and Jeff Haman 2003. The Biscuit Fire: Management Options for Forest Regeneration, Fire and Insect Risk Reduction and Timber Salvage, July 8, 2003. Oregon State University, College of Forestry, Corvallis, OR: 63 pp.

Sharp, Edward F. 1897. "Portion of West Boundary and Portion of Subdivisions of T. 13 S, R. 8 W.: Surveying Contract No. 661, dated May 21, 1896," IN: County Surveyor's Record, Benton County, Oregon, Government Field Notes, Vol. 3. Corvallis, OR: 413-455.

Shea, Keith R. 1963. Conk Rot in Old-Growth Douglas Fir on the South Pillsbury Tract, Millicoma Tree Farm. Bulletin from the Timberland Division, Weyerhaeuser Company, Tacoma, WA: 15 pp.

Shinn, Dean A. 1977. Man and the Land: An Ecological History of Fire and Grazing on Eastern Oregon Rangelands. MA thesis, Oregon State University, Corvallis, OR: 92 pp.

Silverstein, Michael 1990. "Chinookans of the Lower Columbia," IN: Handbook of North American Indians, Volume 7: Northwest Coast (edited by Wayne Suttles). William C. Sturtevant, series ed., Smithsonian Institute, Washington, DC: 533-546.

Smyth, Arthur V. 2000. Millicoma: Biography of a Pacific Northwestern Forest. Forest History Society, Durham, NC: 145 pp.

Smyth, Arthur V. 2001. "Millicoma: Biography of a Pacific Northwestern Forest," Forest History Today. Spring/Fall: 53-56.

Smyth, Arthur V. 2003. Personal communications. Retired Weyerhaeuser Timber Company forester and lobbyist, Millicoma Forest historian.

Snavely, P. D. and N. S. MacLeod 1981. "The Oregon and Washington Coast Ranges," IN: Forest Soils of the Douglas-fir Region. Paul E. Heilman, Harry W. Anderson, and David M. Baumgartner (eds.), Washington State University Cooperative Extension Service, Pullman, WA: 298 pp.

Speck, Gordon 1970. Northwest Explorations. Binfords & Mort, Publishers, Portland, OR: 394 pp.

Spencer, Omar C. 1950. The Story of Sauvies Island. Oregon Historical Society, Binfords & Mort, Publishers, Portland, OR: 134 pp.

Sperlin, O. B. (ed.) 1931. The Brackenridge Journal for the Oregon Country. University of Washington Press, Seattle, WA: 70 pp.

Spies, Thomas A. and Jerry F. Franklin 1988. "Old-growth and Forest Dynamics in the Douglas-fir Region of Western Oregon and Washington," Natural Areas Journal. Vol. 8: 190-201.

Sprague, F. LeRoy and Henry P. Hansen 1946. "Forest Succession in the McDonald Forest,

Willamette Valley, Oregon," Northwest Science. Vol. XX, No. 4: 89-98.

Spurr, Steven H. 1948. Aerial Photographs in Forestry. The Ronald Press Company, New York, NY: 345 pp.

Spurr, Steven H. 1964. Forest Ecology. The Ronald Press Company, New York, NY: 352 pp.

Starker, Thurman J. 1939. "From The Silvicultural Notebook," The Annual Cruise. The Forestry Club, Oregon State Agricultural College, Corvallis, OR. Vol. XX: 47-48.

Stewart, Lowell O. 1935. Public Land Surveys: History, Instructions, Methods. Collegiate Press, Inc., Ames IA: 202 pp.

Stewart, Omer C. 2002. Forgotten Fires: Native Americans and the Transient Wilderness. University of Oklahoma Press, Norman, OK: 364 pp.

Stout, Benjamin B. 1952. Species Distribution and Soils in the Harvard Forest. Harvard Forest Bulletin No. 24, Harvard Forest, Petersham, MA: 29 pp.

Stout, Benjamin B. (ed.) 1981. Forests in the Here and Now: A Collection of Writings of Hugh Miller Raup, Bullard Professor of Forestry, Emeritus, Harvard University. The Montana Forest and Experiment Station, Missoula, MT: 131 pp.

Stout, Benjamin B. 2003. Personal communications. Retired former Dean of University of Montana College of Forestry.

Strong, Emory 1960. Stone Age on the Columbia. Binfords & Mort Publishing, Portland, OR.

Suttles, Wayne and Kenneth Ames 1997. "Pre-European History," IN: The Rain Forests of Home: Profile of a North American Bioregion. Peter K. Schoonmaker, Bettina von Hagen, and Edward C. Wolf (eds.), Island Press, Washington, DC: 255-274.

Swanson, Fred 2003. Personal communications. Geologist, Pacific Northwest Research Station.

Tappeiner, John C., David Huffman, David Marshall, Thomas A. Spies and John D. Bailey 1997. "Density, Ages, and Growth Rates in Old-Growth and Young-Growth Forests in Coastal Oregon," Canadian Journal of Forest Research, Vol. 27: 638-648.

Taylor, George and Raymond R. Hatton 1999. The Oregon Weather Book: A State of Extremes. Oregon State University Press, Corvallis, OR: 242 pp.

Taylor, George H. and Chris Hannan 1999. The Climate of Oregon: From Rain Forest to Desert. Oregon State University Press, Corvallis, OR: 211 pp.

Taylor, George 2003. Personal communications. Oregon State Climatologist, Oregon Climate Service, Oregon State University.

Teensma, Peter D. A. 1987. Fire History and Regimes of the Central Western Cascades of Oregon. PhD dissertation, University of Oregon, Eugene, OR: 188 pp.

Teensma, Peter D. A., John T. Rienstra, and Mark A. Yeiter 1991. Preliminary Reconstruction and Analysis of Change in Forest Stand Age Classes of the Oregon Coast Range From 1850 to 1940. Technical Note T/N OR-9, USDI Bureau of Land Management, Oregon State Office, Portland, OR: 9 pp. w/4 maps.

Thilenius, John F. 1964. Synecology of the White oak *(Quercus garryanna)* Woodlands of the Willamette Valley. PhD dissertation, Oregon State University, Corvallis, OR: 151 pp.

Thilenius, John F. 1968. "The *Quercus Garryanna* Forests of the Willamette Valley, Oregon," Ecology. Vol. 49, No. 6: 1124-1133.

Thompson, Gilbert and A.J. Johnson 1900. Map of the State of Oregon Showing the Classification of Lands and Forests. Professional Paper No. 4, United States Geological Survey. Government Printing Office, Washington, D.C. Scale, 1 in. : 10 mi. Size, 30 in. x 37 in. (Gannett 1902).

Thompson, Lucy (Che-na-wah Weitch-ah-wah) 1991. To the American Indian: Reminiscences of a Yurok Woman. Heyday Books, Berkeley, CA: 292 pp.

Thwaites, Reuben G. (ed.) 1959a. Original Journals of the Lewis and Clark Expedition, 1804-1806, Volume Three. Antiquarian Press Ltd., New York, NY: 363 pp.

Thwaites, Reuben G. (ed.) 1959b. Original Journals of the Lewis and Clark Expedition, 1804-1806, Volume Four. Antiquarian Press Ltd., New York, NY: 372 pp.

Tillamook Headlight-Herald 1947. "Forestry Warns Against Fern Fires," Headlight-Herald. Tillamook, OR, February 20: pg. 1.

Todt, Donn L. and Nan Hannon 1998. "Plant Food Resource Ranking on the Upper Klamath River of Oregon and California: A Methodology with Archaeological Applications," Journal of Ethnobiology. Vol. 18, No. 2: 273-308.

Towle, Jerry C. 1974. Woodland in the Willamette Valley: An Historical Geography. PhD. dissertation. University of Oregon, Eugene, OR: 159 pp. w/maps.

Towle, Jerry C. 1982. "Changing Geography of the Willamette Valley Woodland," Oregon Historical Quarterly. Vol. 83, No. 1: 66-88.

Turner, N. J. 1991. "Burning Mountain Sides for Better Crops: Aboriginal Landscape Burning in British Columbia," Archaeology in Montana. Vol. 32, No. 2: 57-73.

UCAR (University Corporation for Atmospheric Research) 2000. Wildfires. URL: http://www.ucar.edu/communications/factsheets/wildfires.html (September 30, 2003): 3 pg.

US Coast Survey 1859. "Coast of Oregon," IN: Report of the Superintendent of the Coast Survey Showing the Progress of the Survey During the Year 1858. Executive Document No., 33, 35th Congress, 2nd Session, James B. Steedman, Printer. Washington, DC: 373-386.

USDA Siuslaw National Forest ca. 1988. Cascade Head Experimental Forest: Current Studies. Hebo Ranger District. Anonymous, un-dated tour guide of research plots w/tables, had-drawn maps, and annotations, in possession of the author: 6 pp.

US Public Law 88-577 1964. An Act to Establish a National Wilderness Preservation System for the Permanent Good of the Whole People, and for Other Purposes. Public Law 88-577, 88th Congress, S. 4. (September 3). URL: http://www.fs.fed.us/outernet/htnf/wildacta.htm (September 18, 2003).

Vale, Thomas R. (ed.) 2002. Fire, Native Peoples, and the Natural Landscape. Island Press, Washington, DC: 315 pp.

Vansina, Jan 1996. "Oral Tradition and Historical Methodology," IN: Oral History, An Interdisciplinary Anthology. David K. Dunaway and Willa K. Baum (eds.), Alta Mira Press, Walnut Creek, CA: 121-125.

Vaughn, Warren N. 1923. Early Settlement of Tillamook County. Unpublished memoirs based on journal entries. Handwritten ca. 1890 and typewritten in 1923, on file with the Tillamook Historical Society, Tillamook, OR: 83 pp.

Venell, Rosetta 2003. Personal communications. Owner, Venell Farms, Inc., Corvallis, OR.

Wade, Dale, George Custer, Jim Thorsen, Paul Kaskey, John Kush, Bill Twomey, and Doug Voltolino 1998. "Reintroduction of Fire into Fire-Dependent Ecosystems: Some Southern Examples," IN: Fire in Ecosystem Management: Shifting the Paradigm from Suppression to Prescription. Teresa L. Pruden and Leonard A Brennan (eds.), Tall Timbers Ecology Conference proceedings, No. 20, Tall Timbers Research Station, Tallahassee, FL: 94-98.

Wakefield, Rex 1988. Personal communications. Former Supervisor of the USDA Siuslaw National Forest.

Walling, A. G. 1884. History of Southern Oregon, Comprising Jackson, Josephine, Douglas, Curry, and Coos Counties, Compiled from the Most Authentic Sources. Printing and Lithographing House of A. G. Walling,
Portland, OR: 545 pp.

Walstad, John D. and Kenneth W. Seidel 1990. "Use and Benefits of Prescribed Fire in Reforestation," IN: Natural and Prescribed Fire in Pacific Northwest Forests. John D. Walstad, Steven R. Radosevich, and David V. sandberg (eds.), Oregon State University Press, Corvallis, OR: 67-79.

Warre, Henry J. 1976. Overland to Oregon in 1845. M.O.M. Printing, Ottowa, Canada: 149 pp. (1: 2)

Wasson, George B. 2000. "Oral Traditions of the Earliest Humans on the Oregon Coast: A Comparison of Ethnohistorical and Archaeological Records," Changing Landscapes. Proceedings of the 3rd Annual Coquille Cultural Preservation Conference, 1999, Robert J. Losey (ed.), Coquille Indian Tribe, North Bend, OR: 19-32.

Webster, Kimball 1853. "Third Standard Parallel South, thru R. 6 W. and West and South Boundaries of T. 14 S., R. 6 W.: Contract No. 37, dated September 19, 1853," IN: Benton County, Oregon Surveyor's Record, Government Field Notes, Vol. 3. Corvallis, OR: 95, 98-101.

Weisberg, Peter J. and Frederick J. Swanson 2003. "Regional Synchroneity in Fire Regimes of Western Oregon and Washington, USA," Forest Ecology and Management. Vol. 172: 17-28.

Weyerhaeuser Timber Company 1947. Coos Bay Growth and Yield Study. Unpublished report prepared by Forestry Department for Weyerhaeuser Timber Company, Tacoma, WA: 31 pp.

White, C. Albert 1980. Durability of Bearing Trees. Cadastral Survey Training Staff, USDI Bureau of Land Management, Portland, OR: 106 pp.

White, C. Albert n.d. A History of the Rectangular Survey System. USDI Bureau of Land Management. US Government Printing Office, Washington, DC: 774 pp.

White, Richard 1999. “Indian Land Use and Environmental Change, Island County, Washington: A Case Study,” IN: Indians, Fire and the Land in the Pacific Northwest. Robert Boyd (ed.), Oregon State University Press, Corvallis, OR: 36-49.

Whitlock, Cathy and Margaret A. Knox 2002. "Prehistoric Burning in the Pacific Northwest: Human Versus Climatic Influences," IN: Fire, Native Peoples, and the Natural Landscape. Thomas R. Vale (ed.), Island Press, Washington, DC: 195-231.

Whitlow II, Leonard 1988. Grande Ronde Indian Reservation, Yamhill County, Oregon: 1860 Census, 1888 Census, 1901 Family Register (Family Groups, Births 1902-1937, Deaths 1902-1937). Published by Leonard Whitlow II, Portland, OR: 42 pp.

Whitlow II, Leonard 2003. Personal communication. Authority on genealogy and demographics of Siletz tribal families from 1856 to the present.

Whitney, Gordon G. 1994. From Coastal Wilderness to Fruit Plain: A History of Environmental Change in Temperate North America, 1500 to the Present. Cambridge University Press, New York, NY: 451 pp.

Whitney, Gordon G. and Joseph P. DeCant 2001. "Government Land Office Surveys and Other Early Land Surveys," IN: The Historical Ecology Handbook: A Restorationist's Guide to Reference Ecosystems. Dave Egan and Evelyn A. Howell (eds.), Society for Ecological Restoration, Island Press, Washington, DC: 147-172.

Whitney, Lowell W. 2001. Ecological Relationships Between Columbian White-tailed and Black-tailed Deer in southwest Oregon. MS thesis, Oregon State University, Corvallis, OR: 106 pp.

Wilkes, Charles 1845a. Narrative of the United States Exploring Expedition During the Years 1838, 1839, 1840, 1841, 1842. Lea & Blanchard, Philadelphia, PA. Vol. IV: 539 pp.

Wilkes, Charles 1845b. Narrative of the United States Exploring Expedition During the Years 1838, 1839, 1840, 1841, 1842. Lea & Blanchard, Philadelphia, PA. Vol. V: 558 pp.

Williams, Gerald W. 2000. “Reintroducing Indian-Type Fire: Implications for Land Managers,” Fire Management Today. Vol. 60, No. 3: 40-48.

Williams, Gerald W. 2002. "Aboriginal Use of Fire: Are There Any 'Natural' Plant Communities?" IN: Wilderness and Political Ecology: Aboriginal Influences and the Original State of Nature. Charles E. Kay and Randy T. Simmons (eds.), The University of Utah Press, Salt Lake City, UT: 179-214.

Williams, Gerald W. 2003. References on the American Indian Use of Fire in Ecosystems. USDA Forest Service, Washington DC: 94 pp.

Williams, John H. 1912. The Guardians of the Columbia. John H. Williams, Tacoma, WA: 144 pp.

Williams, Michael 1992. Americans and Their Forests. Cambridge University Press, New York, NY: 599 pp.

Wilson, Mark V., David E. Hibbs, and Edward R. Alverson 1991. "Native Plants, Native Ecosystems, and Native Landscapes," IN: Kalmiopsis. Paper No. 2713, a publication of the Restoration and Conservation Biology Project, Oregon State University Forestry Research Laboratory, Oregon State University and Oregon Department of Agriculture, Corvallis, OR: 13-17.

Winkler, Carol and Tim Bailey 2002. Restoring the Cultural Landscape at Jim’s Creek: Challenges to Preserving a Spirit of Place. Paper presented at the 55th Annual Northwest Anthropological Conference, April 11-13, 2002. USDA Willamette National Forest, Westfir, OR: 13 pp.

Winter, Linda, Linda B. Brubaker, Jerry F. Franklin, Eric A. Miller, and Donald Q. DeWitt 2002. "Initiation of an Old-growth Douglas-fir Stand in the Pacific Northwest: A Reconstruction from Tree-ring Records," Canadian Journal of Forest Research. Vol. 32: 1039-1056.

Worona, Marc A. and Cathy Whitlock 1995. "Late Quaternary Vegetation and Climate History Near Little Lake, Central Coast Range, Oregon," Geological Society of America Bulletin. Vol. 107, No. 7: 867-876.

Younker, Jason, Mark A. Tveskov, and David G. Lewis (eds.) 2000. Changing Landscapes: “Telling Our Stories.” Proceedings of the Fourth Annual Coquille Cultural Preservation Conference, 2000, The Coquille Indian Tribe, North Bend, OR: 122 pp.

Zaffino, Matt 2003. Personal communication. Chief Meteorologist, KGW-TV, Portland, Oregon.

Zenk, Henry B. 1990a. "Kalapuyans," IN: Handbook of North American Indians, Volume 7: Northwest Coast (edited by Wayne Suttles). William C. Sturtevant, series ed., Smithsonian Institute, Washington, DC: 547-553.

Zenk, Henry B. 1990b. "Alseans," IN: Handbook of North American Indians, Volume 7: Northwest Coast (edited by Wayne Suttles). William C. Sturtevant, series ed., Smithsonian Institute, Washington, DC: 568-571.

Zenk, Henry B. 1990c. "Siuslawans and Coosans," IN: Handbook of North American Indians, Volume 7: Northwest Coast (edited by Wayne Suttles). William C. Sturtevant, series ed., Smithsonian Institute, Washington, DC: 572-579.

Zobel, Donald B. 2002. "Ecosystem Use by Indigenous People in An Oregon Coastal Landscape," Northwest Science, Vol. 76, No. 4: 304-314.

Zybach, Bob and Eric Badzinski 1998. Siletz River, 1939: Farms, Clearcuts, Old-Growth, and Indian Prairies. Mapped and indexed July, 1939 aerial photograph coverage of Siletz River basin in western Oregon. Internet report posted August 27, 1998 for Oregon Websites and Watersheds Project, Inc., Philomath, OR. URL: www.ORWW.org/Reports/R/SZ/AP-INTRO.html (June 16, 2003).

Zybach, Bob, Mack Barrington and Thomas Downey 1995. "Converting Historical Information to GIS: Political Boundaries of the Douglas-Fir Region, 1788 to 1995," Journal of Forestry. Vol. 93, No. 5: 15-20.

Zybach, Bob and Patrick Maeder 1996. Draft Report on 1996 Inventory of 1955-1977 Oregon Timber Tax Maps. NW Maps Co., Corvallis, OR: 89 pp.

Zybach, Bob 1982. "Reforestation: A Look Into the Future of Contract Logging," The Log. Associated Oregon Loggers, Inc., Salem, OR (November): 26-28, 37-39.

Zybach, Bob 1983. "Renewed Resources: The Reforestation of the Tillamook Burn (1948-1983)," A.R.C. Quarterly. Associated Reforestation Contractors, Inc., Salem, OR (Fall): 13-17.

Zybach, Bob 1992. Alsea River Drainage, Oregon: Sources and Uses of Historical Information for the Drift Creek and Lobster Valley Areas. Coastal Oregon Productivity Enhancement (COPE) Program, USDA Pacific Northwest Research Station, Corvallis, OR: 67 pp.

Zybach, Bob 1994. Siletz Gorge Forest History: Methods and Recommendations for Using Northwest Forest History Information with GIS Applications. Center for Environmental Studies, Santa Barbara, CA: 77 pp. w/24 appendices.

Zybach, Bob 1999. Using Oral Histories to Document Changing Forest Cover Patterns: Soap Creek Valley, Oregon, 1500-1999. MAIS thesis, Oregon State University, Corvallis, OR: 321 pp.

Zybach, Bob 2002a. "Clearcutting and Fire: We Need to Know More About the Historical Link Between Logging Practices and Catastrophic Forest Blazes," The Oregonian. Portland, OR (July 29).

Zybach, Bob 2002b. "The Alseya Valley Prairie Complex, ca. 1850: Native Landscapes in Western GLO Surveys," IN: Changing Landscapes: "Sustaining Traditions." Proceedings of the 5th and 6th Annual Coquille Cultural Preservation Conferences, 2001 and 2002, Don Ivy and R. Scott Byram (eds.), Coquille Indian Tribe, North Bend, OR: 161-188.

# Appendices

# LIST OF APPENDICES

# APPENDIX A.

## Oregon Coast Range Plants and Animals, ca. 1750

This appendix consists of three tables. Table A.01 is a list of native bearing trees (White 1980), shrubs, and other understory vascular plants noted by General Land Office surveyors in Alsea Valley (see Table 3.04; Appendix D; Zybach 2002b: 178).

Table A.02 is a list of native Coast Range food plants used by precontact Indian families (see Table 3.06; Lake and Zybach: In Review; Zybach 2002b: 174-176). The "Fire" column notes whether plants benefit from regular disturbance by burning (XX), or whether they have simply adapted to such disturbances (X).

Table A.03 is a list of native Coast Range animals used for food by precontact Indian families (see Table 3.07; Lake and Zybach: In Review). The "Fire" column notes whether plants benefit from regular disturbance by burning (XX), whether they have adapted to such disturbances (X), or whether regular burning has no apparent effect on their existence (0).

Table A.01 Native trees and other plants noted by GLO surveyors.

| **GLO Bearing Trees** | ***Latin Name*** | **GLO Surveyor Notes** |
|---|---|---|
| Alder (Red) | *Alnus rubra* | |
| Ash (Oregon) | *Fraxinus latifolia* | |
| Cherry, Wild | *Prunus emarginata* | "Cherry" |
| Chinquapin | *Chrysolepis chrysophylla* | |
| Chittam | *Rhamnus purshiana* | [Cascara buckthorn] |
| (Cottonwood, Black) | *Populus balsamifera* | "Balm" [of Gilead] |
| Crabapple (Wild) | *Pyrus fusca* | |
| Dogwood | *Cornus nuttalli* | |
| (Douglas-fir) | *Pseudotsuga menziesii* | "Fir" |
| (Fir, "True") | *Abies, spp.* | "Spruce" [?] |
| (Fir, Grand) | *Abies grandis* | |
| (Fir, Noble) | *Abies nobilis* | |
| Fir, White | *Abies concolor* | [?] |
| Hemlock (Western) | *Tsuga heterophylla* | |
| (Madrone) | *Arbutus menziesii* | " Mountain Balm" [?] |
| Maple (Bigleaf) | *Acer macrophyllum* | |
| Myrtle | *Myrica californica* | [?] |
| Oak (White) | *Quercas garryanna* | |
| (Redcedar) | *Thuja plicata* | "Cedar" |
| Spruce (Sitka) | *Picea sitchensis* | [True fir?] |
| Willow | *Salix, spp.* | |
| Yew | *Taxus brevifolia* | |
| **Other Plants** | | |
| Arrowwood | *Holodiscus discolor* | [Ocean spray] |
| Blackberry (Wild) | *Rubus ursinus* | |
| (Brackenfern) | *Pteridium aquilinum* | "Fern" |
| Camas | *Camassia, spp.* | |
| (Chokecherry) | *Prunus virginiana* | "Cherry" |
| Devil's ("War") Club | *Oplopanax horridum* | "Devil's Walking Stick" |
| Ferns | *Polystichum, spp.* | [And others] |
| Gooseberry | *Ribes divaricatum* | |
| Grasses | *Poaceae, spp.* | [And others] |
| Hazel | *Corylus cornuta* | [Filbert] |
| Huckleberry | *Vaccinium, spp.* | |
| Laurel | *Kalmia occidentalist (?)* | [Madrone?] |
| Lilac | *Ceanothus, spp. (?)* | [Buckbrush?] |
| Oregon Grape | *Berberis aquifolium* | "Wild grape" |
| Pea (Wild, "Indian") | *Vicia americana* | [Also *Lathyrus*, spp.] |
| Raspberry (Wild) | *Rubus leucodermis* | [Blackcap] |
| Rhododendron (Western) | *Rhododendron macrophyllum* | |
| Salal | *Gaultheria shallon* | |
| Salmonberry | *Rubus spectabilis* | |
| Strawberry (Wild) | *Fragaria, spp.* | |
| Thimbleberry | *Rubus parviflorus* | |
| Vine Maple | *Acer circinatum* | |

Table A.02 Native food plants of the Oregon Coast Range.

| Food Type | Food Name | Fire | Latin Name |
|---|---|---|---|
| **Berries** | Blackberry | XX | *Rubus ursinus* |
| | Black cap | XX | *Rubus leucodermis* |
| | Elderberry | XX | *Sambus spp.* |
| | Gooseberry | XX | *Ribes divaricatum* |
| | Huckleberry | XX | *Vaccinium, spp.* |
| | Kinnikinik | XX | *Arctostaphylos* |
| | Oregon Grape | XX | *Berberis aquifolium* |
| | Salal | XX | *Gaultheria shallon* |
| | Salmonberry | XX | *Rubus spectabilis* |
| | Strawberry | XX | *Fragaria, spp.* |
| | Thimbleberry | XX | *Rubus parviflorus* |
| **Bulbs** | Camas | XX | *Camassia, spp.* |
| | Lily, Chocolate | XX | *Fritillaria lanceolata* |
| | Lily, Tiger | XX | *Lilium columbianum* |
| | Onion | XX | *Allium, spp.* |
| | Wapato | X | *Sagittaria latifolia* |
| **Fruits** | Crabapple | X | *Pyrus fusca* |
| | Cranberries | X | |
| | Cherry, Wild | XX | *Prunus emarginata* |
| | Chokecherry | XX | *Prunus virginiana* |
| | Indian plum | XX | *Oemleria cerasiformis* |
| | Manzanita | XX | |
| | Rosehips | XX | *Rosa spp.* |
| **Grains** | Grass seed | XX | *Poaceae, spp.* |
| | Indian peas | XX | *Vicia Americana* |
| | Sunflower | XX | *Wyethia angustifolia ?* |
| | Tarweed | XX | *Madia, spp.* |
| | Vetch | XX | *Vicia spp.* |
| **Greens** | Dock | XX | *Rumex occidentalis* |
| | Miner's Lettuce | XX | *Claytonia spp.* |
| | Nettles | XX | *Urtica dioica* |
| | Seaweed | X | *Porphyra spp.* |
| | Sweet Clover | X | *Trifolium spp.* |
| **Mushrooms** | Chicken-in-the-woods | | *Laetiporus sulphureaus* |
| | Matsutake | | *Tricholoma spp.* |
| | Morrels | XX | *Morchella spp.* |
| | Puffballs | XX | *Calvatia spp* |
| | Shaggy Manes | | *Coprinus comatus* |
| **Nuts** | Acorns | XX | *Quercas garryanna* |
| | Filberts | XX | *Corylus cornuta* |
| | Myrtle nuts | XX | *Myrica californica* |
| | Pine nuts | X | *Pinus, spp.* |
| **Roots** | Brackenfern | XX | *Pteridium aquilinum* |
| | Mountain carrot | XX | *Lomatum spp.* |
| | Yampah | XX | *Perideridia gairdneri* |
| **Stalks** | Cat-tail | X | *Typha latifolia* |
| | Fiddleheads | XX | *Polystichum, spp.* |
| | Skunk cabbage | X | *Lysichiton americanum* |
| | Thistle (Edible) | XX | |

Table A.03 Native food animals of the Oregon Coast Range.

| Food Type | Food Name | Fire | Latin Name |
|---|---|---|---|
| **Crustaceans** | Crabs, Dungeness | O | *Cancer magister* |
| | Crawdads | X | *Astacus pacifastacus* |
| | Shrimp | O | |
| **Fish** | Eels, Lamprey | X | *Lampetra tridentate* |
| | Eulachon | O | *Thaleichthys pacificus* |
| | Flounder | X | *Platichthys spp.* |
| | Herring, Pacific | O | *Clupea harengus* |
| | Perch | O | *Rhacochilus/Embiotoca* |
| | Rockfish | O | *Sebastes spp.* |
| | Salmon, Chinook | X | *Oncorhynchus tshawytscha* |
| | Salmon, Coho | X | *Oncorhynchus kisutch* |
| | Salmon, Dog | X | *Oncorhynchus keta* |
| | Salmon, Sockeye | X | *Oncorhynchus nerka* |
| | Smelt | X | *Spirinchus spp.* |
| | Sole | O | *Pleuronectidae* |
| | Sturgeon | X | *Acipenser spp.* |
| | Suckers | X | *Catostomus spp* |
| | Trout, Cutthroat | X | *Oncorhynchus clarkii-clarkii* |
| | Trout, Rainbow | X | *Oncorhynchus mykiss* |
| | Trout, Steelhead | X | *Oncorhynchus mykis* |
| **Fowl** | Condor, California | XX | |
| | Doves | XX | |
| | Ducks | XX | *Anas spp.; Aythya spp* |
| | Eagle, Bald | | |
| | Grouse, Ruffed | XX | *Bonansa umbellus* |
| | Geese | XX | *Branta spp.* |
| | Swans | XX | *Cygnus spp.* |
| | Quail, California | XX | *Callipepla californica* |
| | Quail, Mountain | XX | *Oreortyx pictus* |
| | Pigeon, Band-tailed | XX | *Columba fasciata* |
| **Insects** | Caterpillars | XX | |
| | Grass hoppers | XX | |
| | Stoneflies | X | *Plecoptera* |
| | Yellow jackets (larvae) | XX | |
| **Red Meat** | Bear, Black | XX | *Ursus americanus* |
| | Bear, Grizzly | XX | |
| | Beaver | X | *Castor canadensis* |
| | Boomer | XX | *Aplodontia rufa* |
| | Deer, Blacktail | XX | *Odocoileus hemionus* |
| | Deer, Whitetail | XX | *Odocoileus virginiana* |
| | Elk | XX | *Cervus elephus* |
| | Gray Diggers | XX | |
| | Hares | XX | *Lepus americanus* |
| | Rabbits | XX | |
| | Seals | O | *Phoca vitulina* |
| | Sea Lions | O | *Zalophus californianu* |
| | Squirrels, Gray | XX | |
| | Whale, Grey (occasional) | O | *Eschrichtius gibbosus* |
| **Shellfish** | Clams, Butter | X | *Saxidomus giganteus* |
| | Clams, Geoduck | X | *Panopea generosa* |
| | Clams, Freshwater | X | |
| | Clams, Horse | X | *Schizothaerus spp.* |
| | Clams, Razor | O | *Siliqua patula* |
| | Mussels, (saltwater) | O | *Mytilus californianus* |
| | Mussels, freshwater | X | *M. margaritifera* |
| | Oysters (native) | X | *Ostrea lurida* |

# APPENDIX B.

## Oregon Coast Range Landmark Names, 1788 to 1951

Appendix B is a listing of landscape feature names with strong associations to precontact Indian communities and to native landscape plants and patterns in the Oregon Coast Range. The list is comprised of four tables taken almost entirely from two principal references: McArthur's 1982 Fifth Edition of "Oregon Geographic Names," and Pittmon's 1997 "Recreational Atlas of Oregon." Listed names are for landscape features, such peaks, creeks, and rivers, and do not include names of artificial developments, such as towns, post offices, schools, parks, or roads. Some names might appear in more than one table; for example, the name "Camas Prairie" appears on one list as the Indian name for a species of native plant (Table B.01 "camas"), and on another list as a type of plant environment (Table B.03 "prairie").

Table B.01 is a list of Indian-related names given to landscape features, often based on the names of local tribes and nations (see Table 3.01; Map 3.01). Several names relate to Klickitats from eastern Washington, who traveled to western Oregon by horseback during early historical time. These names appear over a wide geographic range, many along documented horse trails frequented by that nation in the 1840s and early 1850s (Zybach 2002b: 170-174). Most other Indian-related names are the simple appellations of either "Indian" or "Squaw." Where known, county names, river basins (see Map 1.03), and Coast Range subregions (see Map 1.01) are also listed, to provide an idea as to the geographical distribution of Indian-related names.

Table B.02 is a list of native plant names (mostly trees, grass, berries, and fern) given to landscape features.

Table B.03 is a list of native plant environments--including prairies, groves, glades, balds, and meadows--given to landscape features.

Table B.04 is a brief list of--usually ephemeral--plant conditions given to landscape features: mostly burns and "dead wood."

Other than Table B.01, which is fairly comprehensive, the listed features in the remaining three tables are only a general sampling of the types and numbers of names dating to early historical time found on the Coast Range. These tables provide a reasonably good indication of the specific locations of a large number of named precontact Indian tribes, and of the species of plants and types of environments associated with each of these named tribes.

Table B.01 American Indian names on the Coast Range landscape.

| Landscape Name | Reference | River Basin | County | N | E | W | S |
|---|---|---|---|---|---|---|---|
| | | | | | | | |
| Alsea Bay | McArthur1982: 13 | Alsea | Lincoln | | | W | |
| Alsea Falls | | Alsea | Benton | | | | |
| Alsea Mountain | | Alsea | Benton | | | W | |
| Alsea River | McArthur 1982: 13 | Alsea | Lincoln | | | W | |
| | | | | | | | |
| Calapooya Creek | McArthur 1982: 102 | Umpqua | Douglas | | | | S |
| Calapooyah Divide | | Umpqua | Lane | | | | S |
| Calapooya Mountains | McArthur 1982: 102 | Willamette | Lane | | E | | |
| Cathlamet Bay | McArthur 1982: 138 | Columbia | Clatsop | N | | | |
| Cathlamet Channel | | Columbia | Clatsop | N | | | |
| Camas Prairie | | Coquille | Douglas | | | | S |
| Camas Prairie | | Siletz | Lincoln | | | W | |
| Camas Valley | McArthur 1982: 103 | Umpqua | Douglas | | | | S |
| Camas Swale | | Willamette | Lane | | E | | |
| Claskanine River | | Youngs | Clatsop | N | | | |
| Clatskanie Mountain | | Clatskanie | Columbia | N | | | |
| Clatskanie River | | Clatskanie | Columbia | N | | | |
| Clatskanie Slough | | Columbia | Columbia | N | | | |
| Clatsop Ridge | | Lewis & Clark | Clatsop | N | | | |
| Clatsop Spit | | Columbia | Clatsop | N | | | |
| Coast Fork Willamette | | Willamette | Lane | | E | | |
| Coos Bay | | Coos | Coos | | | | S |
| Coos Head | | Ocean | Coos | | | | S |
| Coos River | | Coos | Coos | | | | S |
| Coquille River | | Coquille | Coos | | | | S |
| | | | | | | | |
| Digger Creek | | Alsea | Benton | | | W | |
| Digger Mountain | | Alsea | Benton | | | W | |
| | | | | | | | |
| Ecola Creek | | Ocean | Clatsop | N | | | |
| Ecola Point | | Ocean | Clatsop | N | | | |
| Euchre Creek | | Siletz | Lincoln | | | W | |
| Euchre Creek Falls | | Siletz | Lincoln | | | W | |
| Euchre Mountain | | Siletz | Lincoln | | | W | |
| | | | | | | | |
| Indian Beach | | Ocean | Clatsop | N | | | |
| Indian Creek | | | Lane | | | W | |
| Indian Creek | | | Tillamook | | | W | |
| Indian Point | | | Clatsop | N | | | |
| | | | | | | | |
| Kilchis River | | Tillamook | Tillamook | | | W | |
| Klaskani River | | Youngs | Clatsop | N | | | |
| Klickitat Creek | | | Clatsop | N | | | |
| Klickitat Creek | | Alsea | Lincoln | | | W | |
| Klickitat Lake | | Alsea | Lincoln | | | W | |
| Klickitat Mountain | | Yachats | Lane | | | W | |
| Klickitat Ridge | | Yachats | Lane | | | W | |
| Klooshie Creek | | | Clatsop | N | | | |
| | | | | | | | |
| Little Digger Mountain | | Alsea | Lincoln | | | W | |
| Little Luckiamute River | | Luckiamute | Polk | | E | | |

| Landscape Name | Reference | River Basin | County | N | E | W | S |
|---|---|---|---|---|---|---|---|
| Little Nestucca Creek | | Nestucca | Tillamook | | | W | |
| Little Nestucca River | | Nestucca | Tillamook | | | W | |
| Little Yaquina River | | Yaquina | Lincoln | | | W | |
| Long Tom River | | Long Tom | Lane | | E | | |
| Luckiamute Falls | | Luckiamute | Benton | | E | | |
| Luckiamute River | | Luckiamute | Polk | | E | | |
| | | | | | | | |
| Multnomah Island | McArthur 1982: 653 | Columbia | Multnomah | N | | | |
| Multnomah Channel | McArthur 1982: 798 | Columbia | Columbia | N | | | |
| Multnomah River | McArthur 1982: 797 | Columbia | Multnomah | N | | | |
| | | | | | | | |
| Neahkanie Mountain | | Ocean | Tillamook | N | | | |
| Necanicum River | | Necanicum | Clatsop | N | | | |
| Nehalem Bay | | Nehalem | Clatsop | N | | | |
| Nehalem Falls | | Nehalem | | | | | |
| Nehalem River | | Nehalem | Clatsop | N | | | |
| Nehalem Valley (1) | | Nehalem | Clatsop | N | | | |
| Nehalem Valley (2) | | Nehalem | Clatsop | N | | | |
| Neskowin Creek | | | Tillamook | | | W | |
| Neskowin Ridge | | | Tillamook | | | W | |
| Nestucca Bay | | Nestucca | Tillamook | | | W | |
| Nestucca River | | Nestucca | Tillamook | | | W | |
| | | | | | | | |
| Ollala Reservoir | | Yaquina | Lincoln | | | W | |
| Ollala Slough | | Yaquina | Lincoln | | | W | |
| | | | | | | | |
| Shwaeash Creek | | | Clatsop | N | | | |
| Siletz Bay | | Siletz | Lincoln | | | W | |
| Siletz Falls | | Siletz | Lincoln | | | W | |
| Siletz Hill | | Siletz | Lincoln | | | W | |
| Siletz River | | Siletz | Lincoln | | | W | |
| Siltcoos Lake | | Siltcoos | Douglas | | | W | |
| Siltcoos River | | Siltcoos | Douglas | | | W | |
| Siuslaw River | | Siuslaw | Lane | | | W | |
| Skipanon River | | Skipanon | Clatsop | N | | | |
| Skomokawa Channel | | Columbia | Clatsop | N | | | |
| Speelyai Creek | | | Clatsop | N | | | |
| Squaw Creek | | Marys | Benton | | E | | |
| Squaw Creek | | | Clatsop | N | | | |
| Squaw Creek | | Alsea | Lane | | | W | |
| Squaw Creek | | Ocean | Lincoln | | | W | |
| Squaw Island | McArthur 1982: 696 | Ocean | Coos | | | | S |
| | | | | | | | |
| Tahkenitch Lake | | Ocean | Douglas | | | W | |
| Tillamook Bay | | Tillamook | Tillamook | | | W | |
| Tillamook Head | | Ocean | Tillamook | | | W | |
| Tillamook River | | Tillamook | Tillamook | | | W | |
| Tioga Creek | | | Coos | | | | S |
| Tualatin Mountains | | Columbia | Columbia | N | | | |
| Tualatin River | | Tualatin | Washington | | E | | |
| Tumtum Creek | | Marys | Benton | | E | | |
| | | | | | | | |
| Umpqua River | | Umpqua | Douglas | | | | S |
| | | | | | | | |

| Landscape Name | Reference | River Basin | County | N | E | W | S |
|---|---|---|---|---|---|---|---|
| Wapato Lake | McArthur 1982: 773 | Tualatin | Washington | | E | | |
| Wapato Lake (1) | | Gilbert | Columbia | | | | |
| Wapato Lake (2) | | Gilbert | Columbia | | | | |
| Wappatoo Island | McArthur 1982: 652 | Columbia | Columbia | N | | | |
| Willamette River | | Willamette | Multnomah | N | | | |
| Willamette Falls | | Willamette | Clackamas | N | | | |
| | | | | | | | |
| Yachats River | McArthur 1982: 814 | Yachats | Lincoln | | | W | |
| Yakso Falls | McArthur 1982: 814 | Little | Douglas | | | | S |
| Yamhill River | McArthur 1982: 815 | Yamhill | Yamhill | | E | | |
| Yaquina John Point | McArthur 1982: 816 | Ocean | Lincoln | | | W | |
| Yaquina Bay | McArthur 1982: 816 | Yaquina | Lincoln | | | W | |
| Yaquina Falls | | Yaquina | Lincoln | | | W | |
| Yaquina River | McArthur 1982: 816 | Yaquina | Lincoln | | | W | |

Table B.02 Native plant names on the Coast Range landscape.

| Landmark Name | Reference | River Basin | County | N | E | W | S |
|---|---|---|---|---|---|---|---|
| | | | | | | | |
| Alder Creek | | | Columbia | N | | | |
| Alder Creek | | | Lane | | | | |
| Alder Lake | | | Lane | | | | |
| Ash Creek | | | Polk | | E | | |
| Ash Slough | | Gilbert | Multnomah | N | | | |
| Ash Swale | | Willamette | Polk | | E | | |
| Ash Swale Creek | | | Polk | | E | | |
| | | | | | | | |
| Camas Prairie | | Siletz | Lincoln | | | W | |
| Camas Swale | McArthur 1982: 104 | Umpqua | Douglas | | | | S |
| Camas Swale | McArthur 1982: 104 | Willamette | Lane | | E | | |
| Camas Valley | McArthur 1982: 103 | Umpqua | Douglas | | | | S |
| Cedar Butte | | | Tillamook | | | W | |
| Cedar Creek | | Yaquina | Lincoln | | | W | |
| Cedar Creek | | | Columbia | N | | | |
| Cedar Creek | | | Lane | | | | |
| Cedar Creek | | | Polk | | E | | |
| Cedar Creek | | | Tillamook | | | W | |
| Cedar Creek | | | Washington | | E | | |
| Cedar Lake | | | Tillamook | | | W | |
| Clover Ridge | | | Lane | | | | |
| Cottonwood Island | | | Columbia | N | | | |
| | | | | | | | |
| Dogwood Creek | McArthur 1982: 229 | | Clatsop | N | | | |
| | | | | | | | |
| Fern Hill | McArthur 1982: 272 | Columbia | Clatsop | N | | | |
| Fern Hill | | | Columbia | N | | | |
| Fir Creek | | | Lane | | | | |
| Fir Creek | | | Washington | | E | | |
| Firwood | | Clatskanie | Columbia | N | | | |
| | | | | | | | |
| Grass Creek | | | Lane | | | | |
| Grass Creek | | | Lincoln | | | W | |
| Grass Flats Lake | | Gilbert | Columbia | N | | | |
| Grass Lake (1) | | Gilbert | Columbia | N | | | |
| Grass Lake (2) | | Gilbert | Columbia | N | | | |
| Grass Mountain | | Alsea | Benton | | | W | |
| Grassy Creek | | | Lane | | | | |
| Grassy Island | | | Clatsop | N | | | |
| Grassy Lake | | Gilbert | Columbia | N | | | |
| Grassy Lake Creek | | | Clatsop | N | | | |
| | | | | | | | |
| Lily Lake | | Gilbert | Columbia | N | | | |
| | | | | | | | |
| Maple Creek | | | Lane | | | | |
| Maple Hill | | | Columbia | N | | | |
| Myrtle Point | | Coquille | Coos | | | | S |
| | | | | | | | |
| Oak Creek | | Marys | Benton | | | W | |
| Oak Hill | | Willamette | Lane | | | | |

| Landmark Name | Reference | River Basin | County | N | E | W | S |
|---|---|---|---|---|---|---|---|
| Oak Island | | Gilbert | Multnomah | N | | | |
| Oak Point | | Columbia | Columbia | N | | | |
| Olalla Slough | | Yaquina | Lincoln | | | W | |
| Onion Peak | | Ocean | Clatsop | N | | | |
| | | | | | | | |
| Peavine Ridge | | Siletz | Lincoln | | | W | |
| Peavine Ridge | | Alsea | Lincoln | | | W | |
| | | | | | | | |
| Reed Lake (1) | | Gilbert | Columbia | N | | | |
| Reed Lake (2) | | Gilbert | Columbia | N | | | |
| | | | | | | | |
| Salmonberry Creek | | Alsea | Benton | | | W | |
| Salmonberry Lake | | | Columbia | N | | | |
| Salmonberry River | | Salmonberry | Tillamook | | | W | |
| Skunk Cabbage Ridge | | | Columbia | N | | | |
| Sourgrass Lake | | Gilbert | Columbia | N | | | |
| Sourgrass Mountain | | | Lane | | | | |
| | | | | | | | |
| Twin Oaks Lake | | Gilbert | Columbia | N | | | |
| Twin Willows Lake | | Gilbert | Columbia | N | | | |
| Tansy Point | | Columbia | Clatsop | | | | |
| | | | | | | | |
| Wapato Lake | McArthur 1982: 773 | Tualatin | Washington | | E | | |
| Wapato Lake (1) | | Gilbert | Columbia | N | | | |
| Wapato Lake (2) | | Gilbert | Columbia | N | | | |
| Wappatoo Island | McArthur 1982: 652 | | | | | | |
| Willow Bar Island | | | Columbia | N | | | |
| Willow Bar Point | McArthur 1982: 801 | Columbia | Columbia | N | | | |
| Willow Creek | | | Columbia | N | | | |

Table B.03 Plant environment names on the Coast Range landscape.

| Landmark Name | Reference | River Basin | County | N | E | W | S |
|---|---|---|---|---|---|---|---|
| | | | | | | | |
| Ash Swale | | Willamette | Polk | | | W | |
| | | | | | | | |
| Bald Hill | | Marys | Benton | | | W | |
| Bald Mountain | | Willamette | Lane | | E | | |
| Ban Grove Lake | | Gilbert | Columbia | N | | | |
| Big Prairie | | Willamette | Lane | | E | | |
| Brush Creek | | Yaquina | Lincoln | | | W | |
| | | | | | | | |
| Camas Prairie | | Siletz | Lincoln | | | W | |
| Camas Prairie | | Coquille | Douglas | | | | |
| Camas Swale | McArthur 1982: 104 | Umpqua | Douglas | | | | S |
| Camas Swale | McArthur 1982: 104 | Willamette | Lane | | E | | |
| | | | | | | | |
| Enchanted Prairie | | Coquille | Coos | | | | S |
| | | | | | | | |
| Fannys Bottom | | Columbia | Columbia | N | | | |
| Forest Peak | | Luckiamute | Benton | | E | | |
| Forest Lake | | Gilbert | Columbia | N | | | |
| | | | | | | | |
| Glenwood | | | Washington | | | | |
| Grand Prairie | | Willamette | Lane | | E | | |
| | | | | | | | |
| High Prairie | | Willamette | Lane | | | | |
| Halls Prairie | McArthur 1982: 20 | Coquille | Coos | | | | S |
| | | | | | | | |
| Lake in the Woods | | Gilbert | Columbia | N | | | |
| Long Prairie | | Siletz | Lincoln | | | W | |
| | | | | | | | |
| Meadow | | Siuslaw | Lane | | | | |
| Meadow Lake (1) | | Gilbert | Columbia | N | | | |
| Meadow Lake (2) | | Gilbert | Columbia | N | | | |
| Meadow Lake (3) | | Gilbert | Columbia | N | | | |
| Meadow View | | Long Tom | Lane | | E | | |
| | | | | | | | |
| Prairie Channel | | Columbia | Clatsop | N | | | |
| Prairie Mountain | | Alsea | Benton | | | W | |
| Prairie Peak | | Alsea | Benton | | | W | |

Table B.04 Plant condition names on the Coast Range landscape.

| Landmark Name | Reference | River Basin | County | N | E | W | S |
|---|---|---|---|---|---|---|---|
| | | | | | | | |
| Blossom Gulch | | | Coos | | | | S |
| Burnt Creek | | | Coos | | | | S |
| Burnt Mountain | | | Coos | | | | S |
| Burnt Mountain | | | Douglas | | | | S |
| Burnt Ridge | | | Lane | | | | |
| Burnt Timber Mountain | | | Lincoln | | | W | |
| Burnt Top | | | Lane | | | | |
| | | | | | | | |
| Dead Willows Lake | | Gilbert | Columbia | N | | | |
| Deadwood | | Siuslaw | Lane | | | | |
| Deadwood Creek | McArthur 1982: 211 | Siuslaw | Lane | | | | |
| | | | | | | | |
| Green Point | McArthur 1982: 333 | Columbia | Columbia | N | | | |
| Greenleaf Creek | | Siuslaw | Lane | | | | |
| | | | | | | | |
| Timber | | | Washington | | | | |

# APPENDIX C.

## 1910 Smith Interview: "The Big [Yaquina] Fire" of 1849

***And then darkness fell all over the world. The surface of the sun just kept on getting red. The universe was not going to enact a good thing; (a) fire was beginning to approach. Then it got dark all over.***

This is the story of William Smith regarding his family's experiences during the Yaquina Fire of ca. 1849. Smith was an Alsi Indian who told this story, in his native language, to anthropologist Leo Frachtenberg in 1910. Frachtenberg translated the story to English and published it--along with Smith's original Alsi version written in phonetics--in 1920 (Frachtenberg 1920: 213-219). Smith's Alsi name was Clo-ka-tle, which likely translates to "not fat" (Whitlow 2003: personal communication). His memory and attention to detail appear to be excellent, as evidenced by his first hand account of the murder of U. S. Grant (Frachtenberg 1920) by Albert Martin and Abe Logan in 1903. Smith's story, also told in his native tongue and translated to English, closely conforms to newspaper accounts of the event which appeared in the October 23, 1903 Lincoln County Leader and the November 24, 1903 Oregonian (Whitlow 2003: personal communication).

This is the only detailed first-hand account of the Yaquina Fire in existence. The events Smith describes take place a few miles southward of Talbot's 1849 travels (Haskin: 1948), and may well have occurred only a few days or weeks after Talbot's departure from the Alsea River in late August of that year.

I have added some paragraph breaks and fixed a few minor typos for purposes of clarity, without note. Frachetnberg's original comments are enclosed in parenthesis (()). My subsequent editorial comments are enclosed in brackets ([]).

### The Big Fire

We were coming back from Siuslaw, when, long ago, the world was in flames. (The party consisted of) my father and my mother and also my elder brother, and my father's mother and my father's younger brother and his wives--he had three wives--and also one child of one of his wives, and likewise two children of (the other) one of his wives and moreover, three children of (another) one of his wives; (such) only was the number of (the party) [Smith is traveling with his parents, a brother, his paternal grandmother, a paternal uncle, three aunts--all wives of his uncle--

and six cousins, all children of his uncle's wives: in all, two adult men, five adult women and eight children, including at least two older boys, spanning three generations of the same family].

Then it seemed to be getting dark all over. And I was young at that time [Frachtenberg says Smith was about 14 years old; this conforms closely with tribal census records from the 19th century (Whitlow 1988)]. We kept on going. Although the sun stood high, nevertheless it threatened to get dark. Then they kept on saying: "We will not go far anywhere. What on earth is nature going to do?" Thus they would talk. "We will just go down to *Ltowai'sk* [Heceta Head, Tsp. 16 S., Rng. 12 W.] (and) we will build our own fire there." Then, verily, they built their own fire there. And then darkness fell all over the world. The surface of the sun just kept on getting red. The universe was not going to enact a good thing; (a) fire was beginning to approach. Then it got dark all over. The fire seemed to be flying in all directions as soon as darkness enveloped the world. That spoken-of big fire was coming. It became dark all over; the world seemed to be getting red. The fire was falling (all around us). Wherever it would drop (another) fire would start there. The fire seemed to be flying in all directions; its crackling just seemed to make a roaring noise. "We will not got anywhere; we will just stay motionless right here." Then my father kept on saying; "We will never go anywhere. The world is on fire."

Then the fire came to the trail. It was just dark all over; the world just seemed to be getting red. But (it was) not long before some elks were seen coming downstream along that river. Thereupon my father took his gun for them. Then, on his part, my father's younger brother reached for his gun (also). But the elks just stood there motionless. So, they two went there and began to shoot at them, whereupon they two killed one (elk) there. The elk's hair was partially burned and also his legs were partially burned. Then (the elk) was skinned and all his flesh was distributed, whereupon it was carried to the fire.

> ***The people did not remain near the woods. Everybody was staying (near) the ocean on the beach. The fire was flying around just like the birds. It was just dark all over. The sun had disappeared.***

The people did not remain near the woods. Everybody was staying (near) the ocean on the beach. The fire was flying around just like the birds. It was just dark all over. The sun had disappeared. All the hills were on fire. Even the hills (that were near the) sea were burning as soon as the fire arrived at the sea. Everywhere even the blossoms of he highest trees burned down.

"What, indeed, can we do (to help ourselves)? Who is going to come here to tell us (of the conditions in other places)?"--"Yes, (I) wonder if anybody will be so void of sense that he will not (know enough to) go to the water?" Thus they would talk as soon as darkness fell over the world. The crackling of the fire just seemed to rear all over. "Now we are just going to stay (here). (I) wonder how we can go anywhere (else)?"

My grandmother was crying all the time. She was crying for her people. "All my people must have perished in the flames." Her child would there speak to her continually. "Thou shalt not cry all the time (or else) my heart will become small. It is nothing (even if) we two only (myself) and my younger brother have survived." Thus my father was continually speaking to his mother.

> ***All sorts of (animals) were coming to the sea: elks, black bears, and cougars-the hair of all (of them) was just partially burned. My grandmother was singing, "(I) wonder, indeed, what nature is going to do."***

All sorts of (animals) were coming to the sea: elks, black bears, and cougars-the hair of all (of them) was just partially burned. My grandmother was singing, "(I) wonder, indeed, what nature is going to do." When I slept, the fire never came to us. My grandmother would speak thus: "Your (dual) hearts shall not be small. It simply got dark all over. (I) wonder when it is going to get light again. Probably for five nights will the world be in flames."

The crackling of the fire (was heard everywhere). Wherever a log lay on the beach (and) whenever the fire dropped there, it would (instantly) catch fire.

But (at last) the crackling of the fire seemed to be dying out. It seemed to have the appearance of birds. The fire was flying in all directions. The fire seemed to be of such a size. The fire was dropping close to where we were staying. (My mother) was watching the children carefully, she never allowed one (to go away from our camp). The children just (had to) stay together. Then they two [probably Smith's father and uncle] began to talk: "We shall not go anywhere, we have plenty of food." Then they two spoke (again): "What are we two going to do (to help) ourselves? Something bad has happened to the universe. We two will just stay here for a long time. Only after the fire shall have disappeared will we two go to see whence it had started." Thus they two would talk among themselves. Then all would speak thus: "None of us will go anywhere; we will just stay together."

***All the black bears went toward the sea; all kinds of deer went toward the ocean. And also cougars, likewise wolves, and, moreover, foxes and wildcats; the hair of all (of them) was partially burned and also their legs. (Such) of their number (as were) partially burned were coming to the water from the east. "No one shall touch (them); they fared poorly. We two will just leave them alone."***

The fire was just terribly hot. The smell of the smoke made an awful odor all over. (But) not far away it was getting light. Where that trail was leading, the ground had burned entirely. All the black bears went toward the sea; all kinds of deer went toward the ocean. And also cougars, likewise wolves, and, moreover, foxes and wildcats; the hair of all (of them) was partially burned and also their legs. (Such) of their number (as were) partially burned were coming to the water from the east. "No one shall touch (them); they fared poorly. We two will just leave them alone." Thus they two would talk among themselves. "Never did nature act thus," thus my grandmother kept on saying. "No matter how long (back I can remember), nature did not act like that." When her two children heard her (speak thus), they would say: "Now we two will just stay here. When the fire will disappear, at that time will we two depart."

Then they two were counting for how many days darkness prevailed all over. For probably ten days it was dark all over. "Not long (afterward) the fire is going to disappear; then, indeed, we two will go away to-morrow." Now, verily, all (the people) were speaking (thus). "(You two) shall watch yourselves carefully when you two will go now," thus my grandmother would speak. Then after it got dark again the fire disappeared right there. Then they two kept on saying, "Now we two will go tomorrow in the morning to have a look." Then in the morning they two ate. "After we two shall have gone, you shall not go far away. You shall just stay (here) Motionless." Then they two were speaking to their (dual) wives. "We two are going to come back to-morrow"--"You two shall take good care of yourselves," thus said my grandmother.

***Where there was a mountain, that place there did not burn. So they two kept on going on that trail, and they arrived at where there was a place (covered) with grass. And only there did the fire reach.***

Then they started. Where there was a mountain, that place there did not burn [possibly Cape Mountain, Tsp. 17 S., Rng. 12 W.]. So they two kept on going on that trail, and they arrived at where there was a place (covered) with grass. And only there did the fire reach. Then they two ascended. Now it was gradually getting light all over; just a little (light) showed far away. At last they came below, whereupon they two started to walk on that beach. Then they two kept on going along the beach. Everywhere even the blossoms of the highest trees had burned down, (as could be seen) after the water came with them to the beach. Now not long (afterward) they two saw a bear walking along the edge of the water, just partially burned.

At last they two arrived at the mouth of the Siuslaw River. All the pine trees (there) were partially burned [see Map 4.09]. Only ashes (could be seen) all over, because all the pine trees had caught fire.

Then they came to a village; whereupon they two were spoken to. "Did you two survive?"--"Yes, we two survived; we just stayed at *Ltowai'sk* [Heceta Head]. (It is) from there that we two have arrived; we are on our way to have a look."--"Yes, we have fared (here) very poorly. We just stayed close to the sea; we brought all our belongings to the beach. All the people stayed close to the ocean. The people stayed close to the water with (their) canoes (in readiness). Thus we stayed. Even the trees (that) lay in the water caught fire. Thus we did it. Nobody (from here) burned; everybody is well. There were two medicine men who were just dancing every night. For that reason those two medicine men were dancing, because they two wanted to find out (what happened) all over the world, (especially) whence the fire originated. Thus we acted." Thus they two were told by a number (of people).

> ***And (only) two people stayed (there). All (the other) people stayed near the ocean. "We two have come here to have a look (at our home). Nobody was burned: all the people are well. Nature (seems to have been) doing its worst thing. Never (before) did nature act like that."***

Then it kept on clearing off far way, and the fire disappeared again. For probably ten days darkness prevailed all over. Then the two were going to return the next day. No matter how large a place was, nevertheless that place burned down (entirely); the mountains caught fire everywhere. Then they two went back. Then they arrived again at where they two were living. And (only) two people stayed (there). All (the other) people stayed near the ocean. "We two have come here to have a look (at our home). Nobody was burned: all the people are well. Nature (seems to have been) doing its worst thing. Never (before) did nature act like that."

Then they started from there and kept on going back (to the place where we stayed). Then (after a long walk) they said, "We will camp here. "All sorts of things were seen close to the water partly burned, but walking around. Then they camped there (at) *Ts !Aam* [Tenmile Creek, Tsp.

15 S., Rng 12 W.]. Then, as soon as daylight appeared again, they started from there. They had very (heavy) packs. Then they kept on going back. "We will camp at Yahach [a prairie slightly north of today's town of Yachats]." Then, verily (after) the sun set they camped there, at Yahach.

**And when daylight appeared they started out. Then they kept on going along the previously mentioned beach. And (it was) not long before they came back to the Alsea River. Then all settled down at the mouth of the river after they came back (there).**

**And now it comes to an end.**

APPENDIX D.

# "Alseya Valley" to Alsea Valley, 1853-1915: GLO Index

An indexed forest and fire history of Alsea Valley and its headwaters from 1853 to 1914, as recorded by the original General Land Office surveyors, a local historian, and an early timber cruiser.

## Table of Contents



## Introduction: The View from Buck Peak, 1853-1914

This appendix is intended to serve a dual purpose. First, it provides an overview of General Land Office (GLO) survey history and methods; a key source of materials used to research this thesis. Second, it provides an indexed sampling of the types of information gathered by GLO surveyors as they surveyed Alsea Valley, in Benton County, Oregon, and compares it with information gathered and recorded by an historian and a timber cruiser who described portions of the same area during the same general period of time. All of the excerpts are indexed by legal description, as assigned by the original land surveyors and used to now.

The cited surveyors are the same as listed in Table 2.03. The Alsea Valley area is located on Map

2.01 and shown on maps 2.12 and 2.13. The historian is David D. Fagan (1885), who is cited throughout this dissertation. The timber cruiser is J. H. Bagley (1915), whose maps and survey notes are typified by Map 2.04. This appendix is indexed by Tsp. and Rng. (see Map 2.01) and by square-mile sections (see Map 2.02). Entries are arranged chronologically, except when noted.

This appendix contains a large number of representative surveyors' field notes on fires, reforestation, location of plant species, and Indian trails. In keeping with the focus of this thesis, I have left out most references to compass bearings, distances, and other measures of little interest to most non-surveyors, as they tend to detract from the narrative flow of the field notes. Likewise, I have left most references to topography, slope, stream crossing locations, and stream widths out of the narrative because this information changes from time to time and/or is readily available from other sources. These excerpts are arranged by legal description and indexed by location, using Map D.01. To get an idea as to the differences in writing style and perspective provided by the different observers, a sample excerpt follows in which each observer is looking over Alsea Valley from a vantage point on--or very near--the valley's eastern boundary, Buck Peak (see Map D.01):

> *Summit of mountain, overlooking the Alseya Valley, course N. and S. . . A fir 24 ins. . . . A fir, 12 ins. . . A fir, 24 . . . A Maple, 20 ins. . . A hemlock, 20 ins. . . Land, very rough and west slope of mountain very steep and rocky*
> --Dennis Hathorn (GLO Surveyor), Buck Peak, June 5, 1856

> *These boundaries include Alsea valley, a beautiful expanse of country some eight miles long and one wide. Not far from the upper end the two branches of the river unite, causing a widening in the vale to about four miles, forming a level prairie now thickly settled, and surrounded with lofty timber-covered hills. Here were the first settlements made more than thirty years ago when many claims were taken under the provisions of the Donation Law, which, it would appear entitled the pre-emptor to the land not covered with brush, and as no one thought of clearing the ground at that period, it was assumed that everything worth occupying had been appropriated. Of later years, however, many have gone into the woods and demonstrated the fact that such lands may be profitably improved.*
> --David D. Fagan (Historian), 1885: 498

> *Yellow or bastard fir is just fair in quality, is low in clears and there is a large amount of defect. Red fir is young and thrifty but low in clears. Hemlock is young, thrifty and low in clears. Cedar is defective and good for posts only . . . A ground fire has been over nearly all of the section years ago and while it did not*

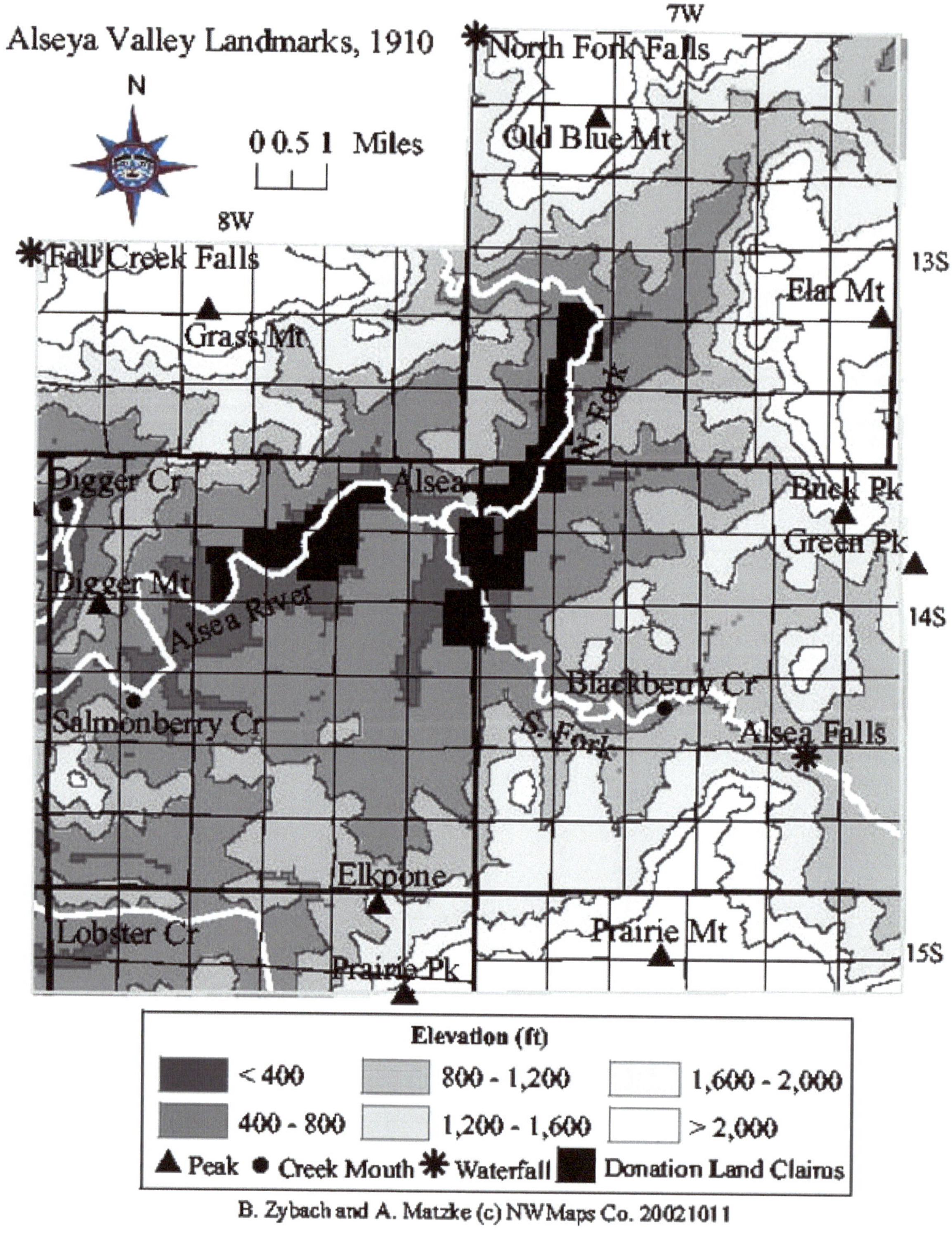


Map D.01 Alsea Valley original land surveys index.

*kill any great amount of timber at the time, a large amount is beginning to show defect. The fire risk is nominal.*
-- J. H. Bagley (Timber Cruiser), Buck Peak, May, 1915

The writers are the nine general land office surveyors listed on Table 2.03, Benton County historian David D. Fagan (1885), and Benton County timber cruiser, J. H. Bagley (1915). The surveyors' names (and the years they worked) were Kimball Webster (1853), Harvey Gordon & Josiah W. Preston (1854), Dennis Hathorn (1856), George Mercer (1865 and/or 1878?), J. M. Dick (1873), Alonzo Gesner (1891), Charles Collier (1891-1893), and Edward Sharp (1897). All of the writers are white males, a trait in common with the primary journalists for the Coast Range (see Table 2.02). Nine of the writers are GLO surveyors, following the same set of specific instructions and using the same specified terms and measures over a period of 44 years, from 1853 until 1897. Near the end of that time, in 1885, while most of the surveyors were still alive, Fagan wrote his detailed history of Benton County. Thirty years later, in 1915, a change in Oregon tax laws caused Benton County to hire Bagley to cruise and map the County's privately owned timberlands (Benton County Commissioners 1914).

Notes are transcribed directly from the references that are cited. Edited portions are clearly marked ( . . . ) and editorial comments are contained in brackets ([]).

**History of the Rectangular Survey, 1780-1850**

In 1784, Thomas Jefferson was made chairman of a committee charged to "ascertain a method of locating and disposing of lands in the Western Territory" in order to help pay national debt incurred fighting the Revolutionary War. Congress subsequently authorized the Public Land Survey (PLS) system of dividing the lands of the US into a series of adjacent rectangles with the Ordinance of May 20, 1785 (Stewart 1935: 2). This system was a radical departure from the existing custom of subdividing land by metes and bounds, which had dominated property surveys in colonial America (Bourdo 1956: 757). The act of February 11, 1805 (enacted coincidentally as Lewis and Clark were preparing to explore and map uncharted lands west of the Western Territory) finalized descriptions of Public Land Survey (PLS) lines and corners as they exist today (Stewart 1935: 12).

The General Land Office (GLO) was established April 25, 1812 for purposes of implementing the act of February 11, 1805 (Stewart 1935: 29-30). Surveys of public lands were to be conducted by independent contractors working under direction of the GLO. In 1849 supervision of the GLO transferred from the Secretary of the Treasury to the newly created Department of the Interior (ibid.: 31). The act of September 27, 1850 authorized the Secretary of the Interior to:

> . . . continue the surveys in Oregon and California, to be made after what is known as the Geodetic method, under such regulations and upon such terms as have been, or may hereafter be prescribed by the Commissioner of the General Land Office , but none other than township lines shall be run where the land is unfit for cultivation.

**General Land Office Survey Methods in Oregon, 1851-1910**

Detailed survey instructions were developed following the act of September 27, 1850 and subsequently issued for the survey of Oregon in 1851 (Moore 1851). These instructions were changed slightly and formalized for the entire US in 1855 (Moore 1855). These latter instructions remained in effect until after 1910, with the result that the entire Pacific Northwest was surveyed by a single prescribed method using only one set of instructions, with only slight modifications from time to time.

By 1855, 12 contract surveyors were working in Oregon and had completed surveys of 1,450 of 1,800 total Donation Land Claims (DLCs) (see Map D.02), in addition to townships and township subdivisions. By 1856, more than 836,000 acres had been surveyed, despite southern Oregon "Indian hostilities", "cloudy weather", and "the mines" that had "hindered progress" on the project (Prosch 1907: 196-197). (The reference to "the mines" was to the lure of instant wealth that had drawn single men and whole families from their homes and jobs in western Oregon to the gold fields of northern California and southwest Oregon, beginning in 1848 (Haskin 1948; Carey 1971: 359, 478-479, 654) and continuing through the Civil War).

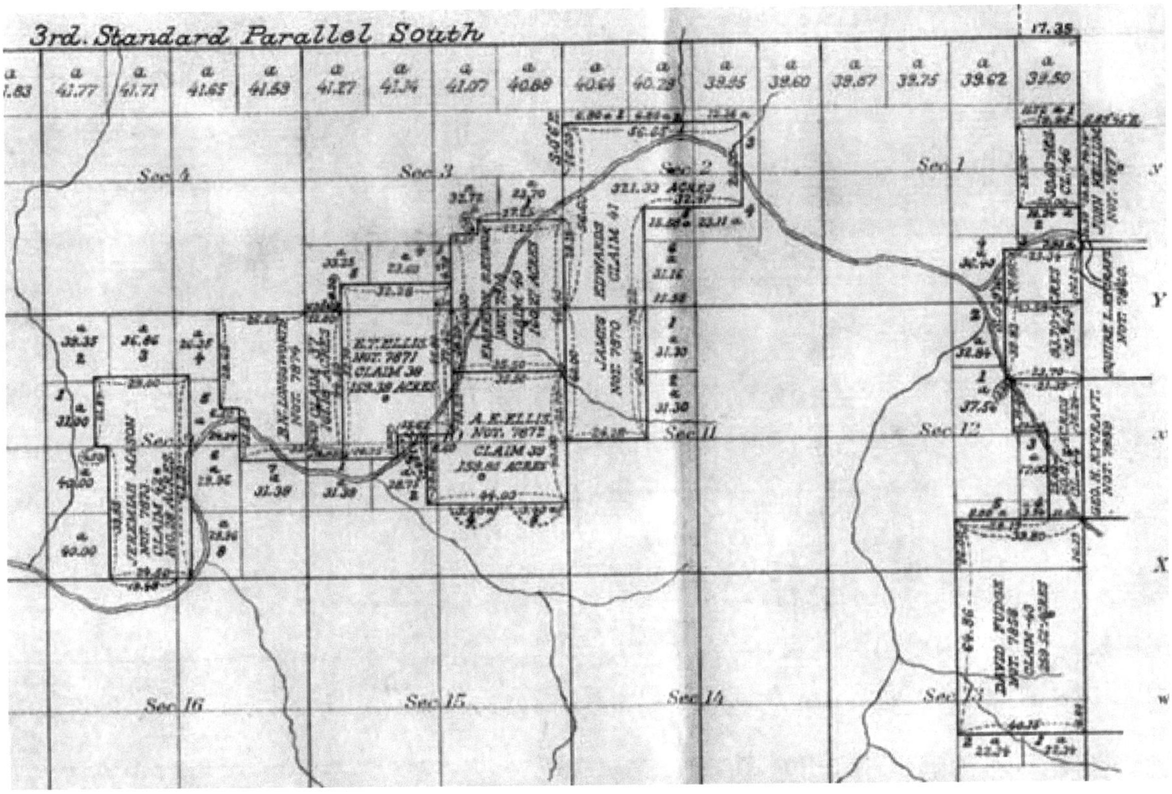


Map D.02 T. 14 S., R. 8 W., GLO Donation Land Claims (DLC), 1856.

The GLO rectangular survey system is based on an systematic arrangement of square miles, or "sections." Each section was located with four corners one mile distance from one another, each marked with four "bearing trees." Bearing trees were established in quadrants: one in each of the four adjacent sections that shared a corner. Halfway between each section corner a "1/4 sec." corner was established with two bearing trees; one on either side of the line. Therefore, each section corner and 1/4 sec. corner located bearing trees in each adjacent section. A section had four corners and four 1/4 sec. corners, so for each forested square mile a total of eight bearing trees could be expected to be established. In instances where trees were absent or too small to be used for surveys, other methods were used to mark corners. Bearing trees were carefully marked, identified as to species and general condition, measured for diameter, and located by distance and compass bearing from each corner. They are protected by strict laws and are the basis for all property and subsequent tax lot descriptions and subdivisions from 1851 to now. When original bearing trees have been destroyed, or are unable to be relocated due to surveyor error or other cause, they are replaced with new bearing trees, so a continuous record of all survey corners and trees is kept from the time of the original survey (1851-1910) until now.

An exact square mile contains 640 acres. Due to the curvature of the earth and occasional surveyor error, a section rarely contains exactly 640 acres. For the same reasons, or in instances where a survey intersects the shoreline of the ocean, a large lake, or wide river, townships sometimes contain less than 36 sections. In such cases, the numbering pattern remains the same as if the township were whole (see Map 2.02). DLCs established by December 1, 1850 could be as many as 640 acres in size; 320 acres for "white settler" or "American half-breed" citizens at least 18 years of age (Carey 1971: 253), and the same amount for his wife, "to be held by her in her own right" (ibid.: 482). This is one of the first federal laws extending equal rights to married women who, by Oregon law at that time, could be as young as 12 years old. Single persons (men and widows older than 18) could claim 320 acres. From December 2, 1850 until 1855, a man and wife could claim 320 acres and a single person 160 acres. All of the DLCs shown on Map D.02 were claimed after 1850, so none are larger than 320 acres. Their relative size indicates whether the claim was made by a single male or a man and wife. DLC surveys were conducted in the same manner as rectangular subdivisions, and were tied into the subdivisions by measure and description, but used a different method for establishing bearing trees at each corner. Because DLCs often had six, eight, or more corners, and because they included trees well-within the boundaries of section lines, these surveys provide us with far more exacting information--particularly near the center of the sections in which they are located--than surveys of sections that do not include DLCs (see maps 3.03 and 3.04).

Map D.03 shows a portion of the GLO map of T. 15 S., R. 7 W., illustrating corrections made in 1940 to the survey of 1897. Note the rectangular sections of the original survey have now been replaced with trapezoids and octagons. This map demonstrates one method by which survey error can be corrected, while still using the original bearing trees. In extreme cases, initial results were voided and entirely new surveys undertaken. Speculation regarding the cause of error in this township is that the "surveyor was lazy" and simply took the "easiest route between corners, rather than the straightest" (Mardis 2002: personal communication).

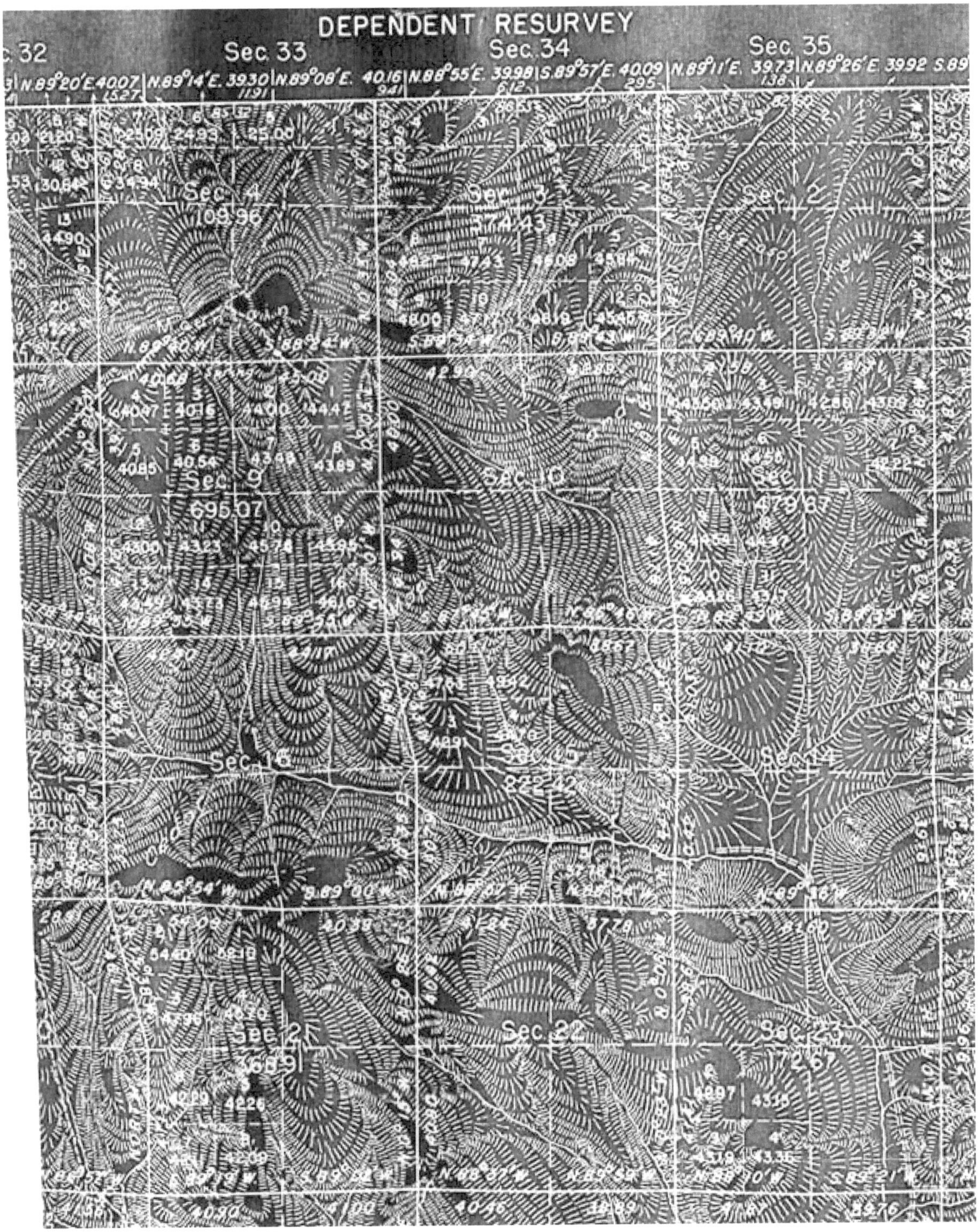


Map D.03 T. 15 S., R. 7 W., GLO errors with corrections, 1897-1940.

Other survey errors noted in the study area were usually noted in the field and corrected at the time (e.g., Gesner 1891a: 244-245; Mercer 1878: 164). Some errors were less troublesome and left uncorrected, as when Collier (1893: 185) reported the diameters of bearing trees had not changed in the 40 years since they were first established and, in one instance, had even decreased an inch in size over a 15 year period.

In addition to surveyor error, fraud and statistical bias have been other causes for concern regarding the quality and usefulness of GLO data. Fraud is a minor problem, although not unknown, with GLO surveys. A well known case that took place in California from 1873 to 1885, the so-called "Benson Frauds" (Stewart 1935: 66), has helped cause some researchers to conclude the practice was far more prevalent than it was. In my own experience, I have examined hundreds of GLO maps and thousands of pages of GLO field notes for Oregon and southwest Washington during the past 30 years (e.g., Braman 1987; Zybach 1992; Zybach 1999) and uncovered only a few minor instances of possible dishonesty: the Sharp (1897) and Gesner (1891a) surveys used in this appendix are about as close as any, and they tend to represent sloppy workmanship rather than outright fraud. In most cases, such poor quality work is usually (as in the instance illustrated by Map D.03) compensated or corrected by others at some later point in time, and has little bearing on the overall quality of the surveys (see Bourdo 1956: 760). The same holds true for surveyor bias. Statistical tests have demonstrated that, while bias certainly exists, it is of little consequence when considered in context of the entire body of work or when considering broad spatial scales (Bourdo 1956:760-767; Schulte and Mladenoff 2001: 7; Whitney and DeCant 2001: 155).

### Tsp. 13 S., Rng 7 W. (13-7): Indian Trail to Alseya

General Description

[Hathorn 1856: 332] The quality of soil in this fractional township along the river is 1st rate; the hilly portions 2nd rate; the timber is principally fir; the open portions are covered with hazel and fern. The unsurveyed portions of the township are too mountainous for settlement or cultivation.

Sec. 1 [N mile between 1 and 6] Top of ridge . . . Top of ridge . . . A maple, 6 ins. . . . A maple, 4 ins. . . . Wells Creek . . . Indian Trail to the Alseya, course SW. Foot of hill, begin to ascend . . . Land, mountainous and unfit for cultivation. Fir, timber (Gordon and Preston 1854: 275)

[N mile between 1 and 6] Top of spur . . . Old corner A fir, 4 ins. . . . marks overgrown. A fir, 8 ins. . . . marks overgrown. A fir, 18 ins. . . . marks overgrown. A yew, 12 ins. . . . marks overgrown. Top of ridge . . . Ravine . . . Top of ridge . . . Bottom of ravine . . . Top of ridge . . . Bottom of ravine . . . Top of ridge . . . A fir, 12 ins. . . . A fir, 12 ins. . . . the bearing trees have probably rotted and slid down the hill . . . Top of a ridge . . . Wells Creek . . . Road from Philomath to Alsea . . . Stream . . . Indian trail, from Willamette Valley to Alsea, course SW. and NE. . . . A red fir, 4 ins. . . . A red fir, 18 ins. . . . A red fir, 5 ins. . . . A red fir, 6 ins. . . . Heavily timbered with fir, cedar and hemlock, and covered with a dense undergrowth of vine maple, salal, rhododendron and devil's war club (Collier 1892: 339-341)

### Tsp. 13 S., Rng. 8 W. (13-8): Grass Mountain

General Description

[Hathorn 1856: 412] This township is all mountainous. Timber principally fir and much of it burnt; none but sec. 36 deemed fit for settlement. There are some burnt openings covered with fern, but no grass land except near the summit of a very high mountain, two miles from the west line and two from the south line of the township. On this there is probably a sec. of grass.

[Collier, September 23, 1891: 407] The greater portion of this township is rough, mountainous and broken. There is much rich bottom land along the Alsea and it's tributaries, which rises in the township. The entire township is covered with heavy timber and dense undergrowth, which has been cleared away in a few spots by the settlers. There are a number of settlers in the township. It should be subdivided.

[Sharp, September 21, 1897: 454-455] This township contains but one variety of land, rough, mountainous and broken. The soil is of one general character and about 3rd rate, productive of much vegetation, vines and timber, owing to the heavy rainfall. It is entirely without grass except in a few open glades on Grass Mountain in about the center of the township. Extensive fires raged over the greater part of this township about forty years ago leaving much dead timber now intermingled with a new growth, in many places being thickets. Much valuable timber exists upon this township, but large quantities have been destroyed by forest fires . . . Grass Mountain, the highest land in the township, is located nearly in the center and from it the water flows north, south, east and west, eventually all in the Alsea River. The township is entirely without settlement, the petitioners praying for the survey are located upon sec. 36 of this township . . . But little, if any, of this township is suitable for settlement. (Sharp, September 21, 1897: 454-455).

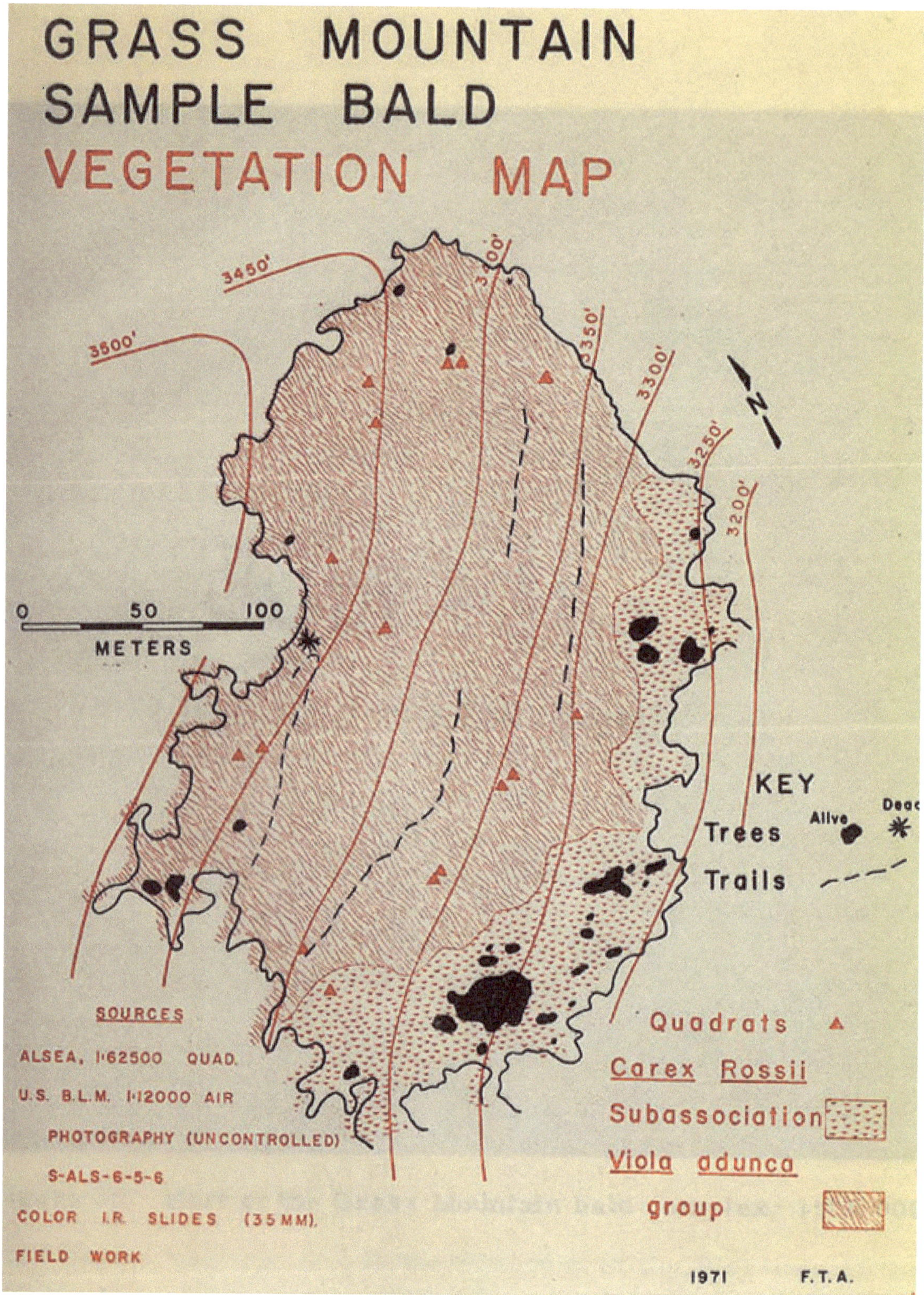


Map D.04 Grass Mountain balds and trails, 1971.

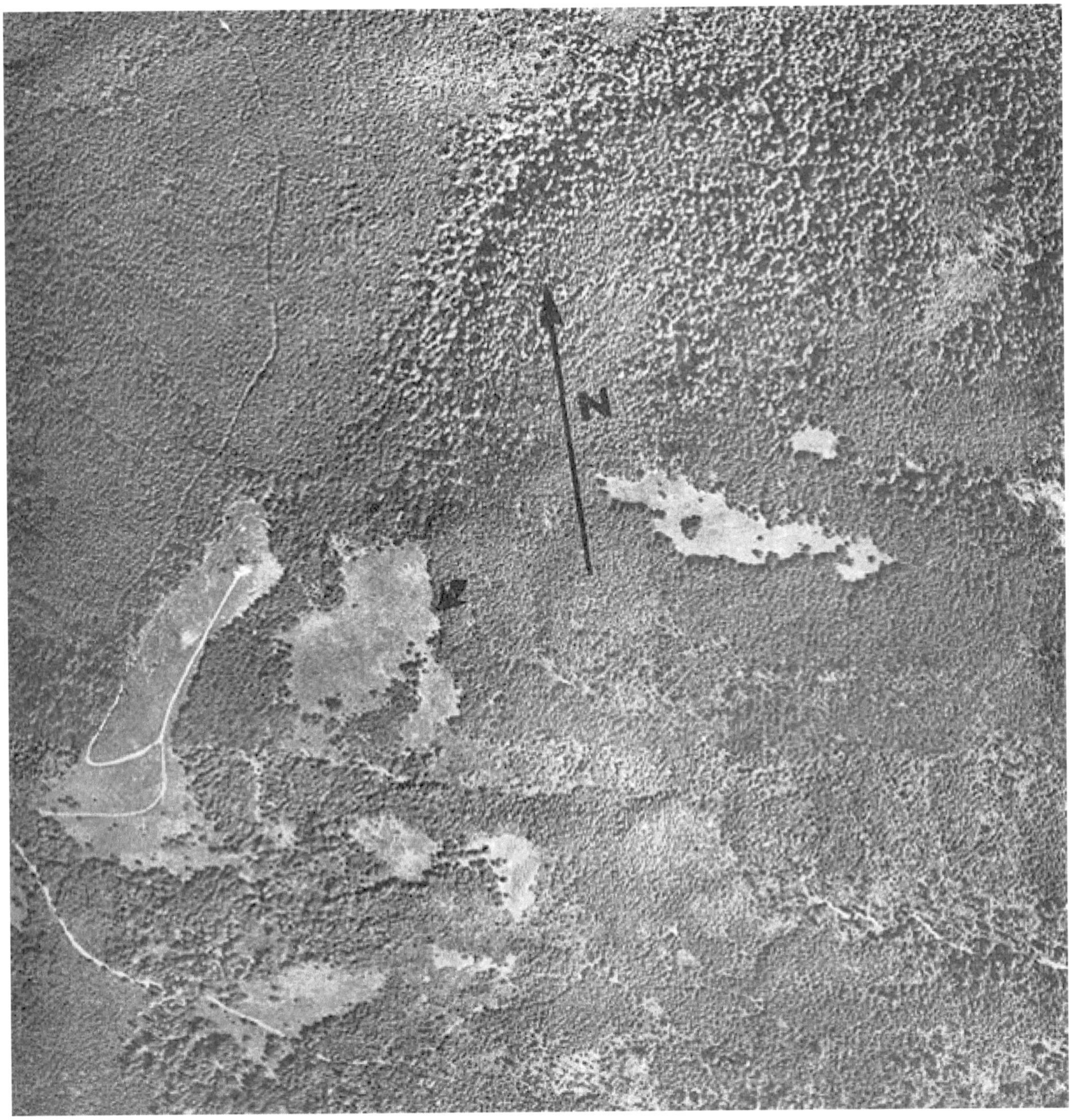


Figure D.01 Aerial photo of Grass Mountain balds and trails, 1966.

**Tsp. 14 S., Rng. 7 W. (14-7): Inmon Mill**

General Description

[Hathorn, c. June 15, 1856: 164] The quality of the soil in this fractional township, along the Alseya and its branches, is excellent and produces abundance of grass. The uplands are 2nd rate soil and covered with a very heavy growth of fern. The timber is principally fir.

[Collier, April 24, 1893: 175] A large part seems to have its timber deadened, offering excellent range for cattle . . . the snow lay just south of our line during the time of the survey. . . There are several settlers scattered through different parts. It should be subdivided.

[Collier May 12, 1893: 228] 43 miles, 77 chs. and 15 lks. of these subdivision lines run over mountainous lands or through timber. . . The usual vegetation of the Coast Range grows strongly everywhere. Wild grass and pea vine grows thick where other growths have been killed. Most of the openings are taken by a rank growth of fern. The whole has been covered by heavy fir, cedar and hemlock forests which has been deadened by fire on about half the area and the slopes exposed afford excellent range for cattle and sheep.

| | |
|---|---|
| **Sec. 1** | [S mile between 1 and 6] A fir, 24 ins. . . . A fir, 18 ins. . . . A fir, 36 ins. . . . Land, all rough mountains. . . Timber, fir, cedar, etc. Undergrowth, hazel, fern, fir, etc. The west 3 1/2 miles of this line is rough mountains of the Coast Range . . . Top of ridge . . . A fir, 28 ins. . . . A fir, 18 ins. . . . A fir , 30 ins. . . . A fir, 30 ins. . . . A fir 40 ins. . . . A fir, 60 ins. . . . A fir, 14 ins.. . . . All rough mountains. . . Timber, fir and cedar, principally dead. Undergrowth, hazel, fern, briars, etc. (Webster October 12, 1853: 98).<br>[W mile between 1 and 36] A sink of earth about 10 feet deep and 10 wide. . . . Summit of ridge. . . A fir 36 ins. . . .A fir 12 ins. . . Summit of ridge. . . A cedar 30 ins. . . . A fir 30 ins. . . Timber, fir cedar and hemlock. Undergrowth, hemlock, fern and salal (Hathorn 1856: 151). |
| **Sec. 2** | [W mile between 2 and 35] Summit of mountain, overlooking the Alseya Valley, course N. and S. . . A fir 24 ins. . . . A fir, 12 ins. . . A fir, 24 . . . A Maple, 20 ins. . . A hemlock, 20 ins. . . . Land, very rough and west slope of mountain very steep and rocky (Hathorn 1856: 151)<br>[S mile between 1 and 2] Along east slope in heavy forest of fir and hemlock. Top of high timbered ridge . . . Thence along steep West slope in heavy timber and thick salal . . . A fir, 24 ins. . . . A fir, 48 ins. . . . Spur from Buck Peak . . . Fir, 60 ins. . . . Ravine . . . Spur . . . Heavily timbered with fir, cedar, and hemlock and covered with dense undergrowth of salal (Collier 1893: 191) |
| **Bagley 1915** | Yellow or bastard fir is just fair in quality, is low in clears and there is a large amount of defect. Red fir is young and thrifty but low in clears. Hemlock is young, thrifty and low in clears. Cedar is defective and good for posts only . . . A ground fire has been over nearly all of the section years ago and while it did not kill any great amount of timber at the time, a large amount is beginning to show defect. The fire risk is nominal. |
| **Sec. 3** | [3/34 W] A fir, 40 ins. . . . Summit of narrow ridge . . . A fir, 36 ins. . . .A fir, 24 ins. . . A cedar, 30 ins. . . .Summit of a ridge . . . A fir, 40 ins. . . .A dogwood 12 ins. . . .Timber, fir, cedar and hemlock. Undergrowth, salal and fern (Hathorn 1856: 152)<br>[2/3 S] A maple, 20 ins. . . . A hemlock, 20 ins. . . . Descend gradually through heavy timber. . . Ascend gradually . . . A fir, 36 ins. . . . A fir, 18 ins. . . .A fir, 18 ins. . . . Ascend rapidly. Top of ridge . . .descend sharply . . . Heavily timbered with fir, cedar and hemlock, with undergrowth of salal, vine maple and arrowwood (Collier 1893: 196) |
| **Sec. 4** | [W mile between 4 and 33] A fir, 36 ins. . . .Summit of ridge . . . A cedar, 48 ins. . . . A vine maple, 9 ins. . . . A fir, 12 ins. . . . A fir, 15 ins. . . .Timber, fir, cedar and hemlock. Undergrowth, vine maple, hazel and fern (Hathorn 1856: 152-153)<br>[S mile between 3 and 4] A fir, 40 ins. . . .mkd. with surveyor's guage but mark mostly burnt off . . . A dogwood, 12 ins. . . .now fallen, mkd. but marks nearly rotted away [exact species and sizes recorded by Hathorn 1856, pg. 152: killed shortly thereafter?]. . . A fir, 5 ins. . . . Descend sharply. Enter bottom. Fence . . . enter meadow . . . Mr. Hutchin's house . . . about 10 chs. dist. Mr. Trout's house . . . about 25 chs. dist. . . . leave meadow and clearing, enter timber and brush . . . Leave bottom; ascend along broken N. slope . . . A maple, 8 ins. . . . A maple, 5 iins. . . . Top of ridge . . . Land . . . bottom . . . partially cleared and seeded to grass . . . Heavily timbered with fir, cedar and |

hemlock and covered with dense undergrowth of same and vine maple, arrowwood, dogwood, salal and maple (Collier 1893: 207-208)

**Boise 1893** Mr. Boise has small clearing and cabin in SW 1/4 (Collier 1893: 229)

**Hutchins 1893** Mr. Hutchins has house, barn and 7 or 8 acres in cultivation or meadow in sec. 4, NE 1/4 (Collier 1893: 229)

**Trout 1893** Mr. Trout has house and barn and some land in cultivation in NE 1/4 (Collier 1893: 229)

**Bagley 1915** Yellow fir is sound and A#1 grade. Red fir is sound and a fair grade. Cedar is sound and a good quality. There is some fire damage as shown in red on the plat [84 acres]. The fire risk is ordinary one.

**<u>Sec. 5</u>** [5/32 W] Enter prairie, course N. and SW. . . . A fir, 30 ins. . . . A fir, 5 ins. . . . Enter river bottom . . . East half timbered same as last mile [4/33 W]; west half prairie (Hathorn 1856: 153)

[4/5 S] A cedar, 15 ins. . . . A cedar 7 ins. . . . A cedar, 16 ins. . . . A cedar 14 ins. . . A red fir, 24 ins. . . . South Fork of Alsea River . . . A fir, 20 ins. . . . An alder, 20 ins. . . . A fir, 14 ins. . . A fir, 12 ins. . . . A fir, 30 ins. . . . A fir, 36 ins. . . . Timber, fir, cedar, hemlock alder, with heavy underbrush (Mercer 1878: 164-165)

**Bagley 1915** Yellow fir is sound and a good quality. Red fir and cedar is a fair quality of timber. There is some fire damage as shown in red on the plat. The fire risk is ordinary one.

**Sec. 6** [W mile between 6 and 31: Error?] Leave prairie and enter timber. . . . Enter North Fork of Alseya River. Leave same . . . Enter prairie . . . Jacob Holgate's house . . . Trail . . . Enter thicket . . . Leave same . . . Leave bottom and commence ascending hill . . . Enter timber . . . A maple, 10 ins. . . . A maple, 4 ins. . . . Enter fern prairie . . . .A maple 20 ins. . . . A maple 20 ins. . . . Land hilly. Timber, principally fir and maple. The high prairie is covered with a thick coat of fern from 6 to 8 ft. high. (Hathorn, June 9, 1856:153-154)

[W mile between 6 and 31: Error?] Trail . . . Enter alder thicket . . . Leave same . . . Enter timber . . . Leave timber and enter fir opening . . . A fir, 40 in. . . . 285 lks. . . . An alder, 10 in. . . . 218 lks. . . . An alder, 9 in. . . 135 lks. . . . An alder, 10 in. . . . 77 lks. . . . No tre convenient NE. Land hilly; about half covered with fir and maple; balance open and all the high portions are covered with a large growth of fern (Hathorn, June 28, 1856: 325-326)

[S mile between 5 and 6] Leave prairie and enter timber . . . A dogwood, 10 ins. . . . A vine maple, 10 ins. . . . Summit of ridge . . . About one-fourth level land . . . Timbered with fir. Under growth maple, hazel, and fern (Hathorn 1856: 161-162)

**Holgate 1856** [#7879: 319.5 A., beginning at NW corner in 13-7-31] A dogwood, 6 ins. . . . A fir, 10 ins. . . . Enter prairie . . . Enter thicket . . .Enter field . . . Enter Holgate and William's house. Leave same. Leave field . . . Trail . . . Enter thicket, Enter prairie, Enter timber . . .Enter prairie . . . A fir, 40 ins. . . . 1028 lks. A fir, 40 ins. . . . 1080 lks. . . . An oak, 12 ins. . . . 233 lks. An oak, 12 ins. . . .177 lks. . . . Summit of ridge . . . Enter timber . . .Enter North Fork of Alseya River. Leave same . . . Enter same Leave same . . . A crab apple, 12 ins. . . .An ash, 30 ins. . . . An ash, 12 ins. . . . A fir, 60 [ins.?] wide . . . 190 lks. . . . Enter prairie or openings . . . An oak, 30 ins. . . .A fir, 36 ins. . . . 820 lks. . . . Leave prairie and enter timber . . . Enter North Fork of Alseya River. Leave same . . . Enter prairie . . . Trail . . . Enter thicket . . . Enter prairie . . . Leave prairie and enter timber . . . To beginning cor. Land about half level prairie and timbered bottoms. . . Timber fir and maple principally. About ten acres in cultivation. Buildings 2nd rate log. (Hathorn 1856a: 469-471)

**Bagley 1915** Yellow fir is a good quality of timber. Red fir and cedar are a coarse grade of timber . . . There has been no recent fire damage and the fire risk is ordinary. The balance of the section is open grass land and farming land.

**Sec. 7** [N mile between 7 and 8] . . Enter openings . . . Enter timber . . . Summit of ridge . . . A fir, 50 ins. . . . 270 lks. A fir, 50 ins. . . . 100 lks. Summit of ridge . . . A fir, 15 ins. . . . 221 lks. A fir, 40 ins. . . . 245 lks. A fir, 48 ins. . . . 127 lks. A fir, 6 ins. . . . 367 lks. Land about the same as last mile [E between 7 and 18] (Hathorn 1856: 160-161)

[E mile between 6 and 7] Leave prairie and enter timber . . . Leave same and enter prairie . . . Enter timber . . . A fir, 30 ins. . . . A fir, 12 ins. . . . Enter openings . . . Summit of ridge . . . Enter timber . . . Enter openings . . . Timber, fir. Undergrowth, maple, hazel, oak and fern (Hathorn 1856: 161)

**Bagley 1915** Yellow or bastard fir is defective, there being considerable stump rot and broken tops. Red and white fir is a young, thrifty growth, no clears. A fire of many years ago has been over the entire section and it has grown up to brush. The present fire risk is nominal.

**Sec. 8** [N mile between 8 and 9] . . . A hemlock, 18 ins. . . . a maple, 12 ins. . . A maple, 24 ins. . . . A maple, 24 ins. . . A fir, 12 ins. . . .A fir, 18 ins. . . . A cedar, 12 ins. . . . Land about the same as last mile [E between 8 and 17] (Hathorn 1856: 162-163)

[E mile between 5 and 8] . . . A fir, 12 ins. . . . A fir, 24 ins. . . .Summit of ridge . . . Land about the same as last two miles [E between 8 and 17; N between 8 and 9] (Hathorn 1856: 163)

**Bagley 1915** Yellow fir is a fair quality of timber. Red fir is a coarse common grade. Cedar is a fair quality. Burned lands are shown on the plat [153 acres]. The fire risk is an ordinar[y] one.

**Sec. 9** [N mile between 9 and 10] Ascend in old burn . . . Enter old burn . . . A dead fir, 30 ins. diam. . . . 37 chs. heavily timbered with fir, cedar and hemlock, remainder in old deadening with scattered groves of fir timber and many old logs and stubs, with patches of dense salal and huckleberry brush (Collier 1893: 205-206)

[W mile between 4 and 9] Fir, 24 ins. . . . A fir, 12 ins. . . . A fir, 10 ins. . . . Enter deadening and slashing in bottom, Boise's house . . . about 7 chs. dist. . . . Heavily timbered with fir, cedar and hemlock with undergrowth of salal, blackberry, hazel, dogwood and vine maple . . . covered with dense undergrowth of weeds and briars and an almost imprenetrable and unsermountable piles of brush of vine maple, etc. (Collier 1893: 217-218)

**Sec. 10** [N mile between 10 and 11] A dead cedar, 20 ins. . . A dead fir, 24, ins. . . . old deadening with logs and stubs, thickly strewn and some thickets of hazel and maple . . . Some scattering groves of green fir on the mile, but most of it is old deadening. Slopes covered with logs, fern and grass (Collier 1893: 193-194)

[W mile between 3 and 10] Top of ridge . . . A fir, 20 ins. . . . A fir, 36 ins. . . . Top of spur . . . Stream . . . Top of spur . . . Heavily timbered with fir, and covered with dense undergrowth of salal and vine maple. The East 29 chs.. partly deadened (Collier 1893: 206-207)

**Sec. 11** Heavily timbered with fir on last 24 chs., scattering live fir and remains of old deadened forest on remainder (Collier 1893: 188)

**Sec. 12** Enter deadening, desc. rapidly through dead timber and brush. . . . Heavily timbered with fir, deadened in some spots (Collier 1893: 187)

**Sec. 14** [W mile between 11 and 14] Ascend gradually over rolling table land in old deadening. . . Top of ridge . . . through belt of live timber. . . . leave live timber . . . A fir, 16 ins. . . . A fir, 30 ins. . . . Old fir logs and stubs, thickly scattered with occasional live fir trees. Covered with good grass and patches of thimbleberry with strong growth of fern, now all flattened down (Collier 1893: 192-193)

**Sec. 15** [N mile between 14 and 15] Top of spur . . . in heavy fir timber . . . Fir, 26 ins. . . . enter opening, old deadening of heavy forest . . . A fir 7 ins. . . . A fir, 36 ins. . . . Road . . . Top of ridge . . . Spring . . . enter dense thicket of alder and willow. . . East branch of Trout Creek . . . leave thicket . . . A dead fir, 36 ins. . . . A dead fir, 24 ins. . . . A dead fir 24 ins. . . . A dead fir 24 ins. . . . heavily timbered with fir, with dense undergrowth of salal, arrowwood and hazel; 56 chs. through deadening over grown in patches with fir and hazel and maple, thick with logs and stubs (Collier 1893: 191-192)

[W mile between 10 and 15] Top of spur . . . A dead fir, 36 ins. . . . A dead fir, 50 ins. . . . Porter's house . . . Top of ridge . . . Some scattering groves of green fir on the mile, but most of it is old deadening. Slopes covered with logs, fern and grass (Collier 1893: 204-205)

**Summers 1893** Mr. Summers has house and barn and about 5 acres in cultivation or meadow in SE 1/4 (Collier 1893: 229)

**Porter 1893** T. W. Porter has house, barn and 2 or 3 acres fenced in NW 1/4 (Collier 1893: 229)

***Sec. 16** [N mile between 15 and 16] . . . Trout Creek. . . Fir, 16 ins. . . . Road to Meyer's . . . A fir, 5 ins. . . . A fir, 5 ins. . . . Road . . . Porter's house . . . Trout Creek . . . A dead fir, 48 ins. . . . dead fir 48, ins. . . live fir, 34, ins. . . . live fir, 30 ins. . . . Heavily timbered with fir, cedar, maple and alder, partly deadened, strewn with big logs, covered in most parts with dense undergrowth of salal, hazel, thimbleberry and dogwood (Collier 1893: 203-204)

**Meyers R. 1893** R. G. Meyers has house, barn, and outhouses and about 15 acres in cultivation and 3/4 acre in orchard in sec. 16, besides a large quantity of range fenced (Collier 1893: 229)

**Smith 1893** Mr. Smith has about 4 acres in cultivation and 1/2 acre in orchard in NW 1/4 of Sec 16 and house and barn in same (Collier 1893: 229)

**Sec. 17** A dead fir, 30 ins. . . . Thence over high open steep grassy S. slope . . . Thence over more gentle descent on grassy S. slope . . . heavily timbered with fir, cedar, maple and alder, with dense undergrowth of salal, arrowwood, hazel, blackberry; remainder open in old burn. (Collier 1893: 228)

**Sec. 18** A fir, 36 ins. . . . A cedar, 15 ins. . . . A fir 8 ins. . . . A cedar, 12 ins. Webster 1853: 99)

[N mile between 17 and 18] Summit of ridge . . . A fir, 36 ins. . . . A fir, 12 ins. . . . Summit of ridge, course E. and W. South trail to the Willamette, course NW. and SE. . . . A maple, 12 ins. . . . A maple, 10 ins. . . . A maple, 10 ins. . . . A cedar, 12 ins. (Hathorn 1856: 159)

**Sec. 19** enter old deadening . . . Heavily timbered with fir, cedar, hemlock, maple and alder, deadened last 59.50 chs., covered with dense undergrowth of cherry, balm, alder, huckleberry, blackberry (Collier 1893: 227)

**Meyers H. 1893** H. C. Meyers has about 5 acres of meadow in sec. 19 in NE 1/4(Collier 1893: 229)

**Sec. 20** [N mile between 20 and 21] Top of rise . . . Rock Creek . . . Top of spur . . . A cedar, 20 ins. . . A cedar, 30 ins. . . . Enter Alsea River ., . . A fir, 50 ins. . . .A dead fir, 24 ins. . . A dead fir, 40 ins. . . . An alder, 5 ins. . . . Heavily timbered with fir, cedar, alder and maple, mostly deadened by fire, covered with dense undergrowth of same and huckleberry, blackberry and salal (Collier 1893: 212-214)

**Hedwig 1893** Mr. Hedwig has 1 acre in cultivation and 6 acres in meadow and house in NW 1/4 (Collier 1893: 229)

**Sec. 21** [N mile between 21 and 22] A dead fir, 40 ins. . . . A fir, 6 ins. . . . . To left bank of Alsea River . . . To right bank . . . ascend on E. slope in heavy live timber. . . A fir, 10 ins. . . . A fir, 4 ins. . . . A fir, 24 ins. . . . A fir, 20 ins. . . .Heavily timbered with fir, cedar, alder and maple; the first 63 chs. having its large timber mostly deadened by old burn, covered with dense undergrowth of huckleberry, salal, arrowwood, cherry and maple (Collier 1893: 201-202)

[W mile between 16 and 21] Top of spur . . . Leave live timber; thence along grassy S. slope . . . A fir, 30 ins. . . . A fir, 18 ins. . . .R. G. Myer's house . . . Enter heavy fir timber . . . 44 chs. heavily timbered with fir and cedar and covered with dense undergrowth of salal, willow, hazel and arrowwood; remainder nearly open with scattering firs. On southern slope, covered with grass and pea vines (Collier 1893: 214)

**Fagan 1885** Ruble's Mills--These mills are located at the upper end of Alsea valley and prove a boon to the country in which they are located. The saw mill was constructed in 1872 and the grist mill in 1873. When first started the latter was fitted with a single pair of burrs and had a capacity of about one hundred bushels in ten hours, while the former had but a single sash saw, capable of producing fifteen hundred feet of lumber per day. In 1884 a new saw mill was built in the forks formed by the Alsea river and Rock creek, and on the opposite side from the old site. This building is fitted with a circular saw, planer, etc., while the lumber manufactured is from the great forests of cedar, fir, alder and maple that abound in the vicinity. There are one hundred and twenty acres attached to the mills.

**Sec. 22** [N mile between 22 and 23] Descend . . . Enter live timber, dense thickets of cherry, vine maple, salal, arrowwood. Enter low wet bottom. South Alsea . . . West's rail fence . . . Enter partial clearing of the underbrush . . . Rail fence . . . leave partial clearing of brush . . . A fir, 18 ins. . . . A fir, 12 ins. . . .Top of rise . . . . Ridge . . . A fir, 18 ins. . . . A fir, 16 ins. . . . A fir, 12 ins. . . . A fir, 30 ins. . . . nearly bare of live timber with old logs and stumps of fir forest; remainder heavily timbered with fir, cedar, hemlock, chinquapin and maple with dense undergrowth in places of salal, vine maple and huckleberry (Collier 1893: 181-182)

[W mile between 15 and 22] Descend along N. side of canyon . . . Enter opening . . . A fir, 10 ins. . . . A fir, 5 ins. . . . Descend sharply into green timber . . . Trout Creek . . . Heavily timbered with fir, cedar and hemlock, with some small glades of opening, with dense undergrowth almost throughout of salal, hazel and arrowwood (Collier 1893: 202-203)

**West 1893** Charles West has house, barn, sheds and about 30 acres in cultivation or meadow in SE 1/4 (Collier 1893: 229)

**Bagley 1915** Yellow or bastard fir is good quality and generally sound. Red fir is a thrifty growth, low in clears. Cedar is only good for posts and shingles . . . This is a railroading proposition down the Alsea river. Fire has been over the entire section except about 25 acres . . . and about 10 acres . . . There is a scattering timber all through the old burn. The present fire risk is nominal. There is a homesteader on the SW 1/4 SE 1/4 . . . All that part of section not reported on is in the old burn and does not carry any timber.

**Sec. 23** [W mile between 14 and 23] . . . Road . . . A fir, 8 ins. . . . 246 lks. . . . A dead fir, 40 ins. . . . 187 lks. . . . Top of steep ascent . . . thence over rounding top of ridge . . . Enter belt of green fir timber. Enter deadening . . . 4 chs. covered with heavy fir forest, remainder through heavy old deadening with patches of undergrowth of salal, fern, etc. and on the high part, good grass (Collier 1893: 183-184)

**Sec. 24** Leave burn and enter timber (Mercer 1878: 167)

**Sec. 25** A fir, 28 ins. . . A fir 18 ins. . . . A fir 15 ins. . . . A fir 24 ins. (Webster 1853: 120)

**Sec. 26** [W mile between 23 and 26] Ascend. Leave live timber. Top of mountain [Name?] . . . Enter bottom with dense underbrush and timber . . . Green Peak Fork of Alsea . . . Cherry, 3 ins. . . . A cherry, 5 ins. . . . A fir, 8 ins. . . . East Fork of Alsea River . . . Leave same. Leave bottom . . . Land, 14 chs. bottom, covered with dense undergrowth and heavy timber; 65.92 chs. mountainous covered with scattering live timber and old logs and stubs of fir, and dense undergrowth of salal, fern, hazel, etc." (Collier 1893: 180-181)

**Sec. 27** [N mile between 26 and 27] Top of rise . . Cedar, 16 ins. . . .Fir, 50 ins. . . . Fir, 50 ins. . . . Leave live timber. Top of ridge . . . A fir, 10 ins. . . . 119 lks. . . . A fir, 36 ins. . . . 250 lks. . . . West's house [Sec. 22] . . . Head of slide and spring . . . Fir, 48 ins. . . . Some croppings of coal . . . A fir, 8 ins. . . . 115 lks. . . . A fir, 4 ins. . . . 120 lks. . . . A fir, 3 ins. . . . 88 lks. . . . A fir, 4 ins. . . . 75 lks. . . . heavily timbered with fir, cedar and hemlock, some groves of fir and some patches of brush on last part (Collier 1893: 178-179)

[W mile between 22 and 27] Cross a little hollow . . . Top of spur . . . West's house . . . West's barn . . . Onto table land . . . A dead cedar, 24 ins. . . A dead cedar, 20 ins. . . . Ridge . . . Top of ridge . . . Timber, mostly deadened, with patches of brush and alder groves along the streams (Collier 1893: 200-201)

**Sec. 28** [N mile between 27 and 28] Descend in old deadening . . . A dead fir, 48 ins. . . . A dead fir, 36 ins. . . . Hole in the ground, 10 ft. in diam., 15 ft. deep, a stream, about 2 lks. wide, observed in the bottom, course W. . . . Ridge . . . An alder, 5 ins. . . . An alder, 10 ins. . . . An alder, 5 ins. . . . A dogwood, 5 ins. . . . through old deadening with many stubs standing, many logs strewn over surface, with occasional thickets of fir, hazel and arrowwood and alder groves along the stream (Collier 1893: 199)

[W mile between 21 and 28] Ascend. Top of ridge . . . Stream [Blackberry Creek] . . . Top of ridge . . . A fir, 7 ins. . . . A fir, 6 ins. . . . Top of ridge . . . In old deadening of heavy fir forest; covered in part with dense undergrowth of cherry, alder, maple and salal, with occasional open grassy slopes (Collier 1893: 212)

**Sec. 29** A dead fir, 72 ins. . . . A dead fir, 60 ins. . . . A dead cedar, 12 ins. . . . A dead cedar, 40 ins. . . .A dead fir, 36 ins. . . A dead fir, 30 ins.. . . In old deadening of heavy fir forest, covered in most parts with dense undergrowth of cherry, balm, alder, maple and fir, with some open grassy patches Dead cedar, 16 ins. . . An alder, 4 ins. . . . An alder, 5 ins. . . A dead fir, 48 ins. . . A dead fir, 24 ins. . . A fir, 48 ins. . . A fir, 48 ins. . . . Scattering live fir timber, but remains of heavy forest (Collier 1893: 211,222- 223)

[W mile between 20 and 29] Descend. Enter bottom . . . Stream [Rock Creek] . . . Asc. Abruptly from bottom. On to spur . . . then up along top of same . . . Head of spur; thence along N. side of mountain. . . A dead fir, 30 ins. . . . A dead fir, 40 ins. . . . Hedwig's house [Sec. 20] . . . about 40 chs. dist. In the Alsea bottom. Top of ridge . . . Spring . . . Timbered with scattering live firs and thick forest of dead firs and cedar, with dense undergrowth on most of the line of cherry, balm, alder, maple, blackberry, huckleberry, etc. (Collier 1893: 223-224)

**Sec. 30** A burnt cedar, 18 ins. . . . a burnt cedar, 20 ins. . . . a burnt fir, 30 ins. . .A burnt yew, 12 ins. . . . Timber, fir and cedar, and mostly burnt (Hathorn 1856a: 158).

**Sec. 31** [N mile between 31 and 32] . . . A fir, 60 ins. . . . A maple, 24 ins. . . . Fir 60 ins. . . . A dead fir, 30 ins. . . A fir, 8 ins. . . A fir, 5 ins. . . . A maple, 4 ins. . . .Heavily timbered with fir, and alder; large timber mostly deadened by old burn, covered in patches with dense undergrowth of salal, cherry, alder and maple (Collier 1893: 219-220)

**Sec. 32** Dead fir timber . . . A fir, 4 ins. . . . A dead fir, 30 ins. . . A dead fir, 36 ins. . . A dead fir, 36 ins. . . . A maple, 4 ins. . . . A cherry, 3 ins. . . . Covered with heavy fir and cedar forests, deadened by old burn, and dense undergrowth of alder, cherry, mountain balm, huckleberry, young fir, and maple (Collier 1893: 209)

**Sec. 33** [N one mile between 33 and 34] Trail to Prairie Mountain, course 750 E., and S. 750 W., on top of ridge of same . . . A fir, 10 ins. . . . A fir, 18 ins. . . . A fir, 20 ins. . . . A fir, 60 ins. . . . Top of Last Chance Ridge . . . enter old deadening . . . A dead fir, 48 ins. . . . A dead fir, 20 ins. . . . A dead fir, 48 ins. . . . A dead fir, 48 ins. . . . Heavily timbered with fir, cedar and hemlock, deadened on last 14 chs., with undergrowth of same and salal (Collier 1893: 197-198)

[28/33 W] Descend steep West slope in old deadening . . . A fir, 5 ins. . . . A fir, 5 ins. . . . A fir, 4 ins. (Collier 1893: 210)

**Sec. 34** Top of Last Chance Ridge . . . enter old deadening . . . Heavily timbered with fir, cedar and hemlock, deadened on last 14 chs. (Collier 1893: 197-198) last 8 chs. in old deadening and fir thicket, remainder covered with heavy timber and dense undergrowth of salal (Collier 1893: 197-198)

**Sec. 35** [W mile between 26 and 35] . . . remains of flume and ditch, starting just above line and running Easterly to Inman's mill (Collier 1893: 178)

**Sec. 36** [30/36 S] A fir, 16 ins. . . . A fir, 15 ins. . . . A fir, 18 ins. . . . Foot of hill . . . A fir 28 ins. . . . A fir, 12 ins. . . . A fir, 12 ins. . . . A cedar, 12 ins. . . . A fir, 40 ins. . . . Timber, fir and cedar. Undergrowth, hazel, fern, fir, etc. (Webster 1853: 100-101)

[25/36 W] South Fork of Alsea River . . . A cedar, 12 ins. . . . A cedar, 14 ins. . . . Same stream . . . Same stream . . . Land rolling; about half good for cultivation. Timber, cedar and hemlock (Mercer 1878: 166)

**Bagley 1915** Yellow fir is sound, fair size and not over ripe. The timber on the SE 1/4 is large and rough, the limbs come low down. The balance of the section is very good grade of timber, old growth and young growth mixed . . . Cedar is scattered along the creeks, is sound and of fairly good quality . . . The timber will have to be taken out by railroad, which can be constructed at a reasonable cost across divide through low pass . . . There is also a low pass through section 8-15-6 . . . No damage by fire. There is danger of fire . . . as here is some very heavy underbrush. Not much fire danger on balance of section as there is very little underbrush except vine maple and alder. The mill on the NE 1/4 NW 1/4 is burned. There are a good many logs cut on the ground. There is a very good 7 room house and a very good large barn on NW 1/4 NW 1/4 which are abandoned . . . There is 7 acres cleared and grass . . . There are also several small garden spots cleared on the old mill site . . . The buildings on the two old mill sites are rough board houses (Bagley, April, 1915)

## Tsp. 14 S., Rng. 8 W. (14-8): Indian Trail to Tidewater

General Description

[Hathorn, July 3, 1856: 278-279] The open portion is covered with a very heavy growth of fern. Timber, being chiefly fir, cedar and maple. There is a plain trail leading from this township to tide water on the Alseya, which is said to be quite passable for horses. The Indians however generally travel it in their canoes from a point near the west line of the township and frequently from near the east line of the same.

[Gesner, June 8, 1891: 313] This township is generally mountainous, rough and broken in the southwest portion . . . It is covered with a dense undergrowth of fir, hemlock, cherry, vine maple, etc. Some excellent fir timber along the western portion of the township . . . There are about thirteen settlers on the portion of the township I subdivided. Some of them having settled there ten or twenty years ago.

**Sec. 1** [S mile between 1 and 6] Enter timber . . . Leave timber and enter prairie on summit of ridge , course NW. and SE. Trail, course NW. and SE. . . . A W. oak, 24 ins. . . . 850 lks. level. An alder, 12 ins. . . . 349 liks. . . . John Kellum's House . . . An Indian Trail . . . Enter thicket . . . Enter prairie . . . Leave same and enter timber, course NW and SE, . . . Indian trail, course E. and W. . . . Enter bottom, course E. and W. Intersect right bank of North Fork of Alseya River . . . Leave river . . . Same river . . . Leave timber and enter prairie . . . A crab apple, 9 ins. . . . 505 lks. An alder, 20 ins. . . . 500 lks. An oak, 18 ins. . . . 330 lks. An oak, 18 ins. . . .383 lks. The summit of a high mountain bears N. 520 W. . . .

Timber, fir and maple. Undergrowth, maple, briers [sic], vines and fern. Prairie covered with a fine growth of grass. (Hathorn 1856: 154-155)
The post corner was rotted away . . . and the marks on the bearing trees burned off (Dick 1873: 237)
A fir, 50 ins. diam., bears S. 130 E., 310 lks., burned down and nearly destroyed (Gesner 1891a: 242)

**Sec. 4** [W mile between 4 and 33] A fir, 30 in. . . . A fir, 50 in. . . . A fir, 30 in. . . . Summit of ridge . . . Summit of ridge . . . The summit of Holgates Peak [Grass Mountain], bears North. This mt. is open on the S. side near the summit and probably contains about a sec. of the prairie . . . A fir 18 in. . . . A fir, 15 in. (Hathorn 1856: 234-235)
[4/33 E mile between 4 and 33] . . . Spur . . . Spur . . . A fir, 6 ins. . . . A fir, 7 ins. . . . Top of steep ascent at top of rocky bluff . . . thence over table land, sloping gently NE., enter green timber. Trail to Prairie Mountain, course N. 500 E. . . . A hemlock, 20 ins. . . . A fir, 40 ins. . . . A yew, 12 ins. . . . A vine maple, 6 ins. . . .Heavily timbered with fir and hemlock; old growth dead on west 44.50 chs., but young growth in thickets taking its place. Dense undergrowth of salal and rhododendron on last half, 80 chs. (Collier 1893: 172-173)

**Sec. 5** [W one mile between 5 and 33] A fir, 40 ins. . . . A cedar, 48 ins. . . . A maple, 9 ins. . . . A fir, 36 ins. . . . Timber, fir, cedar and maple, and mostly killed by fire. Undergrowth, maple, hazel, fern, briars and vines (Hathorn 1856: 235).

**Sec. 6** [E mile between 6 and 31] . . . Top of spur . . . asc. Over a round, grassy ridge . . . Top of same . . . Ravine . . . An alder, 5 ins. . . . A fir, 8 ins. . . . Top of spur . . . land slide . . . Top of ridge . . . A fir, 5 ins. . . . A fir, 5 ins. . . . A dead fir, 48 ins. . . .A fir, 6 ins. . . . timber, burned. Dense underbrush in most parts of mountain; balance huckleberry and maple, with occasional thickets of fir and open patches of fern and grass (Collier 1893: 169-171)
66.00 Indian trail on summit of ridge, course E. and W. (south between 5 and 6, Hathorn 1856, 14-8: 278)
[W one mile between 6 and 36] Timber, principally fir, cedar and maple, and mostly burnt . . .(Hathorn 1856: 235-236)
[S one mile between 1 and 6] Land, hilly and about half fern prairie; balance burnt fir timber" (Hathorn 1856: 236-237)
[S one mile between 1 and 6] Burnt timber strewn with logs (Gesner 1891: 241-244)

**Sec. 8** 51.00 Indian trail, course NW. and SE.
52.50 Summit of ridge, course NW. and SE.
52.50 Leave prairie and enter timber, course NW. and SE. (north between 8 and 9, Hathorn 1856, 14-8: 273)

**Sec. 17** 19.50 A dim Indian trail, course NE. and SW. (north between 16 and 17, Hathorn 14-8: 272)

**Sec. 19** 45.40 Top of ridge and trail, bears E. and W. (Gesner 1891: 307)

**Sec. 20** 2.50 Top of ridge, bears east and West.
3.00 Trail from Alsea to Lobster, bears East and West (between 20 and 21, Gesner 1891, 14-8: 300)

**Sec. 21** Timber, south 40 chs. Mostly burnt timber; some green alder, maple, and dogwood (Gesner 1891: 285).
27.50 Top of ridge and trail, bears NE. and SW.
38.00 Top of ridge and trail, bears NE. and SW.
74.14 Summit of ridge and trail (west between 16 and 21, Gesner 1891: 285)

**Sec. 22** 10.50 A dim Indian trail, course E. and W. (north between secs. 22 and 23, Hathorn 1856, 14-8: 263)
10.75 Summit of narrow ridge, course N. and S.
Indian trail, course N. and S. (between 22 and 27, Hathorn 1856 14-8: 267)

**Sec. 24** [S mile between 19 and 24] Leave prairie and enter timber . . . Enter prairie . . . summit of ridge . . . Enter timber . . . A fir, 18 ins. . . . A fir, 18 ins. . . . Summit of ridge . . . A fir, 20 ins. . . . A fir, 20 ins. . . . A fir, 15 ins. . . . A fir, 18 ins. (Hathorn 1856a: 158)

**Sec. 25** [S mile between 25 and 30] . . . Summit of a ridge . . . A fir, 36 ins. . . . A maple, 15 ins. . . . Summit of a ridge . . . A burnt cedar, 18 ins. . . . a burnt cedar, 20 ins. . . . a burnt fir, 30 ins. . .A burnt yew, 12 ins. . . . Timber, fir and cedar, and mostly burnt. Undergrowth, maple, hazel, fern, salal, Oregon grape, briers [sic], vines, etc. (Hathorn 1856a: 158).
Creek, 3 lks. Wide, course West; enter burnt woods (Gesner 1891a: 238)

**Sec. 26** Left green and entered burnt timber (Gesner 1891: 292)

**Sec. 27** [W mile between 22 and 27] . . . Summit of narrow ridge, course N. and S. Indian trail, course N. and S. . . . A dogwood, 15 ins. . . . A fir, 50 ins. . . . A fir, 40 ins. . . . A fir, 40 ins. diam. . . . A burnt fir 40 ins. . . . A burnt fir, 40 ins. . . . Timber, fir, cedar and maple, mostly burnt. Undergrowth, maple hazel, fern, salal, briers [sic], vines, etc. (Hathorn 1856a: 267)
[W mile between 22 and 27] . . . Top of ridge and old Lobster Trail, bears SW. and NE. . . . A dogwood, 15 ins. . . . A red fir, 50 ins. . . . Post and bearing trees destroyed by fire . . . A fir, 30 ins. . . .

A fir, 24 ins. . . . A fir, 20 ins. . . . A fir, 16 ins. . . . Timber, fir, maple, dogwood and scatter chinquapin; recent fires have destroyed the principal part of the timber, 80 chs. . . . Dense undergrowth of vine maple, fir, arrowwood, wild cherry and salal berry (Gesner 1891: 283-284).

**Sec. 30:** "Timber, fir, cedar, chinquapin and dogwood; all trees dead in the burn, 80 chs." (Gesner 1891a: 240)

**Sec. 31:** "A deserted cabin, about 10 chs. Up the creek, but the brush was so dense it was impossible to see the house from any part of the line" (Gesner 1891a: 250)
"Timber poor, principally dead." (Gesner 1891a: 251).

**Sec. 33** [W mile between 28 and 33]
Descend steep West slope in old deadening. Enter brush . . . A fir, 5 ins. . . . A fir, 5 ins. . . . A fir, 4 ins. . . . covered with dense undergrowth of cherry, mountain balm, alder, maple and fir (Collier 1893: 210)

**Cox 1891** Wagon road from Lobster to Alsea . . . Mr. Joseph Cox's house . . . about 15 chs. . . . A red fir, 12 ins. . . . A red fir, 18 ins. . . . A dead fir, 36 ins. . . . Wagon road (Gesner 1891: 254)

**Sec. 34:** Corner 27, 28, 33, 34: "A dead fir, 40 ins. . . . A red fir, 20 ins. . . . A dead fir, 30 ins. . . . A red fir, 14 ins." (Gesner 1891: 298)
"Timber, mostly burned fir, cedar" (Gesner 1891: 297)

**Sec. 35** "Timber, burnt" (Gesner 1891a: 253)

**Sec. 36:** "Timber, dead fir and cedar; alder along creek" (Gesner 1891a: 241)
Corner 25, 30, 31, 36: "A dead cedar, 18 ins. diam, bears S. 8 o W., 50 lks. Dist. . . . A dead cedar, 20 ins. diam., bears N. 12o E., 117 lks. dist. . . . A large fir (stump) burned to the ground, bears N. 32o W., 98 lks. dist. The end of a burnt yew log, bears S. 60o E., 73 lks. dist." (Gesner 1891a: 240)
Corner 25, 26, 35, 36: "A dead fir, 30 ins. Diam. Bears S. 62 1/2 W., 158 lks . . . A dead fir, 36 ins. Diam. Bears N. 18 1/2 W., 109 lks. . . . A dead fir, 36 ins. Diam. Bears N. 82o E., 187 lks. . . . A dead fir, 30 ins. Diam., bears S. 43 1/2 E. 229 lks. (Gesner 1891: 293).

## Tsp. 15 S., Rng. 7 W. (15-7): Prairie Mountain

General Description

[Sharp ca. July 15, 1897: 411-412] This township is entirely rough, broken and mountainous. There is no bottom land along the creeks . . . This township is covered with a heavy growth of valuable timber, mostly fir with scattering cedar, hemlock, maple and yew, with dense undergrowth and entirely without grass [!]. A few glades of 3 or 4 acres each occur in secs. 3, 4 and 9 and an extensive glade in sec. 7 of about 100 acres. These glades are covered with a dense growth of fern with little grass . . . There are no trails excepting one leading into the northern part of the township. It is impossible to get a 'pack horse' through the mountains.

**Sec. 1** [1/36 ?] I began at said SW. cor. Of sec. 36 where I found part of the post described in original notes, burned or rotted down. (Collier 1892: 168).
41.50 Left timber and enter prairie, bears E. and W.; this prairie extends in an easterly and westerly direction about 1 1/2 miles each way.
55.00 Summit of Blue Mountain bears E. and W., elevation about 150 ft. above 1/4 S. cor[ner].; this is one of the highest mountains in the Coast Range. (Gesner 1891: 418)
31.50 Cattle trail, course S. 75 W. and N. 75 E.
Trail to Prairie Mountain, course 75o E., and S. 75o W., on top of ridge of same (Collier 1893: 197)

33.25 Top of Blue Mountain, a sharp rocky ridge, bears N. 75 E., S. 75 W.; elevation above cor[ner]. To secs. 11, 12, 13 and 14 about 700 ft. (Gesner 1891: 421)
29.25 oak, 12, 8". "Left timber and entered prairie; this prairie extends easterly several miles along the top of Blue Mountain." (Gesner 1891: 424).

**Sec. 2** Yellow fir is good quality and generally sound except in SW 1/4 SW 1/4 and SE 1/4 NE 1/4 where there is considerable defect in the old growth timber. Hemlock is thrifty growth but low in clears.

**Bagley 1915** Cedar is good for posts only. This section will be easy to log. . . . Fire damage as shown on plat [14 acres]. The fire risk is nominal.

**Sec. 4** [W mile between 4 and 33] . . . A fir, 6 ins. . . . A fir, 7 ins. . . . Trail to Prairie Mountain . . . A hemlock, 20 ins. . . . A fir, 40 ins. . . . A yew, 12 ins. . . . A vine maple, 6 ins. . . .Heavily timbered with fir and hemlock; old growth dead on west 44.50 chs., but young growth in thickets taking its place. Dense undergrowth of salala and rhododendron on last half (Collier 1893: 172-173)

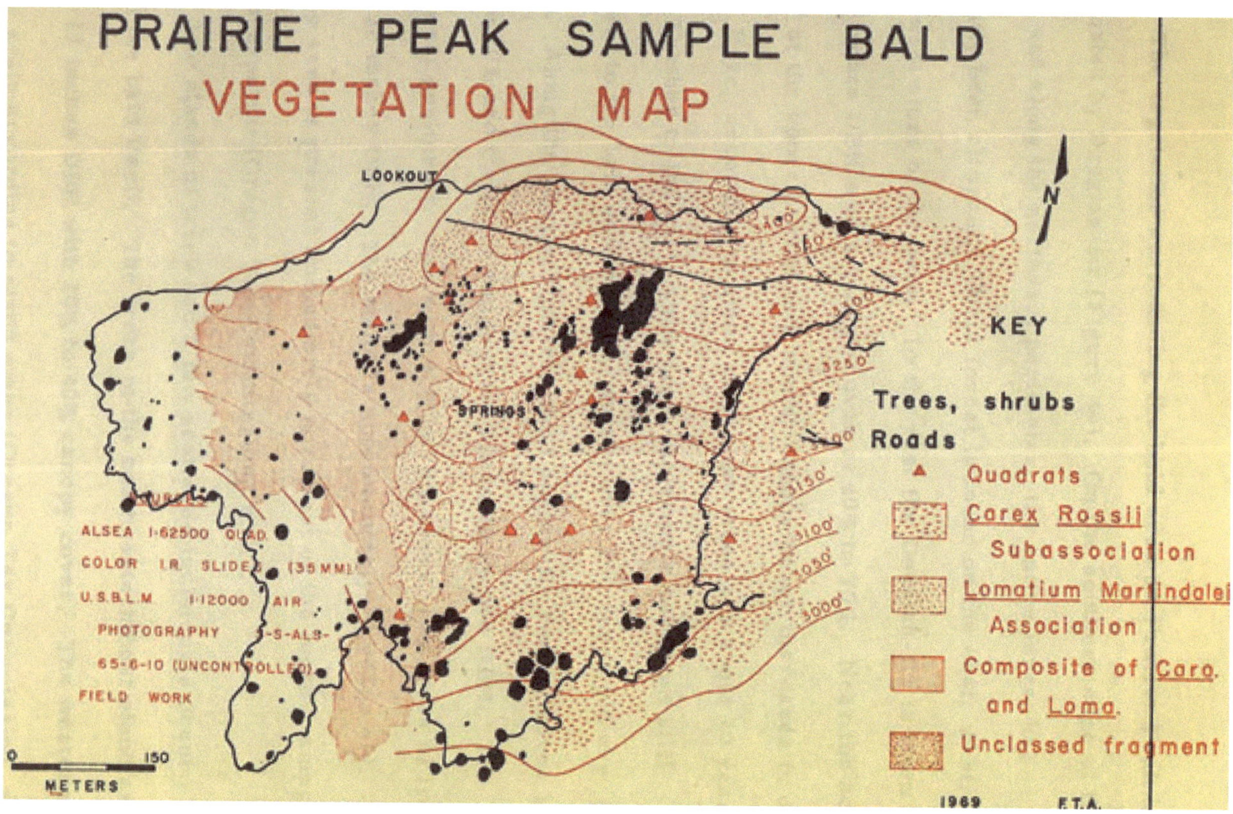


Map D.05 Prairie Peak prairies and trails, 1966.

[Compare the vegetation patterns and trails of Map D.05 and Figure D.02 with GLO survey notes from 1891 to 1897].

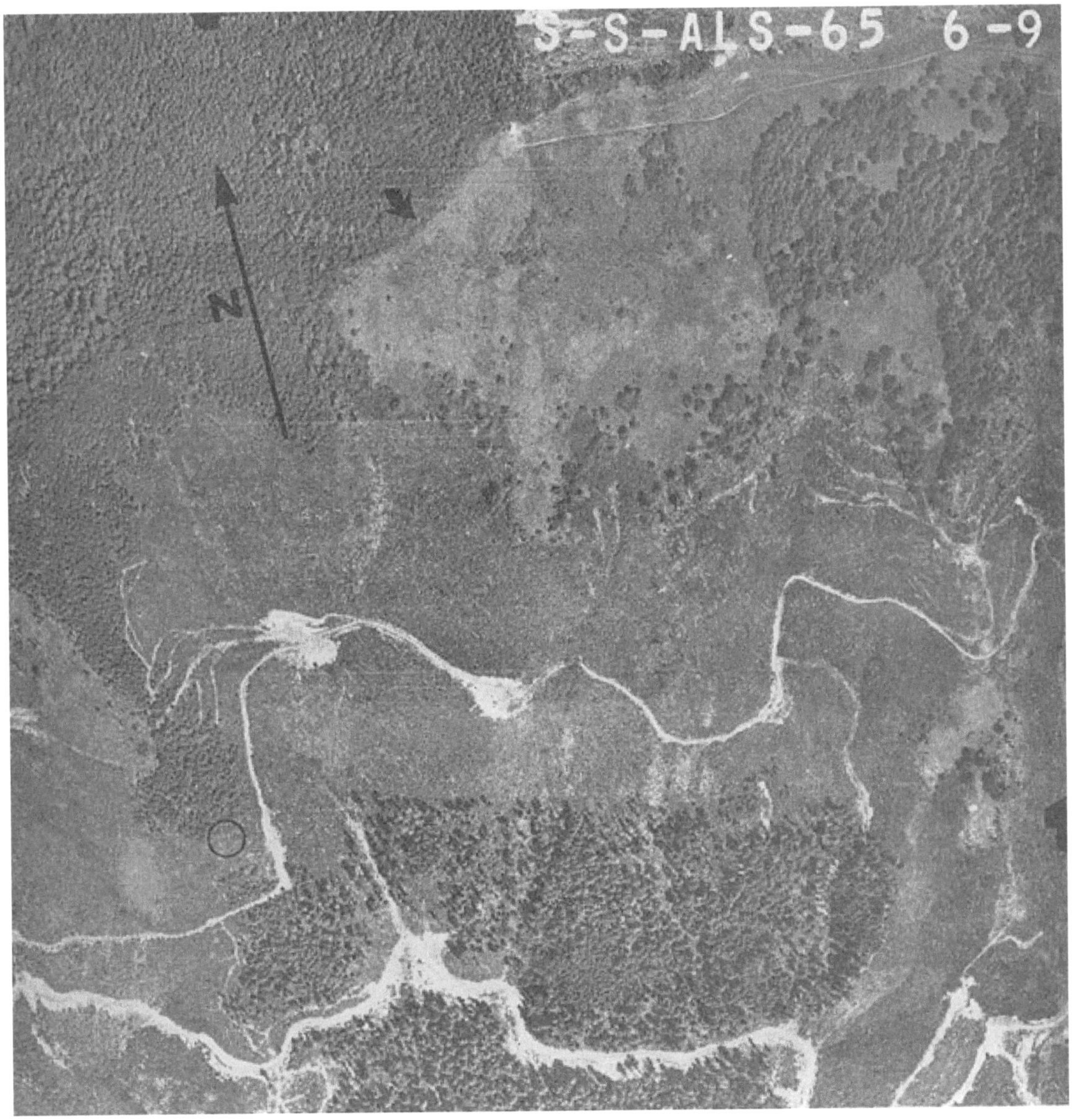


Figure D.02 Aerial photo of Prairie Peak prairies, June 9, 1965.

**Tsp. 15 S., Rng. 8 W. (15-8): Lobster River**

General Description

[David D. Fagan 1885: 498] Conterminous to the valley of the Alsea is a considerable quantity of partially settle country. For eight or ten miles to the southward, until Lobster creek is reached, the land is rolling and productive; while along the valley last-named there is a general settlement, but plenty of room for more.

[Alonzo Gesner, May 29, 1891] The townships on the east [14-7], south [15-8] and west [14-9] are settled with a class of people who intend making homes for themselves.

[Alonzo Gesner, July 28, 1891: 443] This Township is mountainous in character, some portions of it being quite broken. There are about ten settlers along the Lobster River, a tributary of Five Rivers. . . Nearly the entire Township is covered with a dense undergrowth . . . A voting precinct, Post Office and school district are located in this township.

| | |
|---|---|
| **Sec. 1** | [N mile between 1 and 6] A dead fir, 26 ins. . . . A dead fir, 20 ins. . . . [Bummer] Creek . . . Top of ridge . . . A dead fir, 30 ins. . . . foot of Blue Mountain [Prairie Peak], ascend . . . Timber, fir and hemlock, generally burnt wood. Dense undergrowth of hemlock, fir, huckleberry, thimbleberry, salalberry and arrowood (Gesner 1891: 419)<br>[W mile between 1 and 36] [Bummer] Creek . . . pass over spur. A dead fir, 12 ins. . . . Top of ridge . . . Ravine . . . An alder, 10 ins. . . . A red fir, 6 ins. . . . Top of ridge . . . [Swamp] Creek . . . A red fir, 6 ins. . . . . A red fir, 20 ins. . . . A cedar, 30 ins. . . . A red fir, 8 ins. . . . Timber, burned fir and cedar, alder along the streams. Dense undergrowth of vine maple, thimbleberry, cherry, myrtle [?] and hazel (Gesner 1891: 251-252) |
| **Sec. 2** | [W mile between 2 and 35] Top of ridge . . . The top of Elkpone, bears S. 450 E., about 10 chs. dist. which is a round top mountain . . . A dead fir, 48 ins. . . . A red fir, 36 ins. . . .A dead fir, 30 ins. . . .Top of ridge . . . A dead fir, 48 ins. . . . A dead fir, 48 ins. . . . An alder, 15 ins. . . . An alder, 8 ins. . . . Timber, burnt. Dense undergrowth of vine maple, thimbleberry, and wild grape (Gesner 1891: 252-253) |
| **Bagley 1915** | Yellow fir is a fair grad, medium length. Red fir and piling is young sappy timber and has no clears, fit for ties and dimension only. Cedar is a fair grade suitable for shingles . . . No fire damage of recent date and there is not much danger from fire at this time. This [SW] 1/4 section is of very little value as there is very little timber in this locality and the timber is patchy and quite small except a little old timber on the extreme east side. The country on the east for some distance is brushy with small patches of timber and to the north it is all burned. To the west there is a little young timber and a great deal of burn. |
| **Sec. 3** | "Timber, scattering fir, mostly burnt alder along stream, 92 chains" (Gesner 1891: 428).<br>[W mile between 3 and 34] [Meadow] Creek . . . Same creek . . . A small lot . . . Trail . . . Little Lobster . . . A maple, 10 ins. . . . A maple, 10 ins. . . . A maple, 10 ins. . . . A red fir, 50 ins. . . . A red fir, 50 ins. . . . A red fir, 16 ins. . . . Timber, scattering fir, alder and maple. Dense undergrowth of vine maple, hazel, fir, wild grape and thimbleberry (Gesner 1891: 253-254) |
| **Young Mrs. 1891** | (Gesner 1891: 432) |
| **Sec. 4** | "Timber, scattering and burnt fir, cedar and alder along streams" (Gesner 1891: 432)<br>[W mile between 4 and 33] Descend. Wagon road from Lobster to Alsea . . . Mr. Joseph Cox's house [14-8-33]. . . about 15 chs. . . . A red fir, 12 ins. . . . A red fir, 18 ins. . . . A dead fir, 36 ins. . . . Wagon road . . . Same road . . . Same road . . . Mr. Hoge's garden . . . Wagon road . . . near old fir stump . . . Timber, fir and alder, 80 chs. Dense undergrowth of vine maple, hazel and thimbleberry (Gesner 1891: 254-255) |
| **Bagley 1915** | A few scattering yellow fir of good quality. Red fir is a young and rapid growing timber, fit for ties and dimension only at this time. Piling is same kind of timber and will soon make saw logs . . . No recent fire damage but the old fire killed the old timber and it is growing up to young fir and alder. Not much fire risk now. |
| **Sec. 5** | [4/5 S] Top of ridge and divide bet. Little and big Lobster . . . Wagon road and fence . . . A fir, 50 ins. . . . 195 lks. . . . A fir, 60 ins. . . . 242 lks. . . .Wagon road . . . Young's field, fence . . . A. M. Peek's house . . . barn . . . Mr. Young's house . . . barn . . . Mr. Bratton's house . . . barn . . .Mr. |

| | |
|---|---|
| | Young's fence . . . Lobster River . . . river bottom . . . Timber, fir, maple, alder and cedar. Undergrowth of vine maple, cherry, salalberry and thimbleberry (Gesner 1891: 436-437) |
| **Peek A. 1891** | A. M. Peek's house . . . barn (Gesner 1891: 436) |
| **Bratton 1891** | Mr. Bratton's house . . . barn (Gesner 1891: 436) |
| **Young Mr. 1891** | Wagon road . . . Young's field, fence . . . Mr. Young's house . . . barn . . . Mr. Young's fence . . . Lobster River . . . river bottom (Gesner 1891: 436)<br>[W mile between 5 and 32] Creek . . . pass over spur . . . Creek . . . pass over spur . . . Creek . . . A red fir, 24 ins. . . . A red fir, 18 ins. . . . Stream . . . pass over spur . . . Stream . . . pass over spur . . . Stream . . . pass over spur . . . Stream. . . A red fir, 14 ins. . . . A red fir, 16 ins. . . . A red fir, 12 ins. . . . A red fir, 14 ins. . . Timber, fir, alder, cherry and maple. Dense undergrowth of vine maple, cherry, hazel and thimbleberry (Gesner 1891: 255-256) |
| **Sec. 6** | [5/6 S] Ascend. Top of hill and divide bet. Little and big Lobster . . . An alder, 20 ins. . . . A fir, 10 ins. . . .Enter Mr. G. C. Peak's calf pasture, fence . . . Left pasture, entered orchard, fence . . . Entered river bottom. Mr. G. C. Peek's house and Lobster post Office bears . . . 75 lks. . . . Wagon road from Lobster to Alsea . . . Mr. G. C. Peek's barn, bears 150 lks. . . .Entered G. C. Peek's grain field . . . Left grain field and right bank of Lobster River . . . Left Lobster River bottom and foot of hill . . . Top of ridge . . . Timbered with scattering fir, cedar, maple and alder . . . Undergrowth, vine maple, hazel, cherry and salalberry (Gesner 1891: 442-443)<br>[W mile between 6 and 31] . . . A red fir, 32 ins. . . A chittem, 8 ins. . . . A cedar, 30 ins. . . .A small spring . . . Top of ridge . . . Timber, scattering fir, alder, cherry and maple . . . Undergrowth of vine maple, cherry, hazel and salal berry (Gesner 256-257)<br>[S mile between 1 and 6] Descend. A red fir. A cluster of dogwood, about 12 in No. . . . A cherry, 8 ins. . . . A red fir, 10 ins. . . . Top of hill . . . Wagon road . . . Fence . . . Enter Mr. Sapp's yard and orchard. The NE. cor. Of school house in Dist. No. 46 . . . 256 lks.; the NW. cor. Of Mr. John Sapp's house [15-7-7]. . . 318 lks. Fence, leave orchard and yard . . . A maple, 26 ins. . . . An ash, 10 ins. . . . An alder, 5 ins. . . ., An alder, 12 ins. . . . An ash, 8 ins. . . . Timber, fir, maple, alder and dogwood, mostly burned. Dense undergrowth, vine maple, salal, berry, salmon berry, wild cherry and dogwood (Gesner 1891: 415-416) |
| **Peek G. 1891** | Enter Mr. G. C. Peak's calf pasture, fence . . . Left pasture, entered orchard, fence . . . Entered river bottom. Mr. G. C. Peek's house and Lobster post Office bears . . . 75 lks. . . . Wagon road from Lobster to Alsea . . . Mr. G. C. Peek's barn, bears 150 lks. . . .Entered G. C. Peek's grain field . . . Left grain field and right bank of Lobster River (Gesner 1891: 442) |
| **Sec. 7** | [6/7 E] J. Sapp's pasture, fence . . . Lobster River . . . Same river . . . J. Sapp's field, fence . . .An ash, 12 ins. . . . An ash, 36 ins. . . . Timbered with fir, cedar, maple, hazel, salmonberry, gooseberry and salalberry (Gesner 1891: 441) |
| **Sapp 1891** | Wagon road . . .Fence . . . foot of hill. Enter Mr. John Sapp's yard and orchard . . . the NW cor. Of Mr. John Sapp's house bears . . . 318 lks. . . . fence, leave orchard and yard (Gesner 1891: 415). J. Sapp's pasture, fence . . . Lobster River . . . Same river . . . J. Sapp's field, fence (Gesner 1891: 441) |
| **Sec. 8** | "Heavily timbered, south-half mile, green fir, alder, maple and cedar; north-half mile, mostly burned woods." (Gesner 1891: 440) |
| **Sec. 9**<br>**Tucker 1891**<br>**Jenkin 1891**<br>**Morgan 1891**<br>**Smith 1891** | [9/10 N] Mr. Tucker . . . B. W. Jenkin . . . James Morgan<br>(Gesner 1891: 429-430)<br><br>[4/9 W] B. F. Smith (Gesner 1891: 435] |
| **Sec. 11** | "Timber, burned woods, fir and scattering cedar" (Gesner 1891: 423)<br>"Timber, burnt fir and cedar, 80.60 chains" (Gesner 1891: 427) |
| **Sec. 12** | [N mile between 7 and 12] A red fir, 24 in. . . . A red fir, 24 in. . . . A red fir, 24 in. . . . Left timber and enter prairie, bears E. and W.; this prairie extends in an easterly and westerly direction, about 1 1/2 miles each way. Summit of Blue Mountain [Prairie Peak] bears E. and W. . . . this is one of the highest mountains in the Coast Range. Descend. A red fir, 30 in. . . . A hemlock, 20 in. . . . A hemlock, 20 in. . . . A red fir, 32 in. . . . A red fir, 30 in. . . . Timber, red fir and hemlock, 65 chs. Dense undergrowth of fir, hemlock, vine maple, hazel and huckleberry (Gesner 1891: 418) |
| **Sec. 15** | Entered Lobster River bottom, at foot of Blue [Prairie] Mountain Trail. Bears N. 400 W. and S. 400 E. Lobster River, 75 lks. Wide, on line, runs N. 350 W. (Gesner 1891: 429) |

Figure D.03 John Sapp home and family, ca. 1900.

Figure D.03 shows the John Sapp homestead and family. The photograph was taken about 100 years ago, in T. 15 S., R. 8 W., Sec. 6 (Bowen 1990: 34). Alonzo Gesner surveyed this property in July 1891, when the family still lived in a cabin, barely visible near the base of the hill, center background. Details from the survey at that time noted several trails, a wagon road, fences, grain fields, pasture, orchard, yard, and a schoolhouse. Much of the area was "timbered with scattering fir, cedar, maple, alder and dogwood, mostly burned" (note snags in upper left corner). Local native food plants included "hazel, salalberry, salmonberry, gooseberry, and wild cherry" (Gesner 1891: 256-257, 415-416, 442-443). The fenced crops and pasture, yard, and orchard were obviously fashioned from bottomland prairie adjacent to Lobster River (overgrown with brush by the time of this photo), which ran through Sapp's holding.

# APPENDIX E

## Historical Maps vs. Historical Theme Maps, 1850-1940

This appendix provides a brief comparison of the historical maps used in this dissertation (e.g., maps 2.04, 2.05, and 2.06) and a series of thematic maps (Teensma et al: 1991) that have been used extensively over the past 10 years to depict Oregon Coast Range old-growth and catastrophic forest fire patterns. The so-called "Teensma Maps" (named for the senior author of the report that accompanies them) were created to show forest and fire patterns for four points in time: 1850, 1890, 1920, and 1940. They have been used by a wide range of researchers (e.g., Diaz et al 1993; Coultin et al 1996; Botkin 1996: 243-244) to depict patterns of catastrophic fire and to estimate amounts of Coast Range old-growth at different points in time (e.g., Wimberly et al: 2000: 169). Thesc uses can be problematic:

> The small amounts of data available for model parameterization and validation currently limit our ability to accurately estimate historical landscape variability in the Coast Range and other large landscapes . . . Until we can estimate ranges of historical landscape variability more accurately, it will be difficult to substantiate and argument for their use as precise forest-management goals (Wimberly et al 2000: 178)

The Teensma Maps were constructed from many of the same sources used for this dissertation, but are generalized to a far greater degree and contain some basic flaws. For examples, the 1850 map (purportedly based on 1854 GLO survey data, although most of the area wasn't surveyed until the 1870s and later) shows fire patterns that didn't exist until 1868, and the 1940 map fails to show the boundaries of the catastrophic Tillamook Fire of 1939. There are problems with assumed stand ages as well, which are loosely based on tree sizes and contain arithmetical differences between maps (ibid: 1, 7). The following pages contain several comparisons of the Teensma Maps and representative portions of the principal historical maps on which they are based. As with Map 4.03, commentary is kept to a minimum and visual patterns are used to depict basic differences.

Map E.01 shows the four Teensma Maps in relation to the study area of this dissertation. Key problems are that the eastern and southern parts of the Range are not shown, thereby exaggerating the extent of old-growth

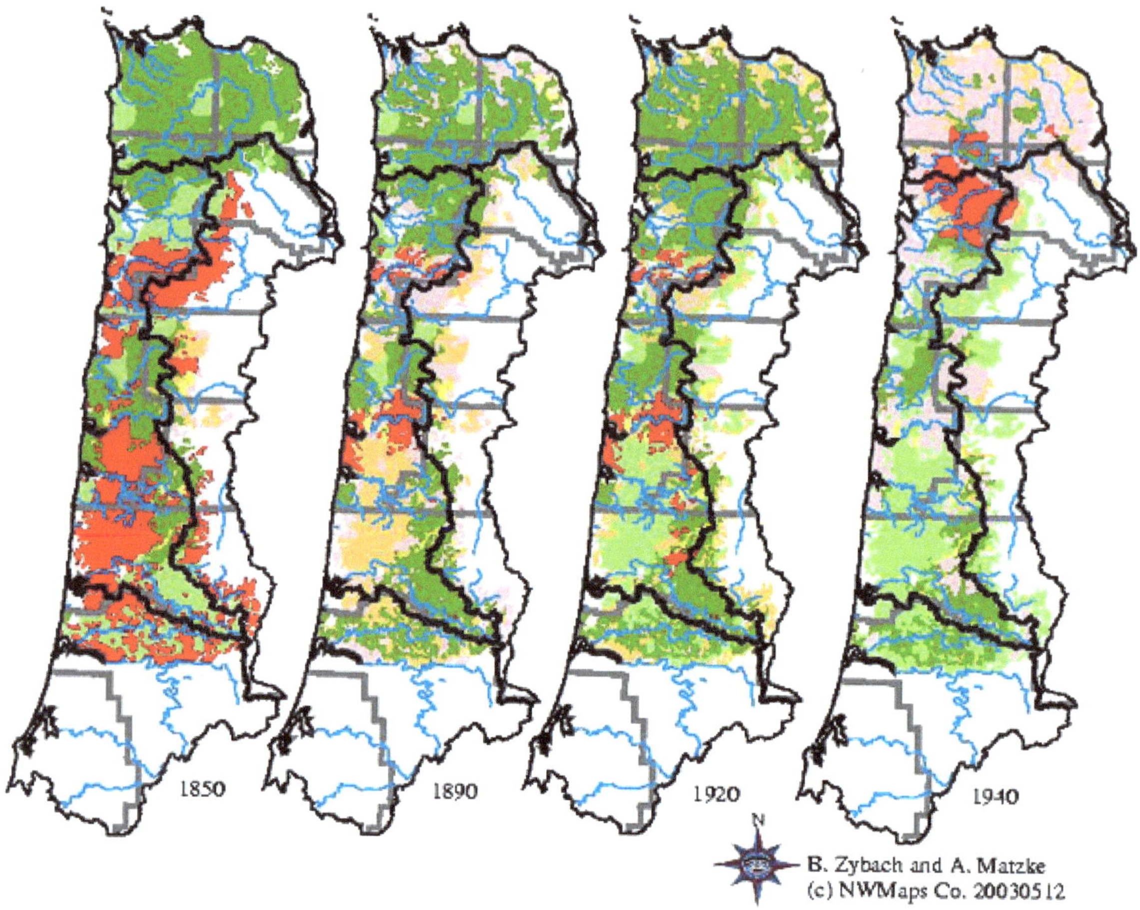


Map E.01 USDI BLM Coast Range fire history maps, 1850-1940.

forests (extensive second-growth stands resulting from the Millicoma and Coos fires in the southern Coast Range are left out) and the percentage of land subjected to catastrophic fire (most of the unburned eastern slope of the Range is also left out). In addition to presenting a bias in favor of old-growth conifer forests over old-growth oak savannahs, the selective bordering of the Coast Range can also present a problem to localized research studies:

> Stand-age analysis of the Coast Range by Teensma et al. (1991) and Juday (1977) did not include the Little Lake area in the large Coast Range fires of the 1800s. No other dendrochronological data for this area are available at present. The record of historical fires in the Little Lake area is therefore limited in comparison to other sedimentary charcoal studies (Clark 1990, Millspaugh and Whitlock 1995) (Long 1996: 17).

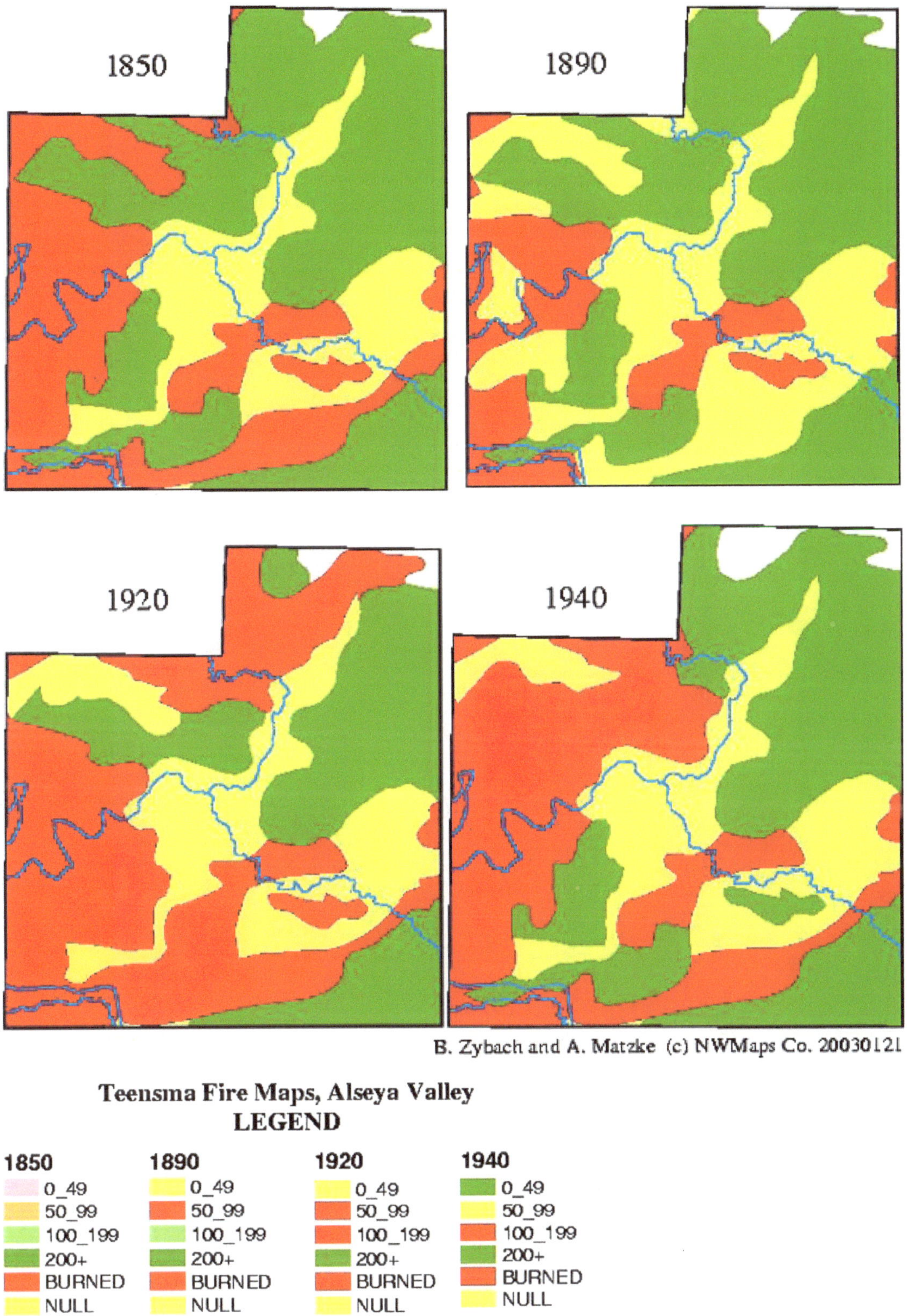


Map E.02 BLM mosaic patterns: Alsea Valley, 1850-1940.

Map E.02 was constructed by extracting the "Alseya Valley" study area from the Teensma Maps and generalizing color codes to highlight the polygons of the parent maps. Two problems are apparent: 1) the patterns do not closely match the patterns of the historical maps on which they are purportedly based (see Map 4.13; Teensma 1991: 1-2), and 2) the polygon boundaries remain virtually static throughout the entire 90-year period they depict. In nature, landscape patterns are dynamic (see Map 6.02) and the boundaries of forested areas and grasslands change constantly and dramatically over time (e.g., figures 3.04 and 3.05). By contrast, the Teensma Maps present a "musical chairs" approach to fire patterns and stand ages; further, the patterns contain arithmetical errors in that polygons are regularly shown to age 50 years or more in the 20 to 40-year periods depicted between maps and major fire events.Map E.03 compares the portion of the coastal strip depicted on the Teensma Maps with actual GLO survey patterns (Christy and Alvorsen 2003) and the historical map patterns used in this dissertation (and cited by Teensma et al 1991: 5-6; also see Map E.04).

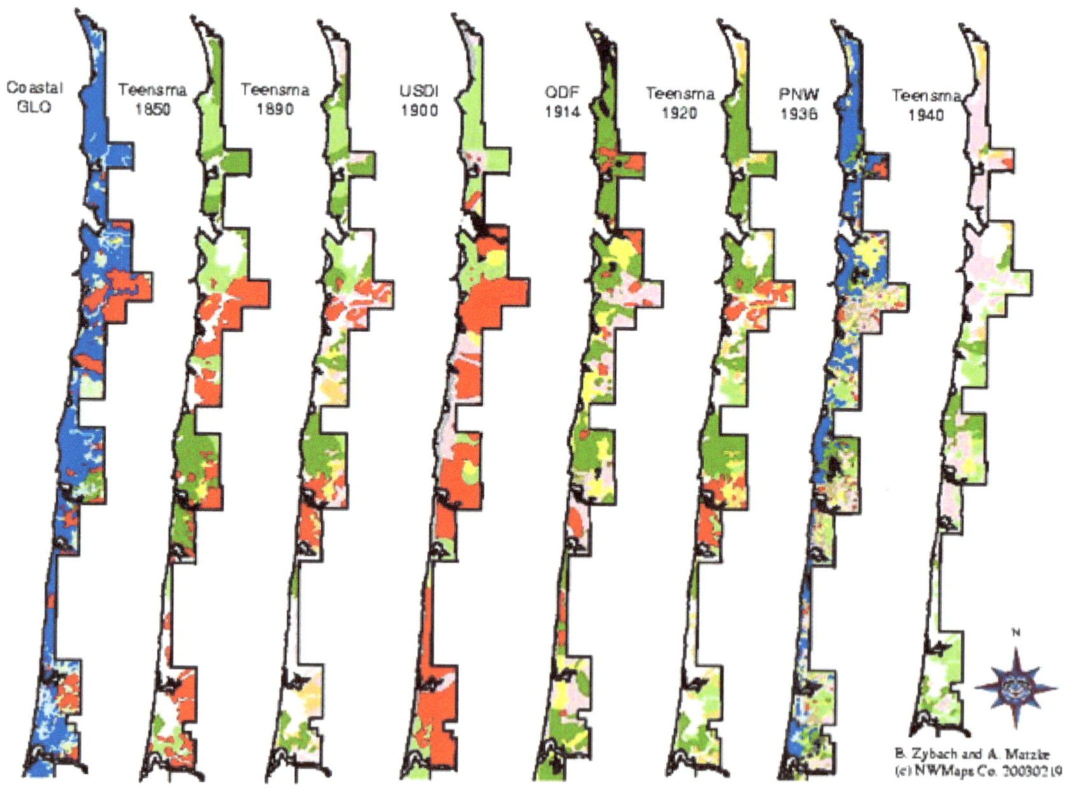


Map E.03 Pacific Ocean coastal strip comparisons, 1850-1940.

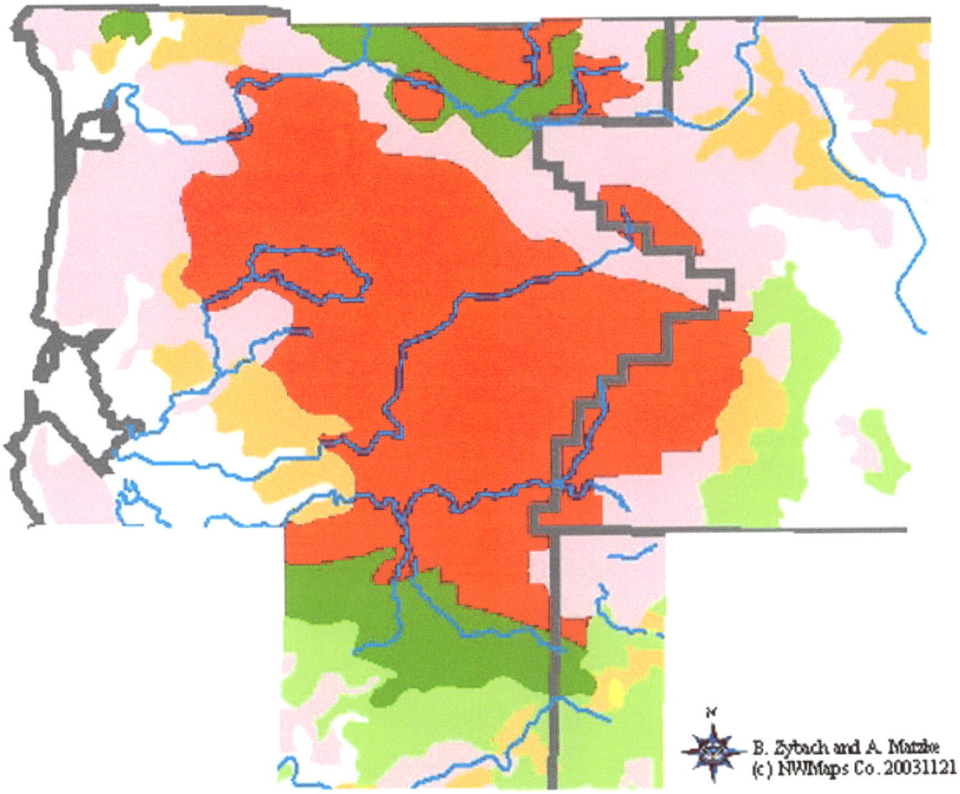


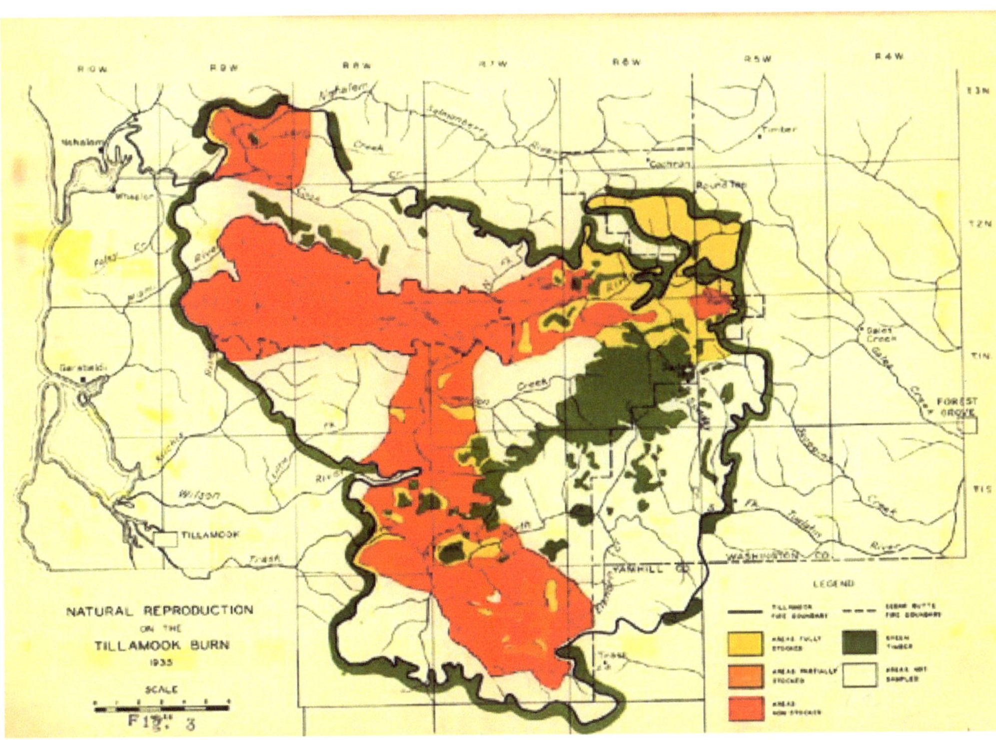


Map E.04 Thematic map vs. historical map: 1933 Tillamook Fire.

Note that the 1850 Teensma pattern conforms fairly well with the GLO pattern, but that the GLO coastal surveys were completed largely between 1870 and 1910. The dates are important because the 1850 pattern assumes catastrophic fire events in precontact time, rather than early historical time. The cause of the fires at those times, then, would likely have been Indians rather than white settlers. Also note: 1) the static nature of the Teensma polygons, and 2) the greater detail contained in the GLO and PNW patterns.

Map E.04 focuses on this latter point (map detail), using the 1933 and 1939 Tillamook fires as an example. The upper pattern is derived from the 1940 Teensma map, while the lower pattern is excerpted from an ODF report (Isaac and Meagher 1936: 20; 1938: 16). Note that the Teensma map fire boundaries are far more general; it contains significantly less information; and it doesn't account for the 1939 Tillamook Fire, which burned nearly 30,000 acres of green timber in addition to 225,000 acres of reburn (Fick an Martin 1992: 89).

In sum, the historical data contains far better information than the thematic data in several regards: 1) the thematic maps contain significant temporal errors; 2) the thematic maps have constricted boundaries, which limit their utility and create a bias toward older forests; 3) the thematic maps depict mechanistic and unlikely patch dynamics through time; and 4) the historical maps are more detailed and contain greater spatial and temporal accuracy than the thematic maps.

APPENDIX F.

## Moravets 1932: Cessation of Indian Burning and Afforestation

The following eyewitness report was made by a forest scientist based in the Douglas-fir Region (including the Oregon Coast Range) in 1932. Moravets' evaluation relies heavily on the testimony of a 90-year old Indian, Charlie Snakelum.

### SECOND GROWTH DOUGLAS FIR FOLLOWS CESSATION OF INDIAN FIRES

By F. L. Moravets, Pacific Northwest For. Exp. Sta.

Whidby Island, once largely deforested by the Indians' repeated "light burning", is now well forested with second growth Douglas fir. It is a striking example of the ability of Douglas fir quickly to reclaim lands long scourged by fire, after the periodic burning ceased. This was noted by the author in the course of type mapping Island County, Washington, in connection with the Forest Survey of the Douglas Fir Region.

This island, the second largest in Continental United States, is approximately 110,000 acres in area. The greater portion of the islands is covered with even-aged stands of second growth Douglas fir, most of which range from 65 to 80 years in age. In practically all of these stands no evidence of previous occupation by old growth Douglas fir was to be found and apparently the areas had been deforested at one time.

This assumption was verified by questioning Charlie Snakelum, the oldest and probably the last survivor of the Indians who occupied the island before the white men came. Old Charlie, over 90 years of age, antedates the second growth stands, and still retaining full possession of his faculties, can plainly recall conditions on the island in his boyhood. Like many aborigines, he has a remarkable memory but, unlike others, he is quite garrulous.

Many of the Puget Sound Indians used the island for a hunting ground and it was also the location of an annual "potlatch" attended by many of the neighboring tribes. Deer were plentiful and large portions of the island were burned over annually to make better hunting. Charlie recalls the time when areas now forested were treeless grass plains. The white settlers located on the better soil from 1850 to 1860 and repeated history by gradually driving the Indians out, stopping the practice of light burning. A scattering stand of relict firs provided seed, and, with the strong prevailing winds disseminating the seed, the denuded areas were soon reclaimed.

Charlie pointed out one area of considerable size slashed by him for a white settler some fifty years ago, which was abandoned after a few years and now supports a dense stand of 40-year old fir. This same condition may be observed on many of the larger islands of Puget Sound, other parts of Washington, and in the Willamette Valley, Oregon.

Practically all of the forest area of these islands was found to be of low site, sites IV and V predominating; consequently, growth is slow, the stands densely stocked, and thinning a slow process. It is not uncommon to find 70-year old trees only four inches in diameter and forty feet high. Stump rot is common in the young stands and what mature trees are found are likely to be defective and wind-skahen. At present the principal use of these stands is fuel wood for lime kilns operating on some of the islands. The young stands can be utilized to some extent for piling.

APPENDIX G.

## Oregon Coast Range Forest Fire History, 1491-2003

This appendix is a table of references that document the forest fire history of the Oregon Coast Range via tree rings, sediment cores, pollen cores, and documented observations. Pre-1491 entries provide possible information on the fuel histories of 1491 and later events. The "Lightning" column denotes proof or assumption of lightning-caused wildfires; the "People" column denotes wildfires caused by people. The "Acres" column denotes the acreage mapped or otherwise measured in conjunction with dated events. It is impossible at the present time to separate Coast Range fire acreage from western Oregon fire acreage for many years due to changes in record-keeping systems (Ballou 2003: personal communications; Fick 2003: personal communications). Fire acreage after 1951 is not limited to the Coast Range (e.g., the 2002 499,000-acre Biscuit Fire in southwest Oregon is included), but is given for western Oregon to emphasize the sharp increase in regional catastrophic wildfires that has occurred since 1987 (Zybach 2002a).

Table G.01 Tree ages and fire years of the Oregon Coast Range

| YEAR | REFERENCE | Lightning | People | Acres |
|---|---|---|---|---|
| **1348** | ODF 1948 | | | |
| **1469** | Connelly and Kertis (1992) | | | |
| **1478** | Impara (1997: 212) | | | |
| **1482** | Teensma (1981) | | | 25,685 |
| **1499** | Connelly & Kertis (1992) | | | |
| **TOTAL** | **<1500** | | | **25,685** |
| **1523** | Connelly & Kertis (1992) | | | |
| **1531** | Impara (1997: 212) | | | |
| **1532** | Teensma (1981) | | | 17,294 |
| | Connelly & Kertis (1992) | | | |
| **1535** | Impara (1997: 212) | | | |
| **1539** | Newton (1973: 1) | | | |
| **1560** | Grantham (1953: 4) | | | |
| | Boyce & Wagg 1953: 8 | | | |
| **1567** | Shea 1963: 4 | | | 28,500 |
| **1585** | Impara 1997: 212 | | | |
| **TOTAL** | **1500-1599** | | | **45,794** |
| **1600** | Grantham 1953: 4 | | | |
| | Boyce & Wagg 1953: 8 | | | |
| **1629** | Impara 1997: 212 | | | |
| **1635** | Newton 1973: 1 | | | |
| **1637** | Impara 1997: 212 | | | |
| **1647** | Sprague & Hansen 1946 | | | |
| **1655** | Impara 1997: 212 | | | |
| **1660** | Thilenius 1968: 1126 | | | |
| **1666** | Impara 1997: 212 | | | |
| **1689** | Impara 1997: 212 | | | |
| **1699** | Impara 1997: 212 | | | |
| **TOTAL** | **1600-1699** | | | |
| **1711** | Thilenius 1968: 1126 | | | |
| **1720** | Weyerhaeuser 1947 | | | |
| **1737** | Impara 1997: 212 | | | |
| **1751** | Impara 1997: 212 | | | |
| **1763** | Long 1996: 35 | | | |
| | Impara 1997: 212 | | | |
| **1765** | Weyerhaeuser 1947 | | Yes | 90,000 |
| **1794** | Impara 1997: 212 | | | |
| **TOTAL** | **1700-1799** | | **Yes** | **90,000** |
| **1826** | Douglas | | Yes | |
| | McLeod | | Yes | |
| **1828** | McLeod | | Yes | |
| | Smith | | Yes | |
| **1832** | Work | | Yes | |
| **1837** | Long (1996: 35) | | | |
| **1841** | Wilkes (1845a; 1845b) | | Yes | |
| **1844** | Applegate | | Yes | |
| **1845** | Chen (1997: 4) | | | 1,502,368 |

| YEAR | REFERENCE | Lightning | People | Acres |
|---|---|---|---|---|
| | Morris (1934b: 319) | | Yes | 380,00 |
| 1846 | Morris | | | |
| | Kirkpatrick (1940: 33) | | Yes | 450,000 |
| 1848 | Albert (1848: 1) | | Yes | |
| | Sprague & Hansen (1946) | | | |
| | Ballou (2002: 67) | | | 300,000 |
| | Chen (1997: 4) | | | 375,592 |
| | Talbot | | | |
| 1849 | Ballou (2002: 67) | | | 800,000 |
| | Morris (1934b: 322) | | | 500,000 |
| | Talbot (Haskin 1948) | | | |
| | Smith (Frachtenberg 1920) | | Yes | |
| 1852 | Impara (1997: 212) | | | |
| 1853 | Ballou (2002: 67) | | | 480,000 |
| | Chen (1997: 4) | | | 321,230 |
| | Kirkpatrick (1940: 33) | | | 320,000 |
| 1868 | Ballou (2002: 67) | | | 300,000 |
| | Chen 1997: 4 | | | 126,021 |
| | Kirkpatrick 1940: 33 | | | 300,000+ |
| | Philips 1988 | | | |
| 1871 | Impara 1997: 212 | | | |
| 1881 | Impara 1997: 212 | | | |
| 1888 | Impara 1997: 212 | | | |
| **TOTAL** | **1800-1899** | | | |
| | Ballou | | | **1,880,000** |
| | Chen | | | **2,325,211** |
| | Impara | | | |
| | Kirkpatrick | | **Yes** | **1,070,000+** |
| | Morris | | **Yes** | **1,005,000+** |
| 1902 | Ballou 2002: 67 | | | 170,000 |
| | Chen 1997: 4 | | | 170,499 |
| 1910 | Kirkpatrick 1940: 35 | No | Yes | 50,556 |
| | Long 1996: 35 | | | |
| 1914 | Ballou 2002: 77 | 147 | 1,107 | 145,971 |
| 1915 | Ballou 2002: 77 | 85 | 1,170 | 109,404 |
| | Forest Log 1994 | | | 109,404 |
| 1917 | Ballou 2002: 77 | 177 | 1,468 | 98,956 |
| | Forest Log 1994 | | | 257,768 |
| 1918 | Ballou 2002: 77 | 139 | 813 | 184,063 |
| | Forest Log 1994 | | | 163,062 |
| 1919 | Ballou 2002: 77 | 181 | 963 | 143,463 |
| | Forest Log 1994 | | | 139,095 |
| **TOTAL** | **1910-1919** | | | |
| | Ballou 2002 | **729** | **5,521** | **681,857** |
| | Forest Log 1994 | | | **669,329** |
| 1920 | Impara 1997: 212 | | | |
| 1922 | Ballou 2002: 77 | 95 | 1,312 | 178,530 |
| | Forest Log 1994 | | | 143,090 |
| | Kirkpatrick 1940: 35 | NO | YES | 8,108 |
| 1924 | Ballou 2002: 77 | 311 | 1,577 | 252,251 |
| | Forest Log 1994 | | | 240,251 |
| 1926 | Ballou 2002: 77 | 150 | 996 | 207,745 |

| YEAR | REFERENCE | Lightning | People | Acres |
|---|---|---|---|---|
| | Forest Log 1994 | | | 207,745 |
| **1928** | Ballou 2002: 77 | 154 | 913 | 103,907 |
| | Forest Log 1994 | | | 103,907 |
| **1929** | Ballou 2002: 77 | 117 | 1,355 | 298,235 |
| | Forest Log 1994 | | | 298,235 |
| | Kirkpatrick 1940: 35 | | | 13,994 |
| **TOTAL** | **1920-1929** | | | |
| | Ballou | **827** | **6,153** | **1,040,668** |
| | Forest Log | | | **993,228** |
| | Impara | | | |
| | Kirkpatrick | **Yes** | **Yes** | **22,102** |
| **1931** | Ballou 2002: 77 | 85 | 1,536 | 188,494 |
| | Forest Log 1994 | | | 175,494 |
| | Kirkpatrick 1940: 35 | | | 5,104 |
| **1932** | Ballou 2002: 77 | 87 | 1,601 | 333,376 |
| | Forest Log 1994: 7 | | | 334,374 |
| **1933** | Ballou 2002: 77 | 239 | 943 | 340,855 |
| | Chen 1997: 4 | | | 261,926 |
| | Forest Log 1994 | | | 340,855 |
| | Kirkpatrick 1940: 35 | | | 267,000 |
| **1936** | Ballou 2002: 77 | 139 | 1,265 | 225,083 |
| | Forest Log 1994 | | | 225,083 |
| | Kirkpatrick 1940: 35 | | | 5,536 |
| | Long 1996: 35 | | | |
| **1938** | Ballou 2002: 77 | 188 | 957 | 79,738 |
| | Forest Log 1994: 7 | | | 79,738 |
| | Kirkpatrick 1940: 35 | | | 11,456 |
| **1939** | Ballou 2002: 77 | 294 | 965 | 308,827 |
| | Chen 1997: 4 | | | 217,448 |
| | Forest Log 1994: 7 | | | 302,049 |
| | Kirkpatrick 1940: 36 | | | 1,758 |
| **TOTAL** | **1931-1939** | | | |
| | Ballou | **1,092** | **7,267** | **1,476,373** |
| | Forest Log | | | **1,457,593** |
| | Kirkpatrick | | | **290,694** |
| **1945** | Ballou 2002: 77 | **333** | **532** | **253,134** |
| | Chen 1997: 4 | | | **180,523** |
| | Forest Log 1994 | | | **122,362** |
| **1951** | Ballou 2002: 77 | **104** | **850** | **125,666** |
| | Forest Log 1994 | | | **125,666** |
| **1967** | Long 1996: 35 | | Yes | |
| **1982** | Long 1996: 35 | | Yes | |
| **1987** | Ballou 2002: 77 | 605 | 976 | 88,907 |
| | Forest Log 1994 | | | 91,441 |
| | Coyle 2002 | 26 | 17 | 88,311 |
| **1996** | Ballou 2002: 67 | | | 118,000 |
| | Coyle 2002 | 9 | 4 | 188,548 |
| **2000** | Coyle 2002 | 4 | 8 | 109,968 |
| **2002** | Ballou 2002: 67 | Yes | | 499,570 |

# APPENDIX H.

## "Alseya Valley" Prairie Relicts, April 14, 2003

This appendix provides photographic evidence of the "cultural legacy" resulting from precontact Indian burning practices in Alsea Valley (see Map 2.13 and Appendix D). The majority of these pictures were taken on April 14, 2003; the day following a series of major lightning storms and heavy thunder showers throughout the local area (Crombie 2003). In addition to recording persistent vegetation patterns of grasslands and forests that date to precontact time, they also provide evidence of how weather events may have changed river uses during those times. The rock shelf that forms the river crossing on the old Indian Trail from the Willamette Valley to Alsea tidewater (see maps 2.11 and 2.12; figures H.01 and H.02) could probably still be safely used by healthy adults--especially with the aid of a crossing rope, log, or weir foot bridge--even in times of heavy rains and swollen streams. Prairies and meadows (e.g., Figure H.03) remained well drained and unflooded. Nathaniel, the kayaker at Clemens Park (Figure H.04), demonstrates the safety and ease in which small canoes could still be used on the river, even following major rainfall events.

Map H.01 (Zybach, Gruen, and Matzke 2003) is an index of six stops made along the "Indian Trail to Tidewater" route mapped in 1856 (see maps 2.11 and 2.12). Each of the six numbered stars represents a series of one to six photographs of that location taken on April 14, 2003. The complete series of photos is included in Appendix I. The vegetation pattern shown on Map H.01 is generally accurate for the 1800-1850 time period, based on GLO survey records (see Map 2.13 and Appendix D; Zybach 2002b).

Each of the six locations and the river crossing was plotted in advance of the photographic survey as suspected major townsites or camping grounds. The stops were planned solely on the basis of historical documentation, rather than known or expected conditions.

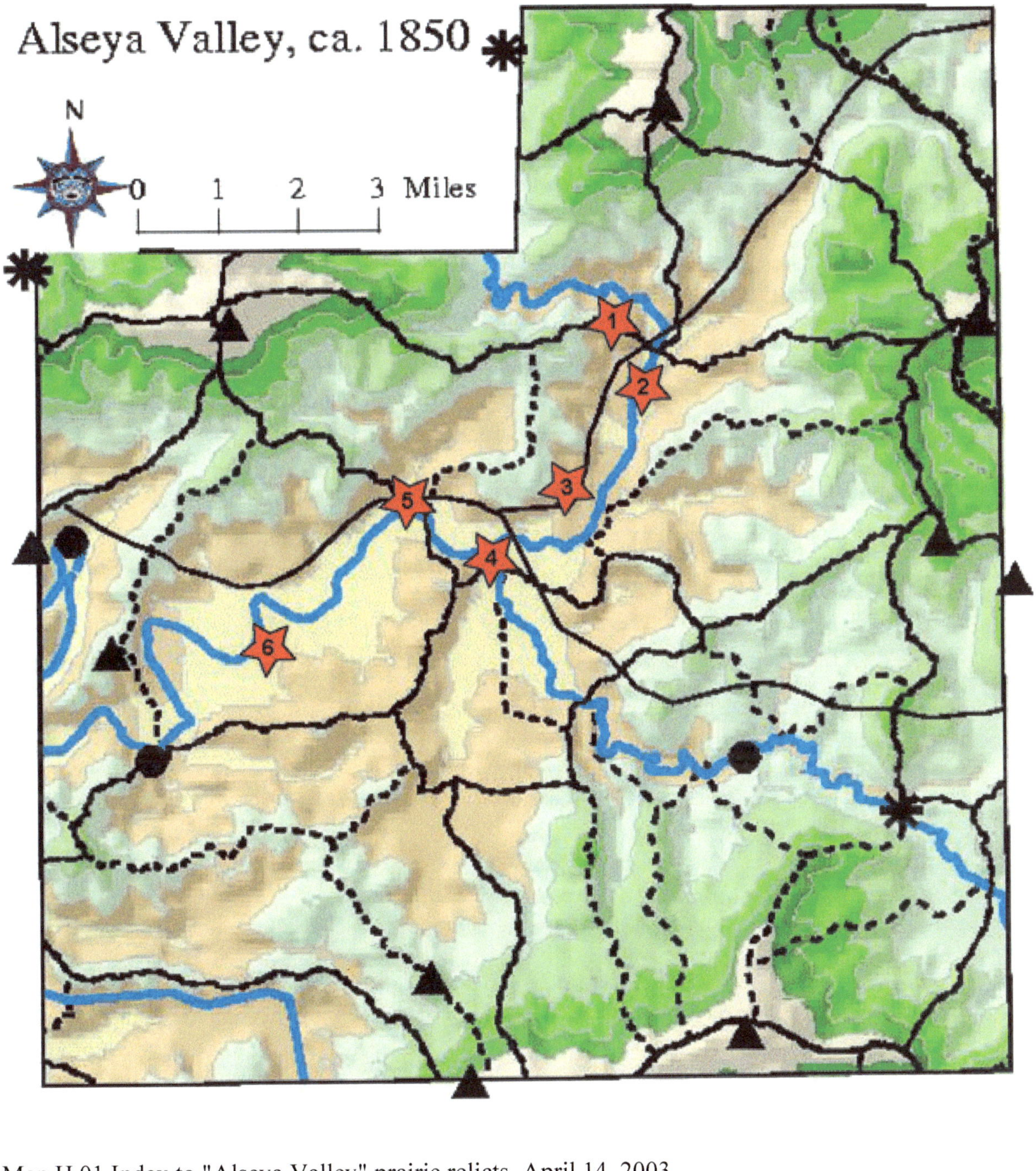


Map H.01 Index to "Alseya Valley" prairie relicts, April 14, 2003.

**Stop 1. North Fork prairie campground and Alsea River crossing.**

People traveling by established trail from the Willamette Valley to ocean-going traffic at Tidewater (the site of an Alsi town) needed only to cross the Alsea River at one location, as shown on Map H.01. Here we found a large, flat rock outcropping that spanned the entire North Fork and that could be safely crossed by an adult, even at high water, as shown on

Figure H.01 (photos by B. Zybach). This crossing is located in Tsp. 13 S., Rng. 7 W., Sec. 28 (see map D.01) and is marked by the bridge on Highway 34, which closely follows the old Indian Trail and a subsequent horse trail and wagon road established in the 1850s. The upper left photo is just above the crossing and could be easily maneuvered with a small canoe. The upper right photo shows the river current as it first encounters the rock outcrop, shown in the lower photo. According to local landowners near the crossing, this location is near the original 1850s Rubles mill site.

Figure H.02 (photo by F. Lake) was taken in October 2003, when the river was at a much lower stage. At that time the river could be crossed with ease without getting wet. (The model is wearing traditional Alsi clothing of a bark skirt, woven hat, and tanned deerskin. She is carrying a basket of ripe blue elderberries gathered a short distance away and a digging stick made of hazel and deer antler).

Figure H.03 (photo by N. Lapham) shows the relict prairie immediately adjacent to the crossing. In addition to the elderberries, the prairie contained oak trees and hazel brush; native food plants dating to precontact time (see appendices A and D). This is the same photo used to illustrate the local trail network (Figure 2.05). Note the openings and older trees along the ridgelines on the right.

Figure H.01 North Fork Alsea Indian Trail crossing, April 14, 2003.

Figure H.02 North Fork Alsea Indian Trail crossing, early fall, 2003.

Figure H.03 North Fork Indian prairie relict, April 14, 2003.

**Stop 2. Clemens Park campground and canoe launch.**

Figure H.03 (photo B. Zybach) shows Nathaniel in a kayak near Clemens Park, indicating the possibility of river travel a day after a major rainstorm. The picnic spot on the north shore indicates the ease in which canoes could be put in and out of the river at this location. Clemens Park is in Tsp. 13 S., Rng. 7 W., Sec. 32 (see Map D.01).

Figure H.04 Canoe campsite, North Fork Alsea River.

**Stop 3. Honey Grove Road riparian meadow, prairie, and river rock.**

Upland prairies, an open grown old-growth Douglas-fir, and a grassy riparian meadow indicate precontact land use patterns at this location. There is also ready availability of river rock for tools such as hammers, wedges, pestles, missiles, etc., as shown by a photo of an easily accessed rock bar in Appendix I. Figure H.05 (photo by B. Zybach) shows a meadow immediately adjacent to the river in Tsp.14 S., Rng. 7 W., Sec. 5 (see Map D.01).

Figure H.05 Riparian prairie relict, North Fork Alsea River.

**Stop 4. Alsea Forks prairie landmark, oak grove and town site.**

Figure H.06 (photo by B. Zybach) was taken near the Alsea High School football field. Does this hill have a name? It is visible from much of Alsea Valley and has to be a local landmark--and most certainly must have been a landmark in precontact time as well. This location is near the forks in the Alsea River, in the heart of the Valley (called *ts! Phahpha* by the Alsi). A significant amount of archaeological evidence near this location indicates that this may have been a major town site in precontact time. The strategic location by both international foot and canoe traffic for trade purposes, the ready availability of a wide range of local seasonal food sources, and the protective location of the valley, deep within the Coast Range, are all likely indicators of long periods of past daily and seasonal use by relatively large numbers of people. This location includes the town of Alsea (see Figure 2.03) in Tsp. 14 S., Rng. 7 W., Sec. 6 (see Map D.01).

Figure H.06 Landmark hill and oak savannah relict, Alsea River Forks.

Figure H.07 (photo by B. Zybach) shows a recently cut three-foot Douglas-fir stump from a tree probably planted adjacent to Alsea High School shortly after World War II. The size indicates the excellent growing condition for conifers, which have been encroaching on old-growth oak near this location during historical time (see Figure 4.01 and Appendix I).

Figure H.07 Douglas-fir stump, Alsea High School.

**5. Mill Creek homesite, campground, and canoe launch.**

This location is near the cabin homesite of "Old Billy," apparently an Alsi Indian who lived here amongst white settlers in the 1850s and perhaps longer (Hathorn 1856b: 262; Rycraft ca.1922). This was probably a good fishing location, was close to town, had easy canoe access, and may have been a major Indian campground, especially after the community of Alsea was established. Figure H.08 shows the mouth of Mill Creek, with a prairie relict on the opposite shore.

Figure H.08 Mouth of Mill Creek.

Figure H.08 (photo by N. Lapham) shows the general location of "Old Billy's homesite, near the Mill Creek crossing on the Indian trail to Tidewater. Historical records say that hundreds of Alsi, Klickitat, Siletz, and Kalapuya Indians would gather near this location for weeks at a time for purposes of gambling, horse racing, and etc. (Rycraft ca. 1922; Phinney 2000). Such gatherings may have occurred during periods of prairie burning, fish runs, berry picking, nut harvesting, and camas digging. Local firewood use during those times must have been enormous. The location is in Tsp. 14 S., Rng. 8 W., Sec. 2 (see Map D.01).

Figure H.09 "Old Billy's" homesite, Alsea Valley.

**Stop 6. The Narrows prairie town site, campground, and river crossing.**

This location is near a confirmed archaeological site and documented Indian campground. Local artifacts include an old-growth oak and filbert orchard and a river crossing on the historic Indian Trail to Lobster Valley. It is also near several known fishing spots. "The Narrows" may have been a secondary town site in Alseya Valley before white settlement. One resident of the 1870s reported: "In 1880 there were not more than 300 whites in the valley and I have seen as many as 1,000 Indians at one time. There was an Indian camp ground at the "narrows" about three miles below Alsea and the Indians used to come and go" (Phinney 2000: 165). The location is in Tsp. 14 S., Rng. 8 W., Sec. 8 (see Map D.01).

Figure H.10 Prairie relict near "The Narrows," in Alsea Valley.

www.ingramcontent.com/pod-product-compliance
Lightning Source LLC
LaVergne TN
LVHW071243210926
848486LV00007B/334